Quality Management in Oil and Gas Projects

Quality Management and Risk Series

Series Editor:
Abdul Razzak Rumane
Senior Engineering Consultant, Kuwait

This new series will include the latest and innovative books related to quality management and risk-related topics. The definition of quality relating to manufacturing, processes, and the service industries is to meet the customer's need, satisfaction, fitness for use, conforms to requirements, and a degree of excellence at an acceptable price, and that of construction projects is fulfillment of owner's needs per defined scope of works within specified schedule and budget to satisfy owner's requirements. With globalization and competitive markets, the emphasis on quality management has increased. Quality has become the most important single factor for the survival and success of any company or organization. The demand for better products and services at the lowest costs has put tremendous pressure on organizations to improve the quality of products, services, and processes, in order to compete in the marketplace. Because of these changes, the ISO 9001 now lists that risk-based thinking must be incorporated into the management system by considering the context of the organization. Quality management and risk management now play an important role in the overall quality management system. This means that books that cover quality need to also cover risk to update practices/processes, tools, and techniques, per ISO 9001. The goal of this new series is to include the books that will meet this need and demand.

Quality Management in Oil and Gas Projects

Abdul Razzak Rumane

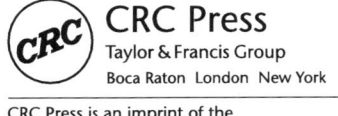

CRC Press
Taylor & Francis Group
Boca Raton London New York

CRC Press is an imprint of the
Taylor & Francis Group, an **informa** business

First edition published 2021
by CRC Press
6000 Broken Sound Parkway NW, Suite 300, Boca Raton, FL 33487-2742

and by CRC Press
2 Park Square, Milton Park, Abingdon, Oxon, OX14 4RN

© 2021 Taylor & Francis Group, LLC

CRC Press is an imprint of Taylor & Francis Group, LLC

The right of Abdul Razzak Rumane to be identified as author of this work has been asserted by him in accordance with sections 77 and 78 of the Copyright, Designs and Patents Act 1988.

Reasonable efforts have been made to publish reliable data and information, but the author and publisher cannot assume responsibility for the validity of all materials or the consequences of their use. The authors and publishers have attempted to trace the copyright holders of all material reproduced in this publication and apologize to copyright holders if permission to publish in this form has not been obtained. If any copyright material has not been acknowledged please write and let us know so we may rectify in any future reprint.

Except as permitted under U.S. Copyright Law, no part of this book may be reprinted, reproduced, transmitted, or utilized in any form by any electronic, mechanical, or other means, now known or hereafter invented, including photocopying, microfilming, and recording, or in any information storage or retrieval system, without written permission from the publishers.

For permission to photocopy or use material electronically from this work, access www.copyright.com or contact the Copyright Clearance Center, Inc. (CCC), 222 Rosewood Drive, Danvers, MA 01923, 978-750-8400. For works that are not available on CCC please contact mpkbookspermissions@tandf.co.uk

Trademark notice: Product or corporate names may be trademarks or registered trademarks and are used only for identification and explanation without intent to infringe.

Library of Congress Cataloging-in-Publication Data

Names: Rumane, Abdul Razzak, author.
Title: Quality tools and techniques in oil and gas projects / Abdul Razzak Rumane.
Other titles: Quality management in oil and gas projects
Description: First edition. | Boca Raton, FL : CRC Press/Taylor & Francis Group, LLC, 2021. | Series: Quality management and risk series | Includes bibliographical references and index.
Identifiers: LCCN 2020041039 (print) | LCCN 2020041040 (ebook) | ISBN 9780367460754 (hardback) | ISBN 9781003145059 (ebook)
Subjects: LCSH: Petroleum engineering. | Total quality management.
Classification: LCC TN870 .R745 2021 (print) | LCC TN870 (ebook) | DDC 665.5068/4—dc23
LC record available at https://lccn.loc.gov/2020041039
LC ebook record available at https://lccn.loc.gov/2020041040

ISBN: 978-0-367-46076-1 (hbk)
ISBN: 978-1-003-14506-6 (ebk)

Typeset in Times
by codeMantra

To

My Parents

For their prayers and love

My prayers are always for my father and my mother who encouraged and inspired me all the times.

I wish they would have been here to see this book and give me blessings.

To

My Wife

I miss my wife, who stood with me all the time during writing of my earlier books.

Contents

Foreword xvii
Preface...xix
Acknowledgments xxiii
Author xxv
Abbreviations...xxvii
Synonyms...xxix

SECTION 1   Quality Tools and Techniques in Oil and Gas Projects

Chapter 1 Overview of Quality..3

    1.1   History of Quality 3
    1.2   Definition of Quality 3
    1.3   Evolution of Quality Management System..............................5
            1.3.1   Quality Inspection..7
            1.3.2   Quality Control..8
            1.3.3   Quality Assurance 10
            1.3.4   Quality Engineering 12
            1.3.5   Quality Management... 12
            1.3.6   Total Quality Management..................................... 14
                      1.3.6.1   Changing Views of Quality 15
                      1.3.6.2   Quality Gurus and Total Quality Management Philosophies.......................... 16
                      1.3.6.3   Principles of Total Quality Management.... 28
    1.4   Definition of Quality for Construction Projects 32
    1.5   Quality Management System 35
            1.5.1   Quality Plan... 36
                      1.5.1.1   Designer's Quality Management Plan 36
                      1.5.1.2   Contractor's Quality Control Plan 37
                      1.5.1.3   Quality Matrix 38
            1.5.2   Quality Assurance 38
            1.5.3   Quality Control.. 40
    1.6   Integrated Quality Management System 42
    1.7   Risk Management 43
            1.7.1   Risk Management Process 45
                      1.7.1.1   Identify Risk 45
                      1.7.1.2   Analyze/Assess Risk 46

		1.7.1.3	Prioritization	49
		1.7.1.4	Plan Risk Response	49
		1.7.1.5	Reduce Risk	51
		1.7.1.6	Monitor and Control Risk	51
	1.7.2	Risks in Construction Projects		51
1.8	Supply Chain Management			52
	1.8.1	Supply Chain Management Process		58
	1.8.2	Supply Chain Stakeholder		59
	1.8.3	Supply Chain Management Processes		60
		1.8.3.1	Product Specification	60
		1.8.3.2	Supplier Selection	60
		1.8.3.3	Purchasing	60
		1.8.3.4	Product Quality	61
		1.8.3.5	Product Handling	61
		1.8.3.6	Improvement	61
	1.8.4	Supply Chain in Construction Projects		62

Chapter 2 Quality Management System 65

2.1	Quality Standards		65
	2.1.1	Importance of Standards	66
2.2	Standards Organizations		67
2.3	International Organization for Standardization		75
2.4	ISO 9000 Quality Management System		76
	2.4.1	Quality System Documentation	78
	2.4.2	Quality Management System Manual	79
		2.4.2.1 Development of Quality Management System	80
2.5	ISO Certification		84
2.6	ISO/TS 29001		95
2.7	Quality Audit		97
	2.7.1	Categories of Auditing	98
	2.7.2	Quality Auditing Process	98

Chapter 3 Overview of Oil and Gas Projects 101

3.1	Oil and Gas Projects		101
	3.1.1	Oil and Gas Project Life Cycle	105
3.2	Project Delivery Systems		110
	3.2.1	Types of Project Delivery Systems	110
		3.2.1.1 Design–build	115
		3.2.1.2 The Turnkey Contract	116
	3.2.2	Types of Contracting/Pricing	116
3.3	Management of Oil and Gas Project		118
	3.3.1	Project Management Processes	121

Contents

Chapter 4 Quality Tools for Oil and Gas Industry 129
    4.1   Introduction 129
    4.2   Categorization of Tools... 129
            4.2.1   Classic Quality Tools 129
                    4.2.1.1   Cause-and-Effect Diagram 130
                    4.2.1.2   Check Sheet 131
                    4.2.1.3   Control Chart 131
                    4.2.1.4   Flowchart 132
                    4.2.1.5   Histogram 132
                    4.2.1.6   Pareto Chart 132
                    4.2.1.7   Pie Chart 133
                    4.2.1.8   Run Chart.. 133
                    4.2.1.9   Scatter Diagram 135
                    4.2.1.10 Stratification... 135
            4.2.2   Management and Planning Tools 135
                    4.2.2.1   Activity Network Diagram 138
                    4.2.2.2   Affinity Diagram 140
                    4.2.2.3   Interrelationship Digraph........................... 143
                    4.2.2.4   Matrix Diagram 145
                    4.2.2.5   Prioritization Matrix 147
                    4.2.2.6   Process Decision Program........................ 147
                    4.2.2.7   Tree Diagram 147
            4.2.3   Process Analysis Tools 147
                    4.2.3.1   Benchmarking... 147
                    4.2.3.2   Cause-and-Effect Diagram 150
                    4.2.3.3   Cost of Quality.. 150
                    4.2.3.4   Critical to Quality 152
                    4.2.3.5   Failure Mode and Effects Analysis 152
                    4.2.3.6   5 Whys Analysis 152
                    4.2.3.7   5W2H 152
                    4.2.3.8   Process Mapping/Flowcharting................. 152
            4.2.4   Process Improvement Tools 152
                    4.2.4.1   Root Cause Analysis................................. 155
                    4.2.4.2   PDCA Cycle.. 155
                    4.2.4.3   SIPOC Analysis.. 155
                    4.2.4.4   Six Sigma DMAIC 158
                    4.2.4.5   Failure Mode and Effects Analysis 158
                    4.2.4.6   Statistical Process Control........................ 158
            4.2.5   Innovation and Creative Tools...................................... 160
                    4.2.5.1   Brainstorming... 160
                    4.2.5.2   Delphi Technique...................................... 162
                    4.2.5.3   5W2H 163
                    4.2.5.4   Mind Mapping 163
                    4.2.5.5   Nominal Group Technique 163
                    4.2.5.6   Six Sigma DMADV 164
                    4.2.5.7   Triz.. 164

		4.2.6	Lean Tools 164

- 4.2.6 Lean Tools 164
  - 4.2.6.1 Cellular Design 164
  - 4.2.6.2 Concurrent Engineering 165
  - 4.2.6.3 5S 166
  - 4.2.6.4 Just in Time 166
  - 4.2.6.5 Kanban 167
  - 4.2.6.6 Kaizan 167
  - 4.2.6.7 Mistake Proofing 167
  - 4.2.6.8 Outsourcing 170
  - 4.2.6.9 Poka-Yoke 171
  - 4.2.6.10 Single Minute Exchange of Die 171
  - 4.2.6.11 Value Stream Mapping 171
  - 4.2.6.12 Visual Management 172
  - 4.2.6.13 Waste Reduction 172
- 4.2.7 Cost of Quality 172
  - 4.2.7.1 Introduction 172
  - 4.2.7.2 Categories of Costs 172
  - 4.2.7.3 Quality Cost in Construction 173
- 4.2.8 Quality Function Deployment 175
- 4.2.9 Six Sigma 177
  - 4.2.9.1 Introduction 177
  - 4.2.9.2 Six Sigma Methodology 178
  - 4.2.9.3 Analytic Tool Sets 182
- 4.2.10 Triz 187
  - 4.2.10.1 TRIZ Methodology 188
  - 4.2.10.2 Application of TRIZ 189
  - 4.2.10.3 TRIZ Process 190

4.3 Tools for Construction Project Development 190
- 4.3.1 Tools for Study Stage 193
- 4.3.2 Tools for Design Stage 201
- 4.3.3 Tools for Bidding and Tendering Stage 204
- 4.3.4 Tools for EPC Stage 204

Chapter 5 HSE in Oil and Gas Projects 213

5.1 Health, Safety, and Environmental Management 213
- 5.1.1 Total Safety Management 216
- 5.1.2 Health, Safety, and Environmental Standards 219
  - 5.1.2.1 ISO 14000 Environmental Management System 219
  - 5.1.2.2 Occupational Health and Safety Assessment Series 18000 220
  - 5.1.2.3 ISO 45000 220

5.2 Health, Safety, and Environmental Management System 231
- 5.2.1 Health, Safety, and Environmental Management Plan 232

5.3	Major Causes of Site Accidents	234
5.4	Preventive Measures	237

SECTION 2 Quality Management Principles, Procedures, Concepts, and Methods in Oil and Gas Projects

Chapter 6 Overview of Quality in Oil and Gas Projects 247

- 6.1 History of Quality 247
- 6.2 Total Quality Management 250
- 6.3 Construction Projects 251
  - 6.3.1 Process-Type Projects 252
  - 6.3.2 Non–Process-Type Projects 252
- 6.4 Life Cycle of Construction Project 253
- 6.5 Quality Definition of Construction Projects 255
  - 6.5.1 Construction Quality and Total Quality Management 261
- 6.6 Quality Principles in Oil and Gas Projects 264
- 6.7 Quality Management System 266

Chapter 7 Selection of Project Teams 271

- 7.1 Project Team 271
- 7.2 Construction Project Stakeholders 274
- 7.3 Selection of Team Member 276
  - 7.3.1 Short-Listing of Team Members 280
    - 7.3.1.1 Bidding Procedure 283
  - 7.3.2 Selection of Design Team 284
    - 7.3.2.1 Request for Proposal 284
  - 7.3.3 Selection of EPC Contractor 290
    - 7.3.3.1 Request for Information 292
    - 7.3.3.2 Request for Qualification/Prequalification Questionnaire 292
    - 7.3.3.3 Request for Quotation 296
  - 7.3.4 Selection of Project Management Consultant 297
  - 7.3.5 Selection of Construction Supervisor 304
  - 7.3.6 Selection of Specialist Consultant 304
  - 7.3.7 Selection of Subcontractor/Subconsultant 308

Chapter 8 Quality Management during Oil and Gas Project Phases 309

- 8.1 Introduction 309
- 8.2 Feasibility Study Phase 312

	8.2.1	Project Initiation	314
	8.2.2	Identification of Need	314
	8.2.3	Feasibility Study	315
	8.2.4	Project Goals and Objectives	317
	8.2.5	Identification of Alternatives/Options	318
		8.2.5.1 Analyze Alternatives	318
		8.2.5.2 Select Preferred Alternative	319
	8.2.6	Finalize Project Delivery and Contracting System	319
	8.2.7	Project Charter	320
8.3	Concept Design Phase		320
	8.3.1	Select Design Team Members	321
	8.3.2	Identify Project Stakeholders	322
		8.3.2.1 Develop Responsibility Matrix	324
	8.3.3	Establish Concept Design Phase Requirements	324
		8.3.3.1 Identify Concept Design Deliverables	324
	8.3.4	Develop Concept Design	326
		8.3.4.1 Collect Data/Information	327
		8.3.4.2 Collect Owner's Requirements	328
		8.3.4.3 Collect Regulatory Requirements	328
	8.3.5	Prepare Concept Design	328
		8.3.5.1 Concept Design Drawings	328
		8.3.5.2 Concept Design Report	329
	8.3.6	Prepare Preliminary Schedule	329
	8.3.7	Estimate Conceptual Cost	331
	8.3.8	Establish Project Quality Requirements	335
		8.3.8.1 Manage Concept Design Quality	339
	8.3.9	Estimate Resources	339
	8.3.10	Identify Project Risks	341
	8.3.11	Identify HSE Requirements	343
	8.3.12	Review Concept Design	343
	8.3.13	Finalize Concept Design	345
	8.3.14	Submit Concept Design Package	345
8.4	FEED Phase		346
	8.4.1	Identify Design Team Members	347
	8.4.2	Identify Project Stakeholders	347
		8.4.2.1 Develop Responsibility Matrix	349
	8.4.3	Establish FEED Requirements	349
		8.4.3.1 Identify FEED Deliverables	349
	8.4.4	Develop FEED	350
		8.4.4.1 Information Collection	350
		8.4.4.2 Data Collection	355
		8.4.4.3 Technical Studies	355
	8.4.5	Prepare FEED	357
		8.4.5.1 FEED Drawings	357
		8.4.5.2 FEED Reports	358
	8.4.6	Prepare Project Schedule/Plans	358

		8.4.7	Evaluate and Estimate Project Cost	358
		8.4.8	Manage FEED Quality	359
		8.4.9	Estimate Resources	361
		8.4.10	Develop Communication Plan	361
		8.4.11	Manage Risks	361
		8.4.12	Manage HSE Requirements	362
		8.4.13	Develop FEED Documents	362
			8.4.13.1 Specifications	362
			8.4.13.2 Material Take Off List	362
			8.4.13.3 Contract Documents	362
		8.4.14	Perform VE	362
			8.4.14.1 Update FEED	366
		8.4.15	Review FEED	366
		8.4.16	Finalize FEED	366
		8.4.17	Develop Tender Documents	366
		8.4.18	Submit FEED Package	371
	8.5	Bidding and Tendering Phase		371
		8.5.1	Organize Tender Documents	372
		8.5.2	Identify Stakeholders/Project Team	372
		8.5.3	Identify Tendering Procedure	373
			8.5.3.1 Define Bidder Selection Procedure	373
			8.5.3.2 Establish Bid Review Procedure	373
		8.5.4	Identify Bidders	373
			8.5.4.1 Prequalify Bidders	373
		8.5.5	Manage Tendering Process	374
			8.5.5.1 Advertise Tender	374
			8.5.5.2 Distribute Tender Documents	374
			8.5.5.3 Conduct Prebid Meeting	374
			8.5.5.4 Submit/Receive Bids	374
		8.5.6	Manage Bidding and Tendering Phase Quality	375
		8.5.7	Manage Risks	376
		8.5.8	Review Bid Documents	377
			8.5.8.1 Evaluate Bids	377
			8.5.8.2 Select Contractor	377
		8.5.9	Award Contract	377
	8.6	Construction Phase		377
		8.6.1	Development of Project Site Facilities	378
			8.6.1.1 Notice to Proceed	380
			8.6.1.2 Kick-off Meeting	381
			8.6.1.3 Mobilization	381
		8.6.2	Identify Stakeholders	384
			8.6.2.1 Identify Owner's Representative/ Project Manager	384
			8.6.2.2 Identify Contractor's Project Team Members	384
			8.6.2.3 Identify PMC Team	390

		8.6.2.4	Identify Supervision Team..........................390
		8.6.2.5	Identify Regulatory Authorities................390
		8.6.2.6	Identify Subcontractors.............................390
		8.6.2.7	Develop Responsibility Matrix.................395
	8.6.3	Establish Construction Phase Requirements............395	
		8.6.3.1	Identify Contract Requirements 396
		8.6.3.2	Develop Construction Phase Scope 397
	8.6.4	Project Planning and Scheduling............................401	
		8.6.4.1	Develop Contractor's Construction Schedule..403
		8.6.4.2	Develop Project S-Curve 405
		8.6.4.3	Develop Contractor's Quality Control Plan 409
	8.6.5	Develop Management Plans....................................409	
		8.6.5.1	Develop Stakeholder Management Plan... 412
		8.6.5.2	Develop Resource Management Plan 412
		8.6.5.3	Develop Communication Management Plan 415
		8.6.5.4	Develop Risk Management Plan............... 418
		8.6.5.5	Develop Contract Management Plan 423
		8.6.5.6	Develop HSE Management Plan...............426
	8.6.6	Execute Project Works 428	
		8.6.6.1	Develop Detailed Engineering 428
		8.6.6.2	Develop Procurement Documents 439
		8.6.6.3	Execute Construction/Installation Works... 446
		8.6.6.4	Monitor and Control Project Works 473
		8.6.6.5	Validate Executed Works..........................494
8.7	Testing, Commissioning, and Handover Phase 495		
	8.7.1	Identify Testing and Start-up Requirements 495	
	8.7.2	Identify Stakeholders..497	
		8.7.2.1	Select Team Members..............................497
		8.7.2.2	Develop Responsibility Matrix.................497
	8.7.3	Develop Scope of Work..497	
	8.7.4	Develop Testing, Commissioning, and Handover Plan 499	
	8.7.5	Execute Testing and Commissioning Works............500	
	8.7.6	Manage Testing and Commissioning Quality..........501	
	8.7.7	Develop Documents 502	
		8.7.7.1	As-Built Drawings 503
		8.7.7.2	Technical Manuals and Documents..........503
		8.7.7.3	Record Books..503
		8.7.7.4	Warranties and Guarantees.......................503
	8.7.8	Monitor Work Progress 504	
	8.7.9	Train Owner's/End User's Personnel 504	
	8.7.10	Handover the Project 504	
		8.7.10.1	Obtain Authorities' Approval 504

			8.7.10.2	Handover Spare Parts 504

		8.7.11	8.7.10.3 Accept/Takeover the Project..................... 505 Close Contract 506
			8.7.11.1 Prepare Punch List 508
			8.7.11.2 Prepare Lesson Learned 508
			8.7.11.3 Issue Substantial Completion Certificate ... 509
			8.7.11.4 Terminate/Demobilize Team Members 509
		8.7.12	Settle Payments 509
		8.7.13	Settle Claims 509

Chapter 9 Operation and Maintenance of Oil and Gas Projects 511

9.1 Operation Management 511
9.2 Maintenance Management 511
    9.2.1 Categories of Maintenance 512
        9.2.1.1 Scheduled Maintenance 512
        9.2.1.2 Preventive Maintenance 513
        9.2.1.3 Corrective or Breakdown Maintenance 513
        9.2.1.4 Predictive Testing and Inspection 513
        9.2.1.5 Routine Maintenance 514
        9.2.1.6 Replacement of Obsolete Items 514
        9.2.1.7 Shutdown Maintenance 514
9.3 Manage Operation and Maintenance Quality 514
    9.3.1 Total Quality Management in Operation
        and Maintenance 515
9.4 Operation and Maintenance Program 515

Bibliography 521

Index 523

Foreword

Today, quality of products and services is what makes the difference and creates the competitive advantage for any organization in any business sector.

Throughout the past decade, the competition for market share and world-class performance was not limited to products and services only but on business sectors and nations level as a result of the globalization that we are living in today. Therefore, we can understand why governments around the world are establishing a national quality programs and awards to encourage and promote the quality movement in every business sector such as healthcare, education, manufacturing, and construction, and in many other sectors.

Of course, the oil and gas industry is one important business sector that played a very important role in our life, and we cannot imagine our life today without it; and that what makes this book of a real value and importance.

Quality and safety are two important success factors for any organization working in oil and gas industry. For us in the Arabian Gulf region, oil and gas industry represents a very important part of our modern history, and our governments are committed to run this heavy and sophisticated industry with the highest quality and excellence standards.

This book was written to be a comprehensive and informative reference in quality management and its applications in oil and gas projects.

I believe it is important reading material and source of knowledge for managers, engineers, and professionals to improve the quality of projects, products, and processes related to this important industry.

The author of this book, Dr. Abdul Razzak Rumane, is a competent and professional quality expert who I know him. I worked with him for years in promoting quality movement in the Middle East region, and I highly appreciate his contribution to the field of quality. It was a pleasure and honor for me to write the foreword of his book.

Dr. Ayed Alamri
President
Saudi Quality Council

PRAISE FOR THIS WORK

As a longtime friend of Dr. Rumane, and knowing his commitment to quality in construction projects, it was only a matter of time for him to add a specific book on quality management for projects in the oil and gas industry. This book compliments his two previous books on construction project quality and auditing in construction projects. His extensive knowledge, experience, and expertise in managing construction projects from a strong quality perspective provide the reader with detailed knowledge and "how to" in the performance of quality management for projects in the oil and gas industry.

This book provides the reader with an in-depth approach on how to prepare a quality management system from an overview of quality in oil and gas projects, through project team selection, quality management during the project phases, and the operation and maintenance of the projects. This book is an important resource as it highlights a detailed project quality management method to use to aid in having a successful project in the oil and gas industry. The details in this book, effectively used as a management tool, will enhance the success of your project. This writing provides a detailed, first-hand perspective on a complete quality management system for projects in the oil and gas industry. This book is truly relevant and has a constructive approach for making management of oil and gas projects highly effective.

Any oil and gas construction project management team, who utilize the knowledge in this book, will develop an effective quality management program that improves on basic construction practices and establishes a true quality constructed project. This book also lists important details to consider for oil and gas construction projects. Also, this book contains tables and figures that support the writings and suggestions put forward in the book. Overall, this book offers detailed insight to oil and gas project quality management that any project management team will find useful and informative. The book is well organized, practical, and clearly written. This book will be a great asset to any project management team working on oil and gas projects.

John F. Mascaro
ASQ Fellow
Construction Quality Systems Consultant

Preface

Quality is a universal phenomenon and a matter of great concern throughout the recorded history of human kind. It was always the determination of the builders and makers of the products to ensure that their products meet the customers' desires. Those were the days where products were of "customized nature".

With the advent of globalization and competitive market, the emphasis on quality management has increased. Quality has become the most important single factor for the survival and success of any company. Customer demands for better products and services at the lowest possible costs have put tremendous pressure on developing and manufacturing organizations to improve the quality of products, services, and processes, in order to compete in the market and improve business results. It become important that construction projects be more qualitative, competitive, and economical to meet owner's expectations.

Construction projects are custom-oriented and custom design having specific requirements set by the customer/owner to be completed within finite duration and budget. Every project has elements that are unique. No two projects are alike. It is always the owner's desire that his project should be unique and better. To a great extent, each project has to be designed and built to serve a specified need. Construction project is a custom than a routine and repetitive business. Construction projects differ from manufacturing or production. Construction projects involve many participants including owners, designers, contractors, and many other professionals from construction-related industries. These participants are both influenced by and depend on each other in addition to "others players" who are part of the construction process. Therefore, construction projects have become more complex and technical, and extensive efforts are required to reduce rework and costs associated with time, materials, engineering, and so forth.

Oil and gas projects are categorized as process-type projects. Quality management in oil and gas projects has specific requirements considering the environmental and safety requirements and hazardous nature of material. Oil and gas projects are capital-intensive and process-oriented. Quality in oil and gas projects is not only the quality of products and equipment used in the construction but also the total management approach to complete the facility as per the scope of works to customer/owner satisfaction and to produce specified quantity and quality and to be completed within specified schedule and within the budget to meet owner's defined purpose.

Quality tools and techniques are very important to develop a comprehensive project quality system. Application of tools and techniques in oil and gas construction projects has a great influence on cost-effectiveness and achieving successful project performance.

Quality in oil and gas projects is achieved through application of various quality control tools and techniques, principles, procedures, concepts, methods, and their applications to various activities/components/subsystems at different phases of the life cycle of an oil and gas project to improve construction process to conveniently manage the project and make them more qualitative, competitive, and economical for operation.

The oil and gas industry has mainly three sectors:

1. Upstream
2. Midstream
3. Downstream

This book mainly discusses quality management in oil and gas projects under downstream sector.

Quality management in construction projects is different to that of manufacturing. Quality in oil and gas construction projects is not only the quality of products and equipment used in the construction but also the total management approach to complete the facility as per the scope of works to customer/owner satisfaction within the budget and to be completed within specified schedule to meet owner's defined purpose.

The life cycle of oil and gas projects is different than civil engineering projects. Turnkey or EPC (engineering, procurement, and construction) type of contract is used mainly for the process type of projects. The book provides, construction professional practitioners, designers (consultant), contractors, quality managers, project managers, construction managers involved in oil and gas projects and other stakeholders in construction industry, all the significant information about usage and applications of various tools and techniques in different phases of oil and gas project focusing through quality management processes to ensure successful completion of project ensuring proper design, detailed engineering, procurement, and construction for making the project most qualitative, competitive, and economical for safe operation and optimized performance.

Apart from managing quality in civil and architectural works of the project, this book also focuses on mechanical works such as piping, processing, storage facilities, and instrumentation and control that are essential parts of the process industry. The products/material/systems used in oil and gas construction projects are expensive, complex, immovable, and long-lived. These are normally produced by other construction-related industries/manufacturers. These industries produce products as per their own quality management practices complying with certain quality standards or against specific requirements for a particular project. The book also discusses procurement procedure for EPC contract. The book also includes managing quality during mobilization, procurement, execution, planning, scheduling, monitoring, control, quality, and testing to achieve the desired results for the project.

For the sake of proper understanding, this book is divided into nine chapters. Each chapter is divided into numbered sections covering quality-related topics which have importance or relevance to understand quality concepts for oil and gas projects.

Chapter 1 is an overview of quality. It discusses definition of quality and evolution of quality management system in different forms such as inspection, quality control, quality assurance, quality management, and total quality management. It also discusses quality definition for construction projects and three elements of quality management system: quality plan, quality assurance, and quality control. A brief about integrated quality management system is also discussed in this chapter. Risk management and supply chain management are also discussed in this chapter.

Preface   xxi

Chapter 2 is about quality management system. It gives a brief introduction to importance of standards and standardization bodies and ISO certification process. It covers brief information about importance of standards and standardization bodies. It also covers the development of QMS and contents of quality manual for EPC contractor.

Chapter 3 provides an overview of oil and gas projects. It discusses project lifecycles of oil and gas projects, which are developed based on systems engineering principles. Different types of project delivery systems and most common project delivery systems that are used in oil and gas project and their contractual relationship are also discussed in this chapter. It also discusses briefly about project management processes for construction projects.

Chapter 4 is about quality tools and their applications in oil and gas projects. It elaborates application of various tools and techniques for specific and appropriate usage at different phases of oil and gas projects starting from project initiation to testing, commissioning, and handover of the project.

Chapter 5 is HSE (Health, Safety, and Environment) in Oil and Gas Projects. This chapter discusses HSE management, total quality management approach to total safety management, different types of ISO standards that are used to monitor HSE activities, HSE management system, and how to develop HSE management plan. It also discusses major causes of construction accidents and the preventive measures to avoid accidents in construction projects.

Chapter 6 is an overview of quality in oil and gas industry. It discusses quality principles for oil and gas projects.

Chapter 7 is about project teams of oil and gas projects such as project management consultant (PMC), designer, and contractor. It discusses selection procedure of project teams.

Chapter 8 details quality management in oil and gas projects and discusses various concepts of quality, principles, methods, tools, and processes, which can be applied from the inception of the project till the issuance of substantial completion certificate. This chapter discusses owner's role while preparing project definition and terms of reference (TOR) for front-end engineering design (FEED) and EPC. It includes information regarding the designer's role and responsibilities to prepare FEED properly specifying important parameters and features of the required processes/services/systems for successful completion of project. Preparation of construction documents and bidding and tendering process are discussed in this chapter. It elaborates different procedures that contractor has to follow during EPC construction process including detailed design, procurement, and construction. This chapter also includes guidelines to contractor about preparation and submission of construction program, logs, product data, and engineering drawings. Various procedures and principles followed during procurement and construction are also discussed. The products/material/systems used in oil and gas construction projects are expensive, complex, immovable, and long-lived. These are normally produced by other construction-related industries/manufacturers. These industries produce products as per their own quality management practices complying with certain quality standards or against specific requirements for a particular project. The owner of the construction project or his representative has no direct control over these companies unless he/his representative/appointed contractor commit to buy their product for use in their facility. These organizations may have their own quality management program. In manufacturing or service industries, the quality management of all in-house manufactured products is performed by manufacturer's own team or under the control of same organization having jurisdiction over their manufacturing plants at different locations. Managing quality of vendor-supplied items/products and subcontracted items has great importance in oil and gas industry. Apart from managing quality in civil and architectural works of the project, it mainly focuses on mechanical works such as piping, processing, storage facilities, and instrumentation and control that are essential parts of the process industry. The chapter also includes managing quality during mobilization, procurement, execution, planning, scheduling, monitoring, control, quality, and testing to achieve the desired results for the project. It also discusses quality system during testing, commissioning, and handover of the project.

Chapter 9 consists of brief information of post-handover activities such as operation and maintenance.

I am certain, this book will meet the requirements of construction professionals in oil and gas industry, quality professionals in oil and gas industry, project owners, students, and academics and satisfy their needs.

Acknowledgments

Share the knowledge with others is the motto of this book.

Many of colleagues and friends extended help while preparing the book by arranging reference material, many thanks to all of them for their support.

I thank publishers and authors, whose writings are included in this book, for extending their support by allowing me to reprint their material.

I thank reviewers, from various professional organizations, for their valuable input to improve my writing. I thank members of ASQ Audit Division, ASQ Design and Construction Division, The Institution of Engineers (India), IEI, Kuwait Chapter, and Kuwait Society of Engineers for their support to bring out this book.

I thank Cindy Renee Carelli, Executive editor of CRC Press; Erin Harris, Senior Editorial assistant; and other staff for their support and contribution to make this construction-related book a reality.

I thank Dr. Ayed Alamri, President, Saudi Quality Council, for his support and nicely worded thought-provoking Foreword.

I thank Mr. John F. Mascaro, former chair, ASQ Design and Construction Division and ASQ Audit Division, for nicely worded praise for this book.

I thank Dr. Adedeji B. Badiru, series editor, CRC Press and Mr. Cliff Moser, former chair, ASQ Design and Construction Division for their best wishes all the time.

I thank Mr. Raymond R. Crawford of Parsons Brinckerhoff and former chair, ASQ Design and Construction Division, for his valuable and timely suggestions during my writing.

I thank Engr. Adel Kharafi, former chairman, Kuwait Society of Engineers, and former president of World Federation of Engineering Organizations (WFEO) for his good wishes all the time. I thank Engr. Ahmad Almershed, former Undersecretary, MSNA, Kuwait, for his good wishes all the time. I thank Engr. Ahmad Alkandari, Minister's Consultant, Kuwait Municipality, for his support and good wishes. I thank Dr. Fadal Safar, former minister of Public Works, Kuwait, member of Board of Directors, Kuwait Petroleum Corporation, for his support and good wishes. I thank Engr. Faisal D. Alatel, chairman, Kuwait Society of Engineers, and president of Federation of Arab Engineers for his support and good wishes. I thank Prof. Mohammed Aichouni, University of Hail, KSA, for his support and good wishes all the time. I thank Dr. Mohammad Ben Salamah, chair, ASQ Kuwait Section, for his support and best wishes. I thank Dr. N. N. Murthy of Jagruti Kiran Consultants, for his support and good wishes. I thank Dr. Othman Alshamrani, Imam Abdulrehman Bin Faisal University, KSA, for his support and good wishes all the time. I thank Ms. Rima, Al Awadhi of Kuwait Oil Company, for her support and good wishes. I thank Engr. Wael, Aljasem of Kuwait Project Management Society, for his support and good wishes. I thank Engr. Yaseen Farraj, former director, Ministry of Public Works, Kuwait, for his support and good wishes.

I thank Engr. Ashraf Hassan Malim of JGC, Engr. Ateeq Mirza of KOC, Engr. Babar Mirza of GOFSCO, Kuwait, Mr. Bashir Ibrahim Parkar of Dar SSH International, Engr. CH. Rama Krushna Chary of Kuwait Oil Company, Engr.

Ganeshan Swaminathan of Dar SSH International, Mr. Noor Mohammed of Strytech A.I. Kuwait, Engr. Syed Fadal Nizar of Kuwait Oil Company, Engr. Ved Prakash Tewari of KAFCO, Engr. D.M. Tripathi of Alghanim International, Mr. Shabbir Chafekar of HOT Engineering and Construction, and Engr. Sivannarayana Edupuganti of Kuwait Oil Company for their valuable input and support.

I extend my thanks to Dr. Ted Coleman, professor and department chair, California State University, San Bernardino, and former chancellor KW University, for his everlasting support.

My special thanks to H.E. Sheikh Rakan Nayef Jaber Al Sabah for his support and good wishes.

I thank my well-wishers whose inspiration made me to complete this book.

Most of the data discussed in this book are from author's practical and professional experience and are accurate to the best of author's knowledge and ability. However, if any discrepancies are observed in the presentation, I would appreciate communicating them to me.

The contribution of my son Ataullah, my daughter Farzeen, and daughter-in-law Masum is worth mentioning here. They encouraged me and helped me in my preparatory work to achieve the final product. I thank my brothers, sisters, and all the family members for their support, encouragement, and good wishes all the time.

Abdul Razzak Rumane

Author

Abdul Razzak Rumane, PhD, is a chartered quality professional-fellow of The Chartered Quality Institute (UK) and a certified consultant engineer in the field of project management and electrical engineering. He obtained a Bachelor of Engineering (Electrical) degree from Marathwada University (now Dr. Babasaheb Ambedkar Marathwada University), India, in 1972 and received his PhD from Kennedy Western University, USA (now Warren National University) in 2005. His dissertation topic was "Quality Engineering Applications in Construction Projects". Dr. Rumane's professional career exceeds 48 years including 10 years in manufacturing industries and over 38 years in construction projects. Presently, he is associated with SIJJEEL Co., Kuwait, as advisor and director, Construction Management.

Dr. Rumane is associated with a number of professional organizations. He is also a Fellow of The Institution of Engineers (India) and has an honorary fellowship with the Chartered Management Association (Hong Kong). He is also a senior member of the Institute of Electrical and Electronics Engineers (USA), a senior member of the American Society for Quality, a member of Kuwait Society of Engineers, a member of SAVE International (The Value Society), and a member of the Project Management Institute. He is also an associate member of the American Society of Civil Engineers, a member of the London Diplomatic Academy, a member of the International Diplomatic Academy, and a member of the Board of Governors of the International Benevolent Research Forum.

As an accomplished engineer, Dr. Rumane has been awarded an honorary doctorate in engineering from The Yorker International University, USA (2007). The World Quality Congress awarded him "Global Award for Excellence in Quality Management and Leadership". The Albert Schweitzer International Foundation honored him with gold medal for "Outstanding contribution in the field of Quality in Construction Projects" and "Outstanding contribution in the field of electrical engineering/consultancy in construction projects in Kuwait". In 2009, he was was selected as one of the Top 100 Engineers of IBC (International Biographical Centre, Cambridge, UK). The European Academy of Informatization also honored him with "World Order of Science-Education-Culture" and a title of "Cavalier", and The Sovereign Order of the Knights of Justice, England, honored him Meritorious Service Medal.

Dr. Rumane has attended many international conferences and has made technical presentations at various conferences. Dr. Rumane is an author of books titled *Quality Management in Construction Projects*, First Edition (2010); *Quality Tools for Managing Construction Projects* (2013); and *Quality Management in Construction Projects*, Second Edition (2017) and an editor of the book titled *Handbook of*

Construction Management: Scope, Schedule, and Cost Control (2016). All of these books are published by CRC Press/Taylor & Francis Group, USA. A book titled *Quality Auditing in Construction Projects: A Handbook* (2019) is published by Routledge/Taylor & Francis Group, UK. His book *Quality Management in Construction Projects* is translated into Korean language.

In 2020, he was the nominating chair of ASQ Kuwait Section for the year 2020. He was secretary ASQ GC in 2019 and secretary ASQ LMC, Kuwait, in 2017 and 2018.

He was honorary chairman of The Institution of Engineers (India), Kuwait Chapter, in 2005–2007, 2013, 2014, and 2016–2017.

Abbreviations

AACE	American Association of Cost Engineers
AAMA	American Architectural Manufacturers Association
ACI	American Concrete Institute
AISC	American Institute of Steel Construction
AMCA	American Composite Manufacturers Association
ANSI	American National Standards Institute
API	American Petroleum Institute
ARI	American Refrigeration Institute
ASCE	American Society of Civil Engineers
ASHRAE	American Society of Heating, Refrigeration and Air-Conditioning Engineers
ASQ	American Society for Quality
ASTM	American Society of Testing Materials
BMS	Building Management System
BREEAM	Building Research Establishment Environmental Assessment
BSI	British Standard Institute
CD	Concept design
CDM	Construction (Design and Management)
CEN	European Committee for Standardization
CIBSE	Chartered Institution of Building Services Engineers
CIE	International Commission on Illumination
CII	Construction Industry Institute
CMAA	Construction Management Association of America
CSC	Construction Specifications, Canada
CSI	Construction Specification Institute
CTI	Cooling tower industry
DIN	Deutsches Institute für Normung
EIA	Environmental impact assessment
EJCDC	Engineering Joint Contract Document Committee
EN	European norms
FEED	Front-end engineering design
FIDIC	Fédération Internationale des Ingénieurs-Conseils
HAZID	Hazard identification
HAZOP	Hazard and operability
HQE	Higher-quality environmental
HSE	Health, safety, and environment
ICE	Institute of Civil Engineers (United Kingdom)
IEC	International Electrotechnical Commission
IEEE	Institute of Electrical and Electronics Engineers
IP	Ingress Protection
ISO	International Organization for Standardization
LEED	Leadership in Energy and Environmental Design

MTO	Material takeoff
NEC	National Electrical Code
NEC	New Engineering Contract
NEMA	National Electrical Manufacturers Association (USA)
NFPA	National Fire Protection Association
NWWDA	National Wood Window and Door Association
OH&S	Occupational health and safety
PHSER	Procedure for Project HSE Review
PMBOK	Project Management Book of Knowledge
PMC	Project management consultant
PMI	Project Management Institute
QMS	Quality management system
QS	Quantity surveyor
RFID	Radio frequency identification
SDI	Steel Door Institute
TIA	Telecommunication Industry Association
UL	Underwriters Laboratories

Synonyms

Consultant	Architect/engineer (A/E), designer, design professionals, designer, consulting engineers, supervision Professional
Contractor	Constructor, builder, EPC contractor
Engineer	Resident project representative
Engineer's representative	Resident engineer
Main contractor	General contractor
Owner	Client, employer
Project charter	Terms of reference (TOR), client brief, definitive project brief
Project manager	Construction manager
Quantity surveyor	Cost estimator, contract attorney, cost engineer, cost and works superintendent

Section 1

Quality Tools and Techniques in Oil and Gas Projects

1 Overview of Quality

1.1 HISTORY OF QUALITY

Quality issues have been of great concern throughout the recorded history of humans. During the New Stone Age, several civilizations emerged, and some 4000–5000 years ago, considerable skills in construction were acquired. The pyramids in Egypt were built approximately 2589–2566 BCE. Hammurabi, the king of Babylonia (1792–1750 BCE), codified the law, according to which, during the Mesopotamian era, builders were responsible for maintaining the quality of buildings and were given the death penalty if any of their construction collapsed and their occupants were killed. The extension of Greek settlements around the Mediterranean after 200 BCE left records showing that temples and theaters were built using marble. India had strict standards for working in gold in the fourth century BCE.

China's recorded quality history can be traced back to earlier than 200 BCE. China had instituted quality control in its handicrafts during the Zhou dynasty between 1100 and 250 BCE. During this period, the handicraft industry was mainly engaged in producing ceremonial artifacts. This industry survived the long succession of dynasties that followed up to 1911 CE.

During the Middle Ages, guilds took the responsibility for quality control upon themselves. Guilds and governments carried out quality control; consumers, of course, carried out informal quality inspection throughout history.

The guilds' involvement in quality was extensive. All craftsmen living in a particular area were required to join the corresponding guild and were responsible for controlling the quality of their own products. If any of the items was found defective, then the craftsman discarded the faulty items. The guilds also initiated punishments for members who turned out shoddy products. They maintained inspections and audits to ensure that artisans followed quality specifications. The guild hierarchy consisted of three categories of workers: apprentice, journeyman, and master. The guilds had established specifications for input materials, manufacturing processes, and finished products, as well as methods of inspection and testing. They were active in managing quality during Middle Ages until the Industrial Revolution marginalized their influence.

1.2 DEFINITION OF QUALITY

Quality has different meanings for different people. The American Society for Quality (ASQ) glossary defines quality as follows:

A subjective term for which each person has his or her own definition. In technical usage, quality can have two meanings:

1. The characteristics of a product or service that bear on its ability to satisfy stated or implied needs.
2. A product or service free of deficiencies.

It further states that it is

- Based on customers' perceptions of a product's design and how well the design matches the original specifications.
- The ability of a product and service to satisfy stated or implied needs.
- Achieved by conforming to established requirements within an organization.

The International Organization for Standardization (ISO), defines quality as "the totality of characteristics of an entity that bears on its ability to satisfy stated or implied needs".

Pyzdek (1999) views that there is no single generally accepted definition of quality. He has quoted five principal approaches to defining quality that have been described by Garvin (1988). These are as follows:

1. *Transcendent*—"Quality cannot be defined, you know what it is." (Persig 1974, p. 213)
2. *Product-based*—"Differences in quality amount to differences in the quantity of some desired ingredient or attribute." (Abbott 1955, pp. 126–127)

    Quality refers to the amounts of the unpriced attributes contained in each unit of the priced attribute. (Leflore 1982, p. 952)

3. *User-based*—"Quality consists of the ability to satisfy wants." (Edwards 1968, p. 37)

    In the final analysis of the market place, the quality of a product depends on how well it fits patterns of consumer preference. (Kuehn and Day 1954, p. 831).

    Quality is fitness for use. (Juran 1974, p. 2-2)

4. *Manufacturing-based*—"Quality (means) conformance to requirements." (Crosby 1979, p. 15)0

    Quality is the degree to which a specific product conforms to a design or specification. (Gilmore 1974, p. 16)

5. *Value-based*—"Quality is the degree of excellence at an acceptable price and the control of variability at an acceptable cost." (Broh 1982, p. 3)

    Quality means best 'for certain customer conditions.' These conditions are (a) the actual use, and (b) the selling price of the product. (Feigenbaum 1991, pp. 1, 25)

The aforementioned definitions can further be summarized under the name of those contributors to the quality movement whose philosophies, methods, and tools have been proved useful in quality practices. They are called the "quality gurus". Their definitions of quality are as follows:

1. *Philip B. Crosby*—Conformance to requirements not as "goodness" nor "elegance".
2. *W. Edward Deming*—Quality should be designed into both product and the process.
3. *Armand V. Feigenbaum*—Best for customer use and selling price.
4. *Kaoru Ishikawa*—Quality of the product as well as after-sales services, quality of management, the company itself, and the human being.
5. *Joseph M. Juran*—Quality is fitness for use.
6. *John S. Oakland*—Quality is meeting customer's requirements.

Based on these definitions, it is possible to evolve a common definition of quality, which is mainly related to the manufacturing, processes, and service industries as follows:

- Meeting the customer's need
- Fitness for use
- Conforming to requirements

1.3 EVOLUTION OF QUALITY MANAGEMENT SYSTEM

The Industrial Revolution began in Europe in the mid-19th century. It gave birth to factories, and the goals of the factories were to increase productivity and reduce costs. Prior to the Industrial Revolution, items were produced by an individual craftsman for individual customers, and it was possible for workers to control the quality of their products. Working conditions then were more conducive to professional pride. Under the factory system, the tasks needed to produce a product were divided among several or many factory workers. Under this system, large groups of workmen were performing similar types of work, and each group was working under the supervision of a foreman who also took on the responsibility of controlling the quality of the work performed. Quality in the factory system was ensured by means of skilled workers, and the quality audit was done by inspectors.

The broad economic result of the factory system was mass production at low costs. The Industrial Revolution changed the situation dramatically with the introduction of a new approach to manufacturing.

The beginning of the 20th century marked the inclusion of process in quality practices. During World War I, the manufacturing process became more complex. Production quality was the responsibility of quality control departments. The introduction of mass production and piecework created quality problems as workmen were interested in earning more money by the production of extra products, which in turn led to bad workmanship. This situation made factories introduce full-time quality inspectors, which marked the real beginning of inspection quality control and thus the introduction of quality control departments headed by superintendents. Walter Shewhart introduced statistical quality control (SQC) in the process. His concept was that quality is not relevant to the finished product but to the process that created the product. His approach to quality was based on continuous monitoring of process variation. The SQC concept freed manufacturers from the time-consuming 100% quality control system because it accepted that variation is tolerable up to certain control limits. Thus, the quality control focus shifted from the end of line to the process.

The systematic approach to quality in industrial manufacturing started during the 1930s when some attention was given to the cost of scrap and rework. With the impact of mass production, which was required during World War II, it became necessary to introduce a more stringent form of quality control. This was instituted by manufacturing units and was identified as SQC. SQC made a significant contribution in that it provided a sampling rather than 100% product inspection. However, SQC was instrumental in exposing the underappreciation of the engineering of product quality.

Harold Kerzner (2001) has given the quality history of the past 100 years:

> During the past 100 years the views of quality have changed dramatically. Prior to World War I, quality was viewed predominantly as inspection, sorting out the good items from the bad. Emphasis was on problem identification. Following World War I and up to the early 1950s, emphasis was still on sorting good items from bad. However, quality control principles were now emerging in the form of
>
> - Statistical and mathematical techniques.
> - Sampling tables.
> - Process control charts.

He further states that, from the early 1950s to the late 1960s, quality control evolved into quality assurance, with its emphasis on problem avoidance rather than problem detection. Additional quality assurance principles emerged, such as:

- The cost of quality
- Zero-defect programs
- Reliability engineering
- Total quality control

Kerzner (2001) has gone further, saying that

> Today, emphasis is being placed on strategic quality management, including such topics as
>
> - Quality is defined by the customer.
> - Quality is linked with profitability on both the market and cost sides.
> - Quality has become a competitive weapon.
> - Quality is now an integral part of the strategic planning process.
> - Quality requires an organization wide commitment. (p. 1087)

Thomas Pyzdek (1999) has stated that, in the past century, quality has moved through four distinct "quality eras": inspection, statistical quality control, quality assurance, and strategic quality management. A fifth era is—emerging—complete integration of quality into the overall business system. Managers in each era were responding to the problems they faced at the time (p. 12).

Overview of Quality

From the foregoing writings and many others on the history of quality, it is evident that the quality system in its different forms has moved through distinct quality eras such as:

1. Quality inspection
2. Quality control
3. Quality assurance
4. Quality engineering
5. Quality management

1.3.1 Quality Inspection

Prior to the Industrial Revolution, items were produced by an individual craftsman, who was responsible for material procurement, production, inspection, and sales. In case any quality problems arose, the customer would take up issues directly with the producer. The Industrial Revolution provided the climate for continuous quality improvement. In the late 19th century, Frederick Taylor's system of scientific management was born. It provided the backup for the early development of quality management through inspection. At the time when goods were produced individually by craftsmen, they inspected their own work at every stage of production and discarded faulty items. When production increased with the development of technology, scientific management was born out of a need for standardization rather than craftsmanship. This approach required each job to be broken down into its component tasks. Individual workers were trained to carry out these limited tasks, making craftsmen redundant in many areas of production. The craftsmen's tasks were divided among many workers. This also resulted in mass production at lower cost, and the concept of standardization started resulting in interchangeability of similar types of bits and pieces of product assemblies. One result of this was a power shift away from workers and toward management.

With this change in the method of production, inspection of the finished product became the norm rather than inspection at every stage. This resulted in wastage because defective goods were not detected early enough in the production process. Wastage added costs that were reflected either in the price paid by the consumer or in reduced profits. Due to the competitive nature of the market, there was pressure on manufacturers to reduce the price for consumers, which in turn required cheaper input prices and lower production costs. In many industries, emphasis was placed on automation to try to reduce the costly mistakes generated by workers. Automation led to greater standardization, with many designs incorporating interchanges of parts. The production of arms for the 1914–1918 war accelerated this process.

An inspection is a specific examination, testing, and formal evaluation exercise and overall appraisal of a process, product, or service to ascertain if it conforms to established requirements. It involves measurements, tests, and gauges applied to certain characteristics in regard to an object or an activity. The results are usually compared with specified requirements and standards for determining whether the item or activity is in line with the target. Inspections are usually nondestructive. Some of the nondestructive methods of inspection are as follows:

- Visual
- Liquid dye penetrant
- Magnetic particle
- Radiography
- Ultrasonic
- Eddy current
- Acoustic emission
- Thermography

The degree to which inspection can be successful is limited by the established requirements. Inspection accuracy depends on:

1. Level of human error
2. Accuracy of the instruments
3. Completeness of the inspection planning

Human errors in inspection are mainly due to:

- Technique errors
- Inadvertent errors
- Conscious errors
- Communication errors

Most construction projects specify that all the contracted works are subject to inspection by the owner/consultant/owner's representative.

1.3.2 Quality Control

The quality control era started at the beginning of the 20th century. The Industrial Revolution had brought about the mechanism and marked the inclusion of process in quality practices. The ASQ termed the quality control era as process orientation that consists of product inspection and SQC.

Thomas Pyzdek (1999) has described the start of the quality control era as follows:

> The Inspection-based approach to quality was challenged by Walter A. Shewhart. In 1931, Shewhart's landmark book *Economic Control of Quality of Manufacturing* introduced the modern era of quality management. In 1924, Shewhart was part of a group working at Western Electric's Inspection Engineering Department of Bell Laboratories. Other members of the group included Harold Dodge, Harry Romig, G.D. Edwards, and Joseph Juran, a veritable "who's who" of the modern quality movement. (p. 13)

Pyzdek further states:

> Quality continued to evolve after World War II. Initially, few commercial forms applied the new, statistical approach. However, those companies that did, achieved spectacular

Overview of Quality

results, and the results were widely reported in the popular and business press. Interest groups, such as the Society of Quality Engineers (1945), began to form around the country. In 1946, the Society of Quality Engineers joined with other groups to form the American Society for Quality (ASQ). In July 1944, the Buffalo Society of Quality Engineers published *Industrial Quality Control*, the first journal devoted to the subject of management discipline. (p. 15)

According to Feigenbaum (1991), the definition of control in industrial terminology is "a process for delegating responsibility and authority for a management activity while retaining the means of assuring satisfactory results". He further states:

> The procedure for meeting the industrial goal is therefore termed quality 'Control,' just as the procedure for measuring production and cost goals are termed, respectively, production 'Control' and cost 'Control.' There are normally four steps in such control:
>
> 1. *Setting Standards.* Determining the required cost–quality, performance–quality, safety–quality, and reliability–quality standards for the products
> 2. *Appraising Conformance.* Comparing the conformance of the manufactured product, or the offered services to the standards
> 3. *Acting When Necessary.* Correcting problems and their causes throughout the full range of those marketing, design, engineering, production, and maintenance factors that influence user's satisfaction
> 4. *Planning of Improvements.* Developing a continuing effort to improve the cost, performance, safety, and reliability standards (p. 10)

Kerzner (2001) describes that

> Quality Control is a collective term for activities and techniques, within the process, that are intended to create specific quality characteristics. Such activities include continually monitoring process, identifying and eliminating problem causes, use of statistical process control to reduce the variability and to increase the efficiency of the process. Quality control certifies that the organization's quality objectives are being met (p. 1099).

Gryna (2001) refers to quality control as the process employed to consistently meet standards. The control process involves observing actual performances, comparing it with some standards, and then taking action if observed performance is significantly different from the standard. Control involves a universal sequence of steps as follows:

1. Choose the control subject, that is, choose what we intend to regulate.
2. Establish measurement.
3. Establish standard of performance, product goal, and process goals.
4. Measure actual performance.
5. Compare actual measured performance against standards.
6. Take action on the difference.

Juran (1999, p. 4.2) defines quality control as a universal managerial process for conducting operation so as to provide stability to prevent adverse change and to "maintain the status quo". To maintain stability, the quality control process evaluates actual performance, compares actual performance with goals, and takes action on the difference.

Chung (1999, p. 4) defines quality control as referring to the activities that are carried out on the production line to prevent or eliminate causes of unsatisfactory performance. In the manufacturing industry, including production of ready-mixed concrete and fabrication of precast units, the major functions of quality control are control of incoming materials, monitoring of the production process, and testing of the finished product.

From the foregoing, quality control can be defined as a process of analyzing data collected through statistical techniques to compare with actual requirements and goals to ensure its compliance with some standards.

Quality control in construction projects is performed at every stage through the use of various control charts, diagrams, checklists, and so on and can be defined as follows:

Checking of executed/installed works to confirm that works have been performed/executed as specified, using specified/approved materials, installation methods and specified references, codes, and standards to meet intended use
Controlling budget
Planning, monitoring, and controlling project schedule

A control chart is a graphical representation of the mathematical model used to detect changes in a parameter of the process. Charting statistical data is a test of the null hypothesis that the process from which the sample came has not changed. A control chart is employed to distinguish between the existence of a stable pattern of variation and the occurrence of an unstable pattern. If an unstable pattern of variation is detected, action may be initiated to discover the cause of the instability. Removal of the assignable cause should permit the process to return to stable state.

There are a variety of methods, tools, and techniques that can be applied for quality control and the improvement process. These are used to create an idea, engender planning, analyze the cause and the process, foster evaluation, and create a wide variety of situations for continuous quality improvement. These tools can also be used during various stages of a construction project.

1.3.3 QUALITY ASSURANCE

Quality assurance is the third era in the quality management system.

The ASQ defines quality assurance as "all the planned and systematic activities implemented within the quality system that can be demonstrated to provide confidence a product or service will fulfill requirements for quality".

The ASQ details this era:

After entering World War II in December 1941, the United States enacted legislation to help gear the civilian economy to military production. At that time, military contracts were typically awarded to manufacturers who submitted the lowest competitive bid. Upon delivery, products were inspected to ensure conformance to requirements.

Overview of Quality

During this period, quality became a means to safety. Unsafe military equipment was clearly unacceptable, and the armed forces inspected virtually every unit of product to ensure that it was safe for operation. This practice required huge inspection forces and caused problems in recruiting and retaining competent inspection personnel. To ease the problems without compromising product safety, the armed forces began to utilize sampling inspection to replace unit-by-unit inspection. With the aid of industry consultants, particularly the Bell Laboratories, they adapted sampling tables and published them in a military standard: Mil-Std-105. The tables were incorporated into the military contracts themselves. In addition to creating military standards, the armed forces helped their suppliers improve their quality by sponsoring training courses in Shewhart's statistical quality control (SQC) techniques. While the training led to quality improvements in some organizations, most companies had little motivation to truly integrate the techniques. As long as government contracts paid the bills, organizations' top priority remained meeting production deadlines. Most SQC programs were terminated once the government's contracts came to an end.

According to ISO 9000 (or BS 5750), quality assurance is "those planned and systematic actions necessary to provide adequate confidence that product or service will satisfy given requirements for quality". The ISO 8402-1994 defines quality assurance as "all the planned and systematic activities implemented within the quality system, and demonstrated as needed, to provide adequate confidence that an entity will fulfill requirements for quality".

The third era of quality management saw the development of quality systems and their application principally to the manufacturing sector. This was due to the impact of the following external environment upon the development take-up of quality systems at this time:

- Growing, and more significantly, maturing populations
- Intensifying competition

These converging trends contributed greatly to the demand for more, cheaper, and better quality products and services. The result was the identification of quality assurance schemes as the only solution to meet this challenge.

Harold Kerzner (2001) has defined quality assurance as the collective term for the formal activities and managerial processes that are planned and undertaken in an attempt to ensure that products and services are delivered at the required quality level. Quality assurance also includes efforts external to these processes that provide information for improving the internal processes. It is the quality assurance function that attempts to ensure that the project scope, cost, and time function are fully integrated (p. 1098).

Brian Thorpe, Peter Sumner, and John Duncan (1996) have described quality assurance as the evolution of QA from techniques of final inspection in the 1930s, followed by quality control, mainly in the manufacturing industries, during the 1940s and 1950s, and then a further extension of controls into the engineering/design phases of these industries during the 1960s.

They further state:

> The term "quality assurance" unfortunately tends to make people think of the finished product and services, whereas it is something far greater; in fact, today's quality system is not something imposed on top of other business systems; it is the system of the business. Our definition of quality assurance is a structural approach to business management and control, which enhances the ability to consistently provide products and services to specification, program and cost. (p. 9)

Quality assurance is the activity of providing evidence to establish confidence among all concerned that quality-related activities are being performed effectively. All these planned or systematic actions are necessary to provide adequate confidence that a product or service will satisfy given requirements for quality.

Quality assurance covers all activities from design, development, production/construction, installation, and servicing to documentation, and also includes regulations of the quality of raw materials, assemblies, products, and components; services related to production; and management, production, and inspection processes.

Quality assurance in construction projects covers all activities performed by the design team, contractor, and quality controller/auditor (supervision staff) to meet owners' objectives as specified and to ensure that the project/facility is fully functional to the satisfaction of the owners/end users.

1.3.4 Quality Engineering

Feigenbaum (1991) defines quality engineering technology as "the body of technical knowledge for formulating policy and for analyzing and planning product quality in order to implement and support that quality system which will yield full customer satisfaction at minimum cost" (p. 234).

Feigenbaum (1991) has further elaborated the entire range of techniques used in quality engineering technology by grouping them under three major headings:

1. *Formulating of quality policy.* Included here are techniques for identifying the quality objectives and quality policy of a particular company as a foundation for quality analysis and systems implementation.
2. *Product-quality analysis.* Techniques for analyzing include those for isolating and identifying the principal actors that relate to the quality of the product in its served market. These factors are then studied for their effects toward producing the desired quality result.
3. *Quality operations planning.* Techniques for implementing the quality system emphasize the development in advance of a proposed course of action and methods for accomplishing the desired quality result. These are the quality planning techniques underlying—and required by—the documentation of key activities of the quality system. (p. 237)

1.3.5 Quality Management

The ASQ glossary defines quality management as "the application of quality management system in managing a process to achieve maximum customer satisfaction at the lowest overall cost to the organization while continuing to improve the process".

Thomas Pyzdek (1999) has described the evolution of the quality management concept:

> The quality assurance perspective suffers from a number of serious shortcomings. Its focus is internal. Specifications are developed by the designers, often with only a vague idea of what customers really want. The scope of quality assurance is generally limited to those activities under the direct control of the organization; important activities such as transportation, storage, installation, and service are typically either ignored or given little attention. Quality assurance pays little or no attention to the competition's offerings. The result is that quality assurance may present a rosy picture, even while quality problems are putting the firm out of business. Such a situation existed in the United States in the latter 1970s.
>
> The approaches taken to achieve the quality edge vary widely among different firms. Some quality leaders pioneer and protect their positions with patents or copyrights. Others focus on relative image or service. Some do a better job of identifying and meeting the needs of special customer segments. And others focus on value-added operations and technologies.
>
> Once a firm obtains a quality advantage, it must continuously work to maintain it. As markets mature, competition erodes any advantage. Quality must be viewed from the customer's perspective, not as conformance to self-imposed requirements. Yet a quality advantage often cannot be obtained only by soliciting customer input, since customers usually are not aware of potential innovations. (p. 19)

According to J.L. Ashford (1989):

> After the second World War the economy of Japan was in ruins. To attain their military objectives, all available resources of capital and of technical manpower had been directed to armaments manufacture, while their civilian economy gained an unenviable reputation for producing poor quality copies of products designed and developed elsewhere. Unless they were able to raise the quality of their products to a level which could compete, and win, in the international marketplace, they stood no chance of becoming a modern industrialized nation.
>
> To learn how to regenerate their industries, they sent a team abroad to study the management practices of other countries and they invited foreign experts to provide advice. Among the latter were two Americans, J.M. Juran and W.E. Deming, who brought a new message, which can be summarized as follows.
>
> 1. The management of quality is crucial to company survival and merits the personal attention and commitment of top management.
> 2. The primary responsibility for quality must lie with those doing the work. Control by inspection is of limited value.
> 3. To enable a production department to accept responsibility for quality, management must establish systems for the control and verification of work, and must educate and indoctrinate the work force in their application.
> 4. The costs of education and training for quality, and other costs which might be incurred, will be repaid many times over by greater output, less waste, a better quality product and higher profits.
>
> These are the basic principles of management concepts which have since become identified under the generic term of quality management. (p. 5)

However, quality actually emerged as a dominant thinking only since World War II, becoming an integral part of overall business system focused on customer satisfaction, and becoming known in recent times as "total quality management (TQM)", with its three constitutive elements:

- Total: organization-wide
- Quality: customer satisfaction
- Management: systems of managing

1.3.6 Total Quality Management

The TQM concept was born following World War II. It was stimulated by the need to compete in the global market where higher quality, lower cost, and more rapid development are essential to market leadership. Today, TQM is considered a fundamental requirement for any organization to compete, let alone lead, in its market. It is a way of planning, organizing, and understanding each activity of the process and removing all the unnecessary steps routinely followed in the organization. TQM is a philosophy that makes quality values the driving force behind leadership, design, planning, and improvement in activities. Table 1.1 summarizes periodical changes in the quality system.

TABLE 1.1
Periodical Changes in Quality System

Period	System
Middle Ages (1200–1799)	Guilds-skilled craftsmen were responsible to control their own products.
Mid-18th century (Industrial Revolution)	Establishment of factories. Increase in productivity. Mass production. Assembly lines. Several workers were responsible to produce a product. Production by skilled workers and quality audit by inspectors.
Early 19th century	Craftsmanship model of production.
Late 19th century (1880s)	Frederick Taylor and "scientific management". Quality management through inspection.
Beginning of 20th century (1920s)	Walter Shewhart introduced statistical process control. Introduction of full-time quality inspection and quality control department. Quality management.
1930s	Introduction of sampling method.
1950s	Introduction of statistical quality process in Japan.
Late 1960s	Introduction of quality assurance.
1970s	Total quality control.
	Quality management.
1980s	Total quality management.
Beginning of the 21st century	Integrated quality management (IQM).

Source: Abdul Razzak Rumane. (2017). *Quality Management in Construction Projects*, Second Edition. Reprinted with permission from Taylor & Francis Group.

1.3.6.1 Changing Views of Quality

The failure to address the culture of an organization is frequently the reason for management initiatives either having limited success or failing altogether. Understanding the culture of the organization and using that knowledge to implement cultural change is an important element of TQM. The culture of good teamwork and cooperation at all levels in an organization is essential to the success of TQM. Table 1.2 describes cultural changes needed in an organization to meet TQM.

Harold Kerzner (2001) states that, with the increased complexity of business, the cost of maintaining a meaningful level of quality had been steadily increasing. In order to reverse this trend, TQM is used to achieve a major competitive advantage. Kerzner has described this phenomenon: "During the past twenty years, there has been a revolution toward improved quality. The improvements have occurred not only in product quality, but also in quality leadership and quality project management". He further states:

The push for higher levels of quality appears to be customer driven. Customers are now demanding

- Higher performance requirements
- Faster product developments
- Higher technology levels
- Materials and processes pushed to the limit
- Lower contractor profit managing
- Fewer defects/rejects (p. 1083)

Gryna (2001, p. 3) has described the changing business conditions that the organization should understand to survive in competitive world market. He says:

TABLE 1.2
Cultural Changes Required to Meet Total Quality Management

From	To
• Inspection orientation	• Defect prevention
• Meet the specification	• Continuous improvement
• Get the product out	• Customer satisfaction
• Individual input	• Cooperative efforts
• Sequential engineering	• Team approach
• Quality control department	• Organizational involvement
• Departmental responsibility	• Management commitment
• Short-term objective	• Long-term vision
• People as cost burden	• Human resources as an asset
• Purchase of products or services on price-alone basis	• Purchase on total cost minimization basis
• Minimum cost suppliers	• Mutually beneficial supplier relationship

Source: Abdul Razzak Rumane. (2017). *Quality Management in Construction Projects*, Second Edition. Reprinted with permission from Taylor & Francis Group.

The prominence of product quality in the public mind has resulted in quality becoming a cardinal priority for most organizations. The identification of quality as a core concern has evolved through a number of changing business conditions. These include the following:

1. Competition
2. The customer-focused organization
3. Higher levels of customer expectation
4. Performance improvement
5. Changes in organization forms
6. Changing workforce
7. Information revolution
8. Electronic commerce

1.3.6.2 Quality Gurus and Total Quality Management Philosophies

The TQM approach was developed immediately after World War II. There are prominent researchers and practitioners whose work has dominated this movement. Their ideas, concepts, and approaches in addressing specific quality issues have become part of the accepted wisdom in TQM, resulting in a major and lasting impact within the field. These persons have become known as "quality gurus". They all emphasize involvement of organizational management in the quality efforts. These philosophers are:

1. Philip B. Crosby
2. W. Edwards Deming
3. Armand V. Feigenbaum
4. Kaoru Ishikawa
5. Joseph M. Juran
6. John S. Oakland
7. Shigeo Shingo
8. Genichi Taguchi

A brief summary of their philosophy and approaches is given next.

1.3.6.2.1 Philip B. Crosby
Crosby's philosophy is seen by many to be encapsulated in his five "Absolute Truths of Quality Management". These are as follows:

1. Quality is defined as conformance to requirement, not as "goodness" or "elegance".
2. There is no such thing as a quality problem.
3. It is always cheaper to do it right the first time.
4. The only performance measurement is the cost of quality.
5. The only performance standard is zero defects.

Overview of Quality

Crosby's perspective on quality has three essential beliefs:

1. A belief in qualification
2. Management leadership
3. Prevention rather than cure

Crosby's principal method is his 14-step program for quality management and is illustrated in Table 1.3. His main emphasis is the quantitative, that is, the performance standard of "zero defects".

1.3.6.2.2 W. Edwards Deming

Deming was perhaps the best known figure associated with the quality field and is considered its founding father. His philosophy is based on four principal methods:

1. The plan–do–check–act (PDCA) cycle
2. Statistical process control (SPC)
3. The 14 principles of transformation
4. The seven-point action plan

1.3.6.2.2.1 PDCA Cycle PDCA cycle is an iterative, four-step management method used for continuous improvement of business processes and products. Once it has been completed, it recommences without ceasing. The approach is seen as reemphasizing the responsibility of management to be actively involved in the organization's quality program. The PDCA cycle is also known as the plan–do–study–act (PDSA) cycle.

TABLE 1.3
Fourteen-Step Quality Program: Philip B. Crosby

Step	Description of Quality Program
Step 1	Establish management commitment
Step 2	Form quality improvement teams
Step 3	Establish quality measurements
Step 4	Evaluate the cost of quality
Step 5	Raise quality awareness
Step 6	Take action to correct problems
Step 7	Zero defects planning
Step 8	Train supervisors and managers
Step 9	Hold a "zero defects" day to establish the attitude and expectation within the company
Step 10	Encourage the setting of goals for improvement
Step 11	Obstacle reporting
Step 12	Recognition for contributors
Step 13	Establish quality councils
Step 14	Do it all over again

Source: Crosby, Philip B. (1979). *Quality Is Free, (Excerpts from the Book).* Reprinted with permission from The McGraw-Hill Companies.

Nancy R. Tague (2005, p. 391) has elaborated on the PDCA cycle as follows:

Description

The PDCA or PDSA cycle consists of a four-step model for carrying out change. Just as a circle has no end, the PDCA cycle should be repeated again and again for continuous improvement. PDCA is a basic model that can be compared to the continuous improvement process, which can be applied on a small scale.

When to use

- As a model for continuous improvement
- When starting a new improvement project
- When developing a new or improved design of process, product, or service
- When defining a repetitive work process
- When planning data collection and analysis in order to verify and prioritize problems or root causes
- When implementing any change

Procedure

1. *Plan.* Recognize an opportunity and plan the change.
2. *Do.* Test the change; carry out a small-scale study.
3. *Check.* Review the test, analyze the results, and identify learnings.
4. *Act.* Take action based on what you learned in the study step. If the change did not work, go through the cycle again with a different plan. If you were successful, incorporate the learning from the test into wider changes. Use what you learned to plan new improvements, beginning the cycle again.

PDCA is mainly used for continuous-process improvement. The PDCA cycle, when used as a process improvement tool for design improvement/design conformance in construction projects to meet owner's requirements, shall indicate the following actions:

Plan: Establish scope.
Do: Develop design.
Check: Review and compare.
Act: Implement comments, take corrective action, and/or release contract documents to construct/build the project/facility.

Figure 1.1 illustrates basic concepts to develop PDCA Cycle and Figure 1.2 illustrates the PDCA cycle model for conformance of construction projects designed to owner requirements/scope of work.

1.3.6.2.2.2 Statistical Process Control SPC is a quantitative approach based on the measurement of process control. Deming believed in the use of SPC charts as the key method for identifying special and common causes and assisting diagnosis of quality problems. His aim was to remove "outliers", that is, quality problems relating to the special causes of failure. This was achieved through training, improved machinery and equipment, and so on. SPC enabled the production process to be brought "under control".

Overview of Quality

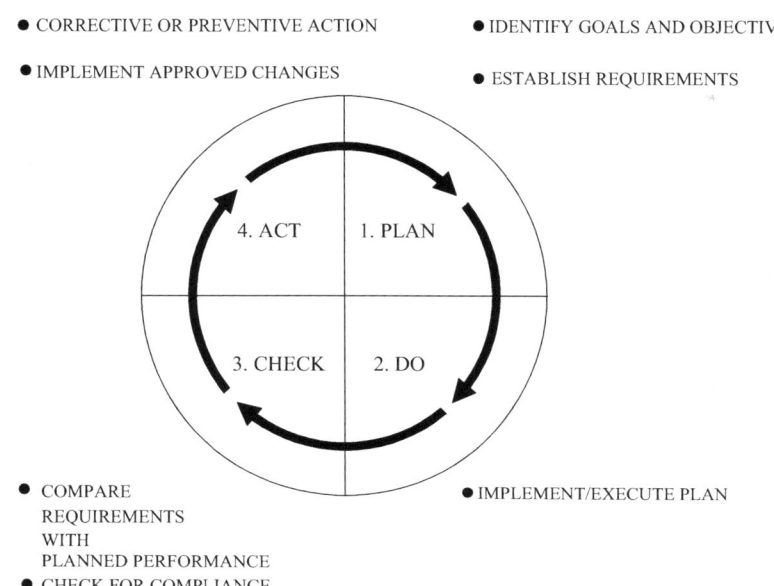

FIGURE 1.1 Development of PDCA cycle. PDCA, plan–do–check–act.

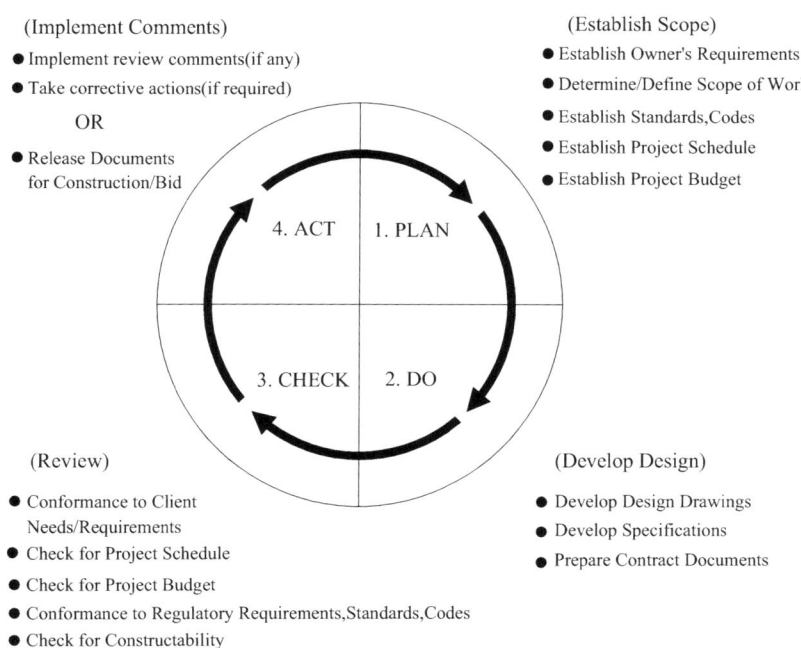

FIGURE 1.2 PDCA cycle for construction projects (design phases). PDCA, PDCA, plan–do–check–act.

The remaining quality problems were considered to be related to common causes, that is, they were inherent in the design of the production process. Eradication of special causes enabled a shift in focus to common causes to further improve quality.

Gryna (2001) describes statistical process control as the application of statistical methods to the measurement and analysis of variation in a process. This technique applies to both in-process parameters and end-process (product) parameters (p. 495). He further states that "a statistical control chart compares process performance data to computed 'statistical control limits,' drawn as limit lines on the chart" (p. 498).

There are two categories of control charts based on the type of data collected. These are as follows:

1. Variable control charts
2. Attributes control charts

Figure 1.3 illustrates categories of control chart.

1.3.6.2.2.3 Principles for Transformation   Deming's 14 principles for transformation are listed in Table 1.4.

1.3.6.2.2.4 The Seven-Point Action Plan   In order to implement the 14 principles, Deming proposed a seven-point action plan. These are listed in Table 1.5.

1.3.6.2.3 Armand V. Feigenbaum

Feigenbaum defines quality as "best for the customer use and selling price", and quality control as an effective method for co-coordinating the quality maintenance and quality improvement efforts at the various groups in an organization, so as to enable production

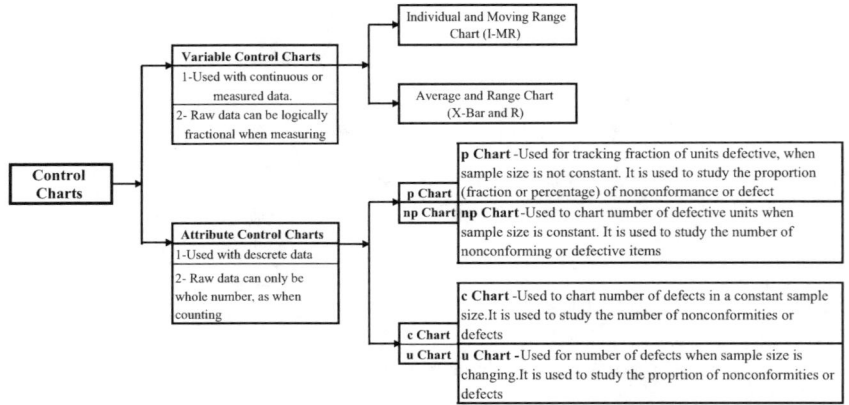

FIGURE 1.3 Categories of control charts. (Abdul Razzak Rumane. (2017). *Quality Tools for Managing Construction Projects*, Second Edition. Reprinted with permission from Taylor & Francis Group.)

TABLE 1.4
Fourteen Principles of Transformation: W. Edwards Deming.

The Quality Gurus

W. Edwards Deming

Principle 1	Create constancy of purpose to improve product and service.
Principle 2	Adopt a new philosophy for the new economic age with management learning what their responsibilities are and by assuming leadership for change.
Principle 3	Cease dependence on mass inspection to achieve quality by building quality into the product.
Principle 4	End awarding business on price. Award business on total cost and move toward single suppliers.
Principle 5	Aim for continuous improvement of the system of production and service to improve productivity and quality and to decrease costs.
Principle 6	Institute training on the job.
Principle 7	Institute leadership with the aim of supervising people to help them to do a better job.
Principle 8	Drive out fear so that everyone can work effectively together for the organization.
Principle 9	Breakdown barriers between departments. Encourage research, design, sales, and production to work together to foresee difficulties in production and use.
Principle 10	Eliminate slogans, exhortations, and numerical targets for the workforce since they are divisory and anyway difficulties belong to the whole system.
Principle 11	Eliminate quotas or work standards and management by objectives or numerical goals; leadership should be substituted instead.
Principle 12	Remove barriers that rob people of their right to pride in their work.
Principle 13	Institute a vigorous education and self-improvement program.
Principle 14	Put everyone in the company to work to accomplish the transformation.

Source: Out of Crisis © (2000) W. Edwards Deming Institute. Reprinted with permission from MIT press.

at the most economical levels that allow for full customer satisfaction. Feigenbaum's philosophy of quality has a four-step approach. These are as follows:

Step 1. Set quality standards
Step 2. Appraise conformance to standards
Step 3. Act when standards are not met
Step 4. Plan to make improvements

1.3.6.2.4 Kaoru Ishikawa

The founding philosophy of Ishikawa approach is "companywide quality control". He has identified 15 effects of companywide quality control. Ishikawa's approach deals with organizational aspects and is supported by the "quality circles" technique and the "seven tools of quality control". Quality circles are Ishikawa's principal method for achieving participation, composed of 5–15 workers from the same

TABLE 1.5
Seven Points Action Plan: Deming

	The Quality Gurus
	W. Edwards Deming
Point 1	Management must agree on the meaning of the quality program, its implications, and the direction to take.
Point 2	Top management must accept and adopt the new philosophy.
Point 3	Top management must communicate the plan and the necessity for it to the people in the organization.
Point 4	Every activity must be recognized as a step in a process and the customers of that process identified. The customers are responsible for the next stage of the process.
Point 5	Each stage must adopt the "Deming" or "Shewhart" cycle—plan, do, check, action—as the basis of quality improvement.
Point 6	Team working must be engendered and encouraged to improve inputs and outputs. Everyone must be enabled to contribute to this process.
Point 7	Construct an organization for quality with the support of knowledgeable statisticians.

Source: Out of Crisis © (2000) W. Edwards Deming Institute. Reprinted with permission from MIT press.

area of achieving, and led by a foreman or supervisor who acts as a group leader to liaison between the workers and the management. The function of quality circles is to identify local problems and recommend the solutions. The aims of quality circles are as follows:

- Contribute to the improvement and development of the enterprise.
- Respect human relations and build a happy workshop offering job satisfaction.
- Deploy human capabilities fully and draw out infinite potential.

Kerzner (2001) states:

> Quality circles are small groups of employees who meet frequently to help resolve company quality problems and provide recommendations to management. Quality circles were initially developed in Japan and only recently have achieved some degree of success in United States. The employees involved in quality circles meet frequently either at someone's home or at the plant before the shift begins. The group identifies problems, analyzes data, recommends solutions, and carries out management-approved changes. The success of quality circles is heavily based upon management's willingness to employ recommendations.

He further states:

> The key elements of quality circles include
>   They are a team effort.
>   - They are completely voluntary.
>   - Employees are trained in group dynamics, motivation, communications, and problem solving.
>   - Members rely upon each other for help.

Overview of Quality

- Management support is achieved but as needed.
- Creativity is encouraged.
- Management listens to recommendations.

And the benefits of quality circles include

- Improved quality of products and services
- Better organizational communications
- Improved worker performance
- Improved morale (p. 1131)

Ishikawa emphasized four points to be considered for the formation of the quality circle. These are as follows:

1. *Voluntarism.* Circles are to be created on voluntary basis, and not by a command from above. Begin circle activities with people who wish to participate.
2. *Self-development.* Circle members must be willing to study.
3. *Mutual development.* Circle members must aspire to expand their horizons and cooperate with other circles.
4. *Eventual total participation.* Circles must establish their ultimate goal full participation of all workers in the same workplace.

The quantitative techniques of the Ishikawa approach are referred as "Ishikawa's Seven Tools of Quality Control", listed in Table 1.6. The approach includes both quantitative and qualitative aspects, which, taken together, focus on achieving companywide quality.

These form a set of pictures of quality, representing in diagrammatic or chart form the quality status of the operation of process being reviewed. Ishikawa considered that all staff should be trained in these techniques as they have a useful role to play in managing quality.

TABLE 1.6
Seven Tools of Quality Control by Kaoru Ishikawa

Sr. No.	Name of Quality Tool	Usage
Tool 1	Pareto charts	Used to identify the most significant cause or problem.
Tool 2	Ishikawa/fishbone diagrams	Charts of cause and effect in processes.
Tool 3	Stratification	Layer charts that place each set of data successively on top of the previous one. How are the data made up?
Tool 4	Check sheets	To provide a record of quality. How often it occurs?
Tool 5	Histogram	Graphs used to display frequency of various ranges of valves of a quantity.
Tool 6	Scatter diagram:	To determine whether there is a correlation between two factors.
Tool 7	Control charts:	A device in statistical process control to determine whether or not the process is stable.

Ishikawa developed a technique of graphically displaying the causes of any quality problem. His method is called by several names, such as the Ishikawa diagram, fishbone diagram, and cause-and-effect diagram. The Ishikawa diagram is essentially an end or goal-oriented picture of a problem situation. It is based around a set of "M" causes such as Manpower (personnel), Machine (plant and equipment), Material (raw material and parts), Method (techniques and technology), Measurement (sampling, instrumentation), and Mother Nature (environment). Figure 1.4 illustrates the Ishikawa "fishbone" diagram.

There are six steps that are used to perform a cause-and-effect analysis. These are as follows:

Step 1. Identify the problem to analyze its technical cause.
Step 2. Select an interdisciplinary brainstorm team.
Step 3. Draw a problem box and prime arrows.
Step 4. Specify major categories contributing to the problem.
Step 5. Identify a defect cause.
Step 6. Identify corrective action and perform the analysis in the same manner as for the cause-and-effect analysis.

Nancy R. Tague (2005, p. 248) has elaborated on the fishbone diagram as follows:

Description

The Fishbone Diagram identifies many possible causes for an effect or problem. It can be used to structure a brainstorming session. It immediately sorts ideas into useful categories.

When to use

- When identifying possible causes of a problem
- When the team's thinking tends to fall into ruts

Procedure

1. Agree on a problem statement (effect). Write it at the center right of the flipchart or whiteboard. Draw a box around it and draw a horizontal arrow running to it.

THE QUALITY GURUS

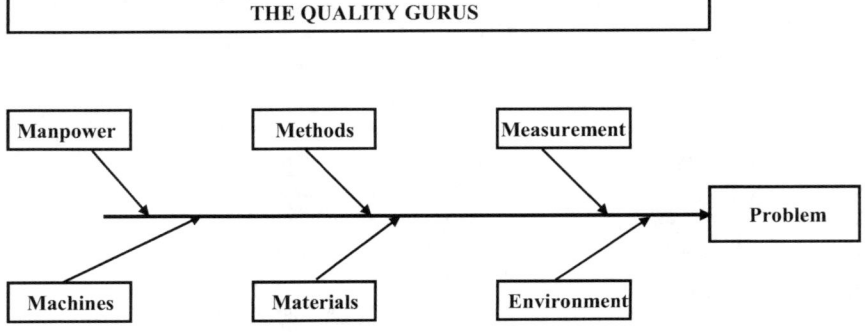

FIGURE 1.4 The Ishikawa "fishbone" diagram.

Overview of Quality

2. Brainstorm the major categories of causes of the problem. If there is difficulty here, use generic headings: method, machines (equipment), people (manpower), materials, measurement, and environment. Write the categories of causes as branches from the main arrow.
3. Brainstorm all the possible causes of the problem. Ask "Why does this happen?" As each idea is given, the facilitator writes it as a branch from the appropriate category. Causes can be written in several places if they relate to several categories or multiple relationships.
4. Ask again, "Why does this happen?" about each cause. Write subcauses branching off the causes. Continue to ask "Why" and generate deeper levels of causes. Layers of branches indicate causal relationships.
5. When the team runs out of ideas, focus attention on places on the fishbone where ideas are few.

Summarizing Ishikawa's approach, it can be seen to contain both quantitative and qualitative aspects, which, taken together, focus on achieving companywide quality.

1.3.6.2.5 Joseph M. Juran

Juran's philosophy is perhaps best summed as "Quality does not happen by accident; it has to be planned".

The emphasis of Juran's work is on planning organizational issues, management's responsibility for quality, and the need to set goals and targets for improvement. Juran's definition of quality is "Fitness for use or purpose". His thinking on quality is an operational framework of three quality processes. These are as follows:

1. Quality planning
2. Quality control
3. Quality improvement

These are best known as Juran's quality trilogy.

Shtub, Bard, and Globerson (1994, p. 284) describe the Juran trilogy as follows:

Quality planning. In preparing to meet organizational goals, the end result should be a process that is capable of meeting those goals under operating conditions. Quality planning might include identifying internal and external customers, determining customer needs, developing a product or service that responds to those needs, establishing goals that meet the needs of customers and suppliers at a minimum cost, and proving that the process is capable of meeting quality goals under operating conditions. A necessary step is for managers to engage cross-functional teams and openly supply data to team members so that they may work together with unity of purpose.

Quality control. At the heart of this process is the collection and analysis of data for the purpose of determining how best to meet project goals under normal operating conditions. One may have to decide on control subjects, units of measurement, standards of performance, and degrees of conformance. To measure the difference between the actual performance before and after the process or system has been modified, the data should be statistically significant and the process or system should be in statistical control. Task forces working on various problems need to establish baseline data so that they can determine if the implemented recommendations are responsible for the observed improvements.

Quality improvement. This process is concerned with breaking through to a new level of performance. The end result is that the particular process or system is obviously at a higher level of quality in delivering either a product or a service.

Juran's approach, like those of his colleagues, stresses the involvement of employees in all phases of a project. The philosophy and procedures require that managers listen to employees and help them rank the processes and systems that need improving.

1.3.6.2.6 John S. Oakland

Oakland's philosophy of quality is "We cannot avoid seeing how quality has developed into the most important competitive weapon, and many organizations have realized that TQM is the [sic] way of managing for the future (Oakland, 1993, Preface)". He gives absolute importance to the pursuit of quality as the cornerstone of organizational success. Figure 1.5 illustrates Oakland's major features in his "Total Quality Management model."

Oakland's view is that "quality starts at the top", with quality parameters inherent in every organizational decision. He offers his own overarching approach for

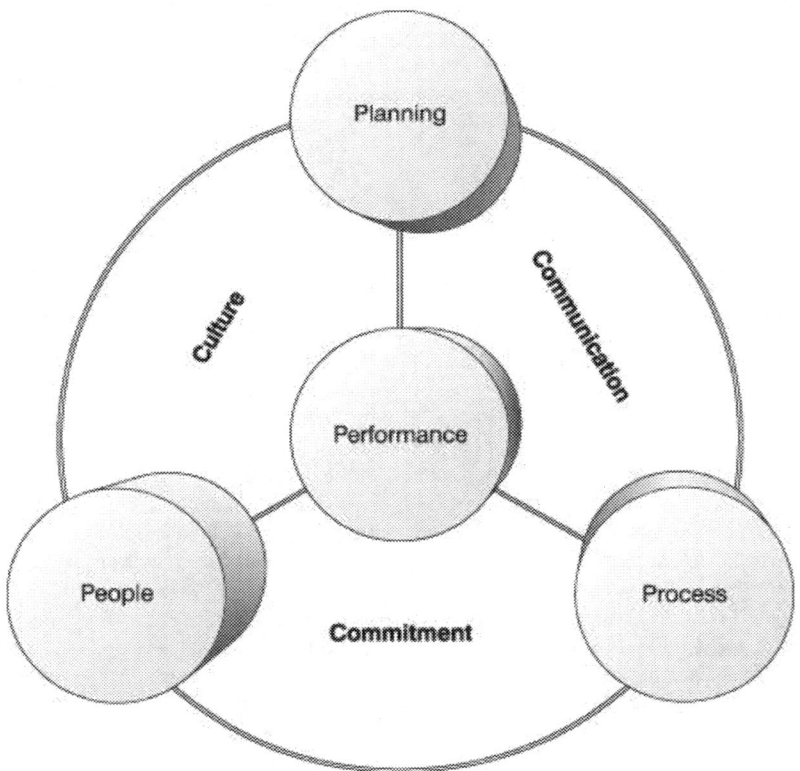

FIGURE 1.5 Total quality management model: John S. Oakland. (John S. Oakland, (2014). TQM. Reprinted with permission from Taylor & Francis Group.)

TQM on the many well-established methods, tools, and techniques for achieving quality and some new insight. The overarching method is his "Ten Points for Senior Management". Table 1.7 illustrates this.

1.3.6.2.7 Shigeo Shingo

Shingo's early philosophy embraced the "scientific management ideas" originated by Fredrick Taylor in the early part of the 20th century. Shingo believed that statistical methods detect error too late in the manufacturing process. He suggested that, instead of detecting errors, it was better to engage in preventative measures aimed at eliminating error sources. Shingo continues to believe in mechanizing the monitoring of error, considering that human assessment was "inconsistent" and prone to error and introduction of controls within a process. He used people to identify underlying causes and produce preventative solutions. Shingo has a clear belief, like Crosby, in a "zero defects" approach. His approach emphasizes zero defects through good engineering and process investigation and rectification.

Shingo is strongly associated with the "just-in-time" manufacturing philosophy. He was the inventor of the single-minute exchange of die system that drastically reduced the equipment setup time from hours to minutes. Just-in-time is an integrated set of activities designed to achieve high-level volume production, with minimal inventories of parts that arrive at the workstation when they are needed.

Shingo is also associated with the Poka-Yoke system to achieve zero defects (failsafe procedures). The Poka-Yoke system includes checklists that (1) prevent workers from making an error that leads to defects before starting or (2) give rapid feedback of abnormalities in the process to the worker in time to correct them.

TABLE 1.7
Ten Points for Senior Management : John S. Oakland

The Quality Gurus	
John S. Oakland	
Point 1	Long-term commitment.
Point 2	Change the culture to "right first time".
Point 3	Train the people to understand the "customer–supplier relationship".
Point 4	Buy products and services on total cost (sic).
Point 5	Recognize that system improvement must be managed.
Point 6	Adopt modern methods of supervision and training and eliminate fear.
Point 7	Eliminate barriers, manage processes, and improve communications and teamwork.
Point 8	Eliminate, arbitrary goals, standards based only on numbers, barriers to pride of workmanship, fiction (use the correct tools to establish facts).
Point 9	Constantly educate and retrain the in-house experts.
Point 10	Utilize a systematic approach to TQM implementation.

TQM, total quality management.
Source: John S. Oakland (2014). *Total Quality Management.* Reprinted with permission from Taylor & Francis Group.

1.3.6.2.8 Genichi Taguchi

Taguchi's two founding ideas of quality work are essentially quantitative. The first is a statistical method to identify and eradicate quality problems. The second rests on designing products and processes to build in quality right from the outset.

Taguchi's prime concern is with customer satisfaction and with the potential for "loss of reputation and goodwill" associated with failure to meet customer expectation. Such a failure, he considered, would lead the customer to buy elsewhere in future, damaging the prospects of the company, its employees, and society. He saw that loss not only occurred when a product was outside its specification but also when it varied from its target value. Taguchi recognized the organization as "open system", interacting with its environment. The principal tools and techniques were espoused by Taguchi center on the concept of continuous improvement and eradicating, as far as possible, potential causes of "non-quality" at the outset. His concept of product development has three stages:

1. System design stage
2. Parameter stage
3. Tolerance design stage

The first stage is concerned with system design reasoning involving both product and process. This framework is carried on to the second stage—parameter design. The third stage, tolerance design, enables the recognition of factors that may significantly affect the variability of the product.

1.3.6.2.9 Summary of Philosophies

Although there are differences in certain areas among these philosophers, all of them generally advocate the same steps. Their emphasis is on customers' satisfaction, management leadership, teamwork, continuous improvement, and minimizing defects.

Based on these, the common features of their philosophies can be summarized as follows:

1. Quality is conformance to the customer's defined needs.
2. Senior management is responsible for quality.
3. Institute continuous improvement of process, product, and services through the application of various tools and procedures to achieve higher level of quality.
4. Establish performance measurement standards to avoid defects.
5. Take a team approach by involving every member of the organization.
6. Provide training and education to everyone in the organization.
7. Establish leadership to help employees perform a better job.

Thus, their concept of quality forms the basic tenets of TQM.

1.3.6.3 Principles of Total Quality Management

Juran describe TQM in terms of the Juran Trilogy, which involves quality planning, quality control, and quality improvement.

In the JUSE's (Japanese Union of Scientists and Engineers) view, as mentioned by Juran and Godfrey (1999, p. 14.3),

Overview of Quality

TQM is a management approach that strives for the following in any business environment:

- Under strong top management leadership established clear mid and long-term vision and strategies.
- Properly utilize the concepts, values and scientific methods of TQM.
- Regard human resources and information as vital organizational infrastructures.
- Under an appropriate management system, effectively operate a quality assurance system and other cross-functional management systems such as cost, delivery, environment and safety.
- Supported by fundamental organizational powers, such as core technology, speed and vitality, ensure sound relationship with customers, employees, society suppliers and stockholders.
- Continuously release corporate objectives in the form of achieving an organization's mission, building an organization with a respectable presence and continuously securing profits.

As per ASQ Quality Glossary:

Total Quality Management (TQM) is a term used to describe a management approach to quality improvement. Since then TQM has taken many meanings. Simply put, it is a management approach to long-term success through customer satisfaction. TQM is based nm all members of organization in improving processes, products, services and the culture in which they work. The methods for implementing this approach are found in the teachings of such quality leaders as Philip B. Crosby, W. Edwards Deming, Armand V. Feigenbaum, Kaoru Ishikawa, and Joseph M. Juran.

The ASQ described the history of TQM as follows;

The history of Total Quality Management (TQM) began initially as a term coined by the Naval Air Systems Command to describe its Japanese-style management approach to quality improvement. An umbrella methodology for continually improving the quality of all processes, it draws on a knowledge of the principles and practices of:

- The behavioral sciences
- The analysis of quantitative and nonquantitative data
- Economics theories
- Process analysis

It has further described the evolution of TQM as follows:

- 1920s
  - Some of the seeds of quality management were planted as the principles of scientific management swept through U.S. industry
  - Businesses clearly separated the processes of planning and carrying out the plan, and union opposition arose as workers were deprived of a voice in the conditions and functions of their work
  - The Hawthorne experiments in the late 1920s showed how worker productivity could be impacted by participation
- 1930s
  - Walter Shewhart developed the methods for statistical analysis and control of quality

- 1950s
  - W. Edwards Deming taught methods for statistical analysis and control of quality to Japanese engineers and executives. This can be considered the origin of TQM
  - Joseph M. Juran taught the concepts of controlling quality and managerial breakthrough
  - Armand V. Feigenbaum's book *Total Quality Control,* a forerunner for the present understanding of TQM was published
  - Philip B. Crosby's promotion of zero defects paved the way for quality improvement in many companies
- 1968
  - The Japanese named their approach to total quality *companywide quality control.* It is around this time that the term quality management systems arises
  - Kaoru Ishikawa synthesis of the philosophy contributed to Japan's ascendency as quality leader
- Today
  - TQM is the name for the philosophy of a broad and systemic approach to managing organizational quality
  - Quality standards such as the ISO 9000 series and quality award programs such as the Deming Prize and the Malcolm Baldrige National Quality Award specify principles and processes that comprise TQM.

TQM focuses on participative management and strong operational accountability at the individual contributor level. Total quality involves not just managers but also everyone in the organization in a complete transformation of the prevailing culture. It is a change to the way people do things and relies on trust between managers and staff. TQM is applicable to all kinds of organizations, in both the public and private sectors. It is also applicable to those providing services as well as those involved in producing goods or manufacturing activities.

According to Construction Industry Institute (CII) source document 74, "Total Quality Management is often termed a journey, not a destination". This is because of its nature as a collection of improvement-centered processes and techniques that are performed in a transformed management environment. The concept of "continuous improvement" holds that this environment must prevail for the life of the enterprise and that the methods will routinely be used on a regular, recurring basis. The improvement process never ends; therefore, no true destination is ever reached. Figure 1.6 describes the phases of the TQM journey.

The document further states (p. 1) that from the viewpoint of the individual company, the strategic implications of TQM include the following:

- Survival in an increasingly competitive world
- Better service to the customer
- Enhancement of the organization's "shareholder value"
- Improvement of the overall quality and safety of our facilities
- Reduced project duration and costs
- Better utilization of talents of the people

Overview of Quality

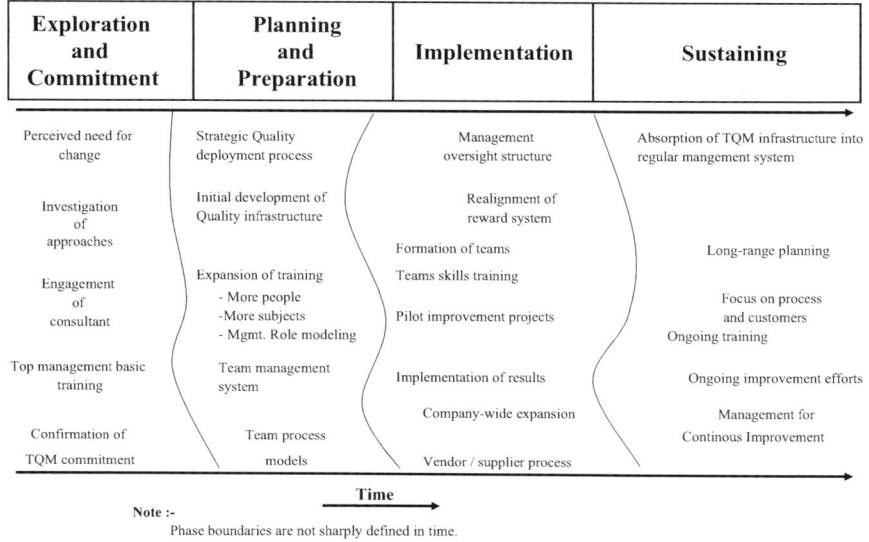

FIGURE 1.6 Phases of the TQM journey. (Courtesy: CII Source document 74. Reprinted with permission from CII, University of Texas.) TQM, total quality management.

Chase, Aquilano, and Jacobs (2001) have defined TQM as

> Managing the entire organization so that it excels on all dimensions of products and services that are important to the customer. This definition is more applicable than another commonly used one—'Conformance to specifications.' Though valid for goods production, the second definition is problematic for many services. Precise specifications for service quality are hard to define and measure. It is possible, however to find out what's important to the customer and then create the kind of organizational culture that motivates and enables the worker to deliver a quality service.

They further state:

> The philosophical elements of TQM stress the operation of the firm using quality as the integrating element. The generic tools consist of (1) various statistical process control (SPC) methods that are used for problem solving and continuous improvement by quality terms, and (2) quality function deployment, which is typically used by managers to drive the voice of the customer into the organization. (p. 260)

The ISO document has listed eight quality management principles on which the quality management system standards of the revised ISO 9000:2000 series are based. These are as follows:

Principle 1—Customer focus
Principle 2—Leadership
Principle 3—Involvement of people
Principle 4—Process approach
Principle 5—System approach to management

Principle 6—Continual improvement
Principle 7—Factual approach to design making
Principle 8—Mutual beneficial supplier relationship

Based on those principles, it can be summarized that TQM is a management philosophy that evolved in Japan after World War II. It places quality as a strategic objective and focuses on continuous improvement of products, processes, services, and cost to compete in the global market by minimizing rework and maximizing profitability to achieve market leadership and customer satisfaction. It is a way of managing people and business processes to meet customer satisfaction. TQM involves everyone in the organization in the effort to increase customer satisfaction and achieve superior performance of the products or services through continuous quality improvement. TQM helps in

- achieving customer satisfaction,
- continuous improvement,
- developing teamwork,
- establishing vision for the employees,
- setting standards and goals for the employees,
- building motivation within the organization, and
- developing corporate culture

TQM is widely accepted as the basis for achieving sustainable competitive advantage. It is required to achieve customer satisfaction and ensure survival in the competitive world because of the global nature of competition, which never rests, and there is no end to product or service improvement.

1.4 DEFINITION OF QUALITY FOR CONSTRUCTION PROJECTS

Quality has different meanings to different people. The definition of quality relating to manufacturing, processes, and service industries is as follows:

- Meeting the customer's need
- Customer satisfaction
- Fitness for use
- Conforming to requirements
- Degree of excellence at an acceptable price

The ISO defines quality as "the totality of characteristics of an entity that bears on its ability to satisfy stated or implied needs".

However, the definition of quality for construction projects is different to that of manufacturing or services industries as the product is not repetitive, but unique piece of work with specific requirements.

Quality in construction project is not only the quality of product and equipment used in the construction of facility but also the total management approach to complete the facility. Quality of construction depends mainly upon the control of construction, which is the primary responsibility of contractor.

Quality in manufacturing passes through series of processes. Material and labor are input through a series of process out of which a product is obtained. The output is monitored by inspection and testing at various stages of production. Any nonconforming product identified is either repaired, reworked, or scrapped, and proper steps are taken to eliminate problem causes. Statistical process control methods are used to reduce the variability and to increase the efficiency of process. In construction projects, the scenario is not the same. If anything goes wrong, the nonconforming work is very difficult to rectify, and remedial actions are sometimes not possible.

Quality management in construction projects is different to that of manufacturing. Quality in construction projects is not only the quality of products and equipment used in the construction but also the total management approach to complete the facility as per the scope of works to customer/owner satisfaction within the budget and to be completed within specified schedule to meet owner's defined purpose. The nature of the contracts between the parties plays a dominant part in the quality system required from the project, and the responsibility for achieving them must therefore be specified in the project documents. The documents include plans, specifications, schedules, bill of quantities, and so on. Quality control in construction typically involves ensuring compliance with minimum standards of material and workmanship in order to ensure the performance of the facility according to the design. These minimum standards are contained in the specification documents. For the purpose of ensuring compliance, random samples and statistical methods are commonly used as the basis for accepting or rejecting work completed and batches of materials. Rejection of a batch is based on nonconformance or violation of the relevant design specifications.

Thus, quality of construction projects can be evolved as follows:

1. Properly defined scope of work.
2. Owner, project manager, design team leader, consultant, and constructor's manager are responsible to implement the quality.
3. Continuous improvement can be achieved at different levels as follows:
   a. Owner—Specify the latest needs.
   b. Designer—Specification to include latest quality materials, products, and equipment.
   c. Constructor—Use latest construction equipment to build the facility.
4. Establishment of performance measures:
   a. Owner—(1) To review and ensure that designer has prepared the contract documents which satisfy his needs, and (2) to check the progress of work to ensure compliance with the contract documents.
   b. Consultant—(1) As a consultant designer to include the owner's requirements explicitly and clearly defined in the contact documents, and (2) as a supervision consultant supervise contractor's work as per contract documents and the specified standards.
   c. Contactor—To construct the facility as specified and use the materials, products, and equipment that satisfy the specified requirements.
5. Team approach—Every member of the project team should know principles of TQM knowing that TQM is collaborative effort and everybody should participate in all the functional areas to improve quality of project works. They should know that it is a collective effort by all the participants to achieve project quality.

6. Training and education—Consultant and contractor should have customized training plans for their management, engineers, supervisors, office staff, technicians, and labors.
7. Establish leadership—Organizational leadership should be established to achieve the specified quality. Encourage and help the staff and labors to understand the quality to be achieved for the project.

These definitions when applied to construction projects relate to the contract specifications or owner/end user requirements to be constructed in such a way that construction of the facility is suitable for owner's use or it meets the owner requirements. Quality in construction is achieved through complex interaction of many participants in the facilities development process.

The quality plan for construction projects is part of the overall project documentation consisting of

1. well-defined specification for all the materials, products, components, and equipment to be used to construct the facility,
2. detailed construction drawings,
3. detailed work procedure,
4. details of the quality standards and codes to be complied,
5. cost of the project,
6. manpower and other resources to be used for the project, and
7. project completion schedule.

Participation involvement of all three parties at different level of construction phases is required to develop quality system and application of quality tools and techniques. With the application of various quality principles, tools, and methods by all the participants at different stages of construction project, rework can be reduced resulting in savings in the project cost and making the project qualitative and economical. This will ensure completion of construction and making the project most qualitative, competitive, and economical.

The authors of *Quality in the Constructed Project* (2012) by the American Society of Civil Engineers (ASCE) have defined as:

> Quality is defined as the delivery of products and services in a manner that meets the reasonable requirements and expectations of the owner, design professional, and constructor, including conformance with contract requirements, prevailing industry standards, and applicable codes, laws, and licensing requirements. Responsibilities refer to the tasks that a participant is expected to perform to accomplish the project objectives as specified by contractual agreement and applicable laws, codes, standards, and regulatory guidelines. Requirements are what each team members expects to achieve or needs to receive during and after their participation in a project.

Chung (1999) states, "Quality may mean different things to different people. Some take it to represent customer satisfaction, others interpret it as compliance with contractual requirements, yet others equate it to attainment of prescribed standards" (p. 3). As regards quality of construction, he further states,

Overview of Quality

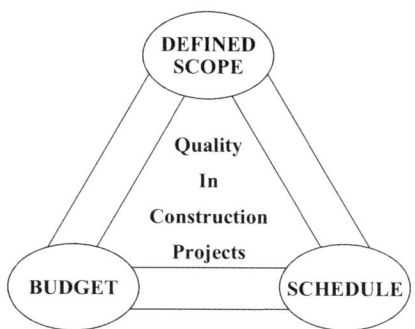

FIGURE 1.7 Construction project quality trilogy.

> Quality of construction is even more difficult to define. First of all, the product is usually not a repetitive unit but a unique piece of work with specific characteristics. Secondly, the needs to be satisfied include not only those of the client but also the expectations of the community into which the completed building will integrate. The construction cost and time of delivery are also important characteristics of quality (p. 3).

The definition of quality for construction projects is different from that of manufacturing or services industries as the product is not repetitive but a unique piece of work with specific requirements. Quality in construction projects is not only the quality of product and equipment used in the construction of a facility but also the total management approach to complete the facility. The quality of construction depends mainly upon the control of construction, which is the primary responsibility of the contractor.

Based on aforementioned, quality of construction projects can be defined as follows: construction project quality is fulfillment of owner's needs as per defined scope of works within a budget and specified schedule to satisfy owner's/user's requirements. The phenomenon of these three components can be called as "construction project quality trilogy" and is illustrated in Figure 1.7

1.5 QUALITY MANAGEMENT SYSTEM

Quality management is an organization-wide approach to understand customer needs and deliver the solutions to fulfill and satisfy the customer. Quality management is managing and implementation of quality system to achieve customer satisfaction at the lowest overall cost to the organization while continuing to improve the process. Quality system is a framework for quality management. It embraces the organization structure, policies, procedures, and processes needed to implement quality management system.

Quality management in construction projects is different to that of manufacturing. Quality in construction projects is not only the quality of products and equipment used in the construction but also the total management approach to complete the facility as per the scope of works to customer/owner satisfaction within the budget and to be completed within specified schedule to meet owner's defined purpose. Quality management in construction addresses both the management of project and the product of the project and all the components of the product. It also involves incorporation of changes or improvements, if needed. Construction Project quality is fulfillment of owner's needs as per defined scope of works within a budget and specified schedule to satisfy owner's/user's requirements.

Quality management system in construction projects mainly consists of the following:

- Quality management planning
- Quality assurance
- Quality control

Each of these processes and activities is to be performed during following main stages of construction project:

1. Study
2. Design
3. Bidding and tendering
4. Construction

1.5.1 Quality Plan

The quality management plan for construction projects is part of the overall project documentation addressing and describing the procedures to manage construction quality and project deliverable. The quality management plan identifies following key components:

- Details of the quality standards and codes to be complied
- Project objectives and project scope of work
- Stakeholders quality requirements
- Regulatory requirements
- Quality matrix for different stages
- Design criteria
- Design procedures
- Detailed construction drawings
- Detailed work procedure
- Well-defined specification for all the materials, products, components, and equipment to be used to construct the facility
- Manpower and other resources to be used for the project
- Inspection and testing procedures
- Quality assurance activities
- Quality control activities
- Defect prevention, corrective action, and rework procedure
- Project completion schedule
- Cost of the project
- Documentation and reporting procedure

1.5.1.1 Designer's Quality Management Plan

Construction projects have involvement of owner, designer (consultant), and contractor. In order to achieve project objectives, both designer and contractor have

Overview of Quality

to develop project quality management plan. The designer's quality management plan shall be based on owner's project objectives, whereas the contractor's plan shall take into consideration requirements of contract documents.

Figure 1.8 illustrates project quality management plan for design stage.

1.5.1.2 Contractor's Quality Control Plan

During construction stage, the contractor prepares contractor's quality control plan (CQCP) based on project-specific requirements. The CQCP is the contractor's everyday

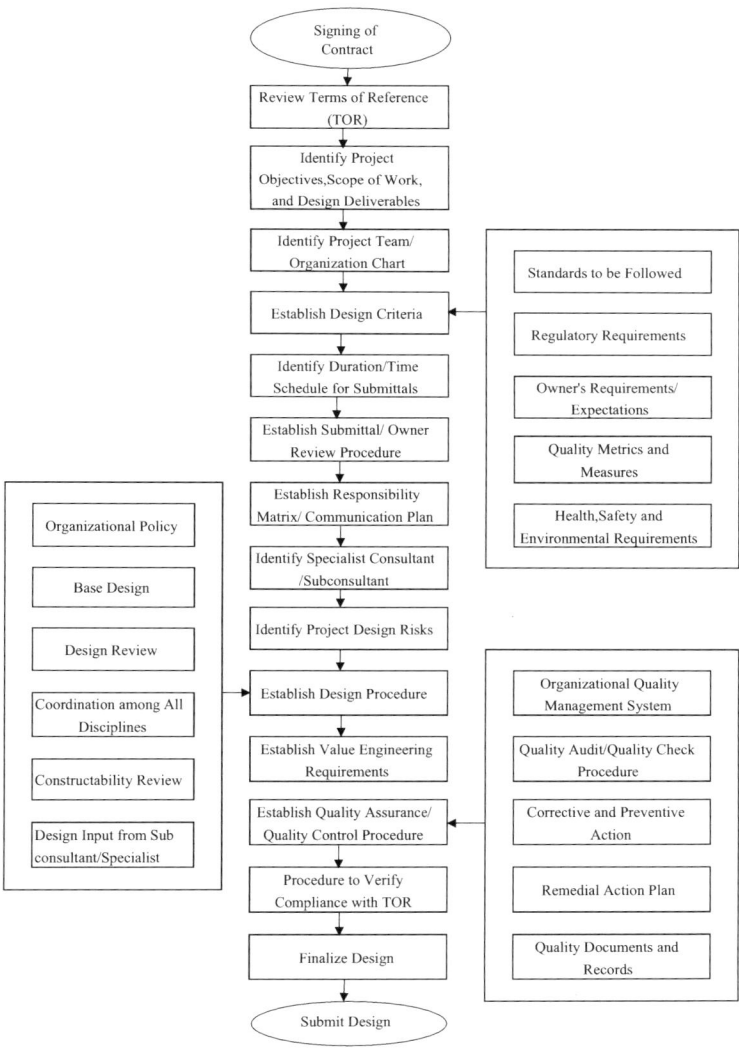

FIGURE 1.8 Project quality management plan for design stage (FEED). (Abdul Razzak Rumane. (2013). *Quality Tools for Managing Construction Projects.* Reprinted with permission from Taylor & Francis Group.) FEED, front-end engineering design.

tool to ensure meeting the performance standards specified in the contract documents. Its contents are drawn from the company's quality system, the contract, and related documents. It is a framework for the contractor's process for achieving quality construction. It is a document setting out the specific quality activities and resources pertaining to a particular contract or project. It is the documentation of contractor's process for delivering the level of construction quality required by the contract. A quality plan is virtually a quality manual tailor-made for the project and is based on contract requirements. Figure 1.9 illustrates process for development of CQCP, and Table 1.8 illustrates table of contents of CQCP.

1.5.1.3 Quality Matrix

In case of construction projects, an organizational framework is established and implemented mainly by three parties: owner, designer/consultant, and contractor.

Table 1.9 illustrates the responsibilities matrix for quality control–related personnel.

1.5.2 Quality Assurance

Quality assurance in construction projects covers all activities performed by design team, contractor, and quality controller/auditor (supervision staff) to meet owner's objectives as specified and to ensure and guarantee that the project/facility is fully functional to the satisfaction of owner/end user. Auditing is part of the quality assurance function.

Quality assurance is the activity for providing evidence to establish confidence, among all concerned, that quality-related activities are being performed effectively. All these planned or systematic actions are necessary to provide adequate confidence that a product or service will satisfy given requirements for quality.

Quality assurance covers all activities from design, development, production/construction, installation, and servicing to documentation and also includes regulations of the quality of raw materials; assemblies, products and components, and services related to production; and management, production, and inspection processes. Following are major activities to be performed for quality assurance of the construction project:

- Confirm that owner's needs and requirements are included in the scope of works (Terms of Reference [TOR]).
- Review and confirm design compliance to TOR.
- Execute works complying with the specified standards and codes.
- Conform to regulatory requirements.
- Execute works as per approved shop drawings.
- Install approved material and equipment on the project.
- Install as per approved method statement or manufacturer's recommendation.
- Coordinate among all the trades.
- Inspect continuously during construction/installation process.
- Identify and correct the deficiencies.
- Submit and review transmittals in time.

Overview of Quality

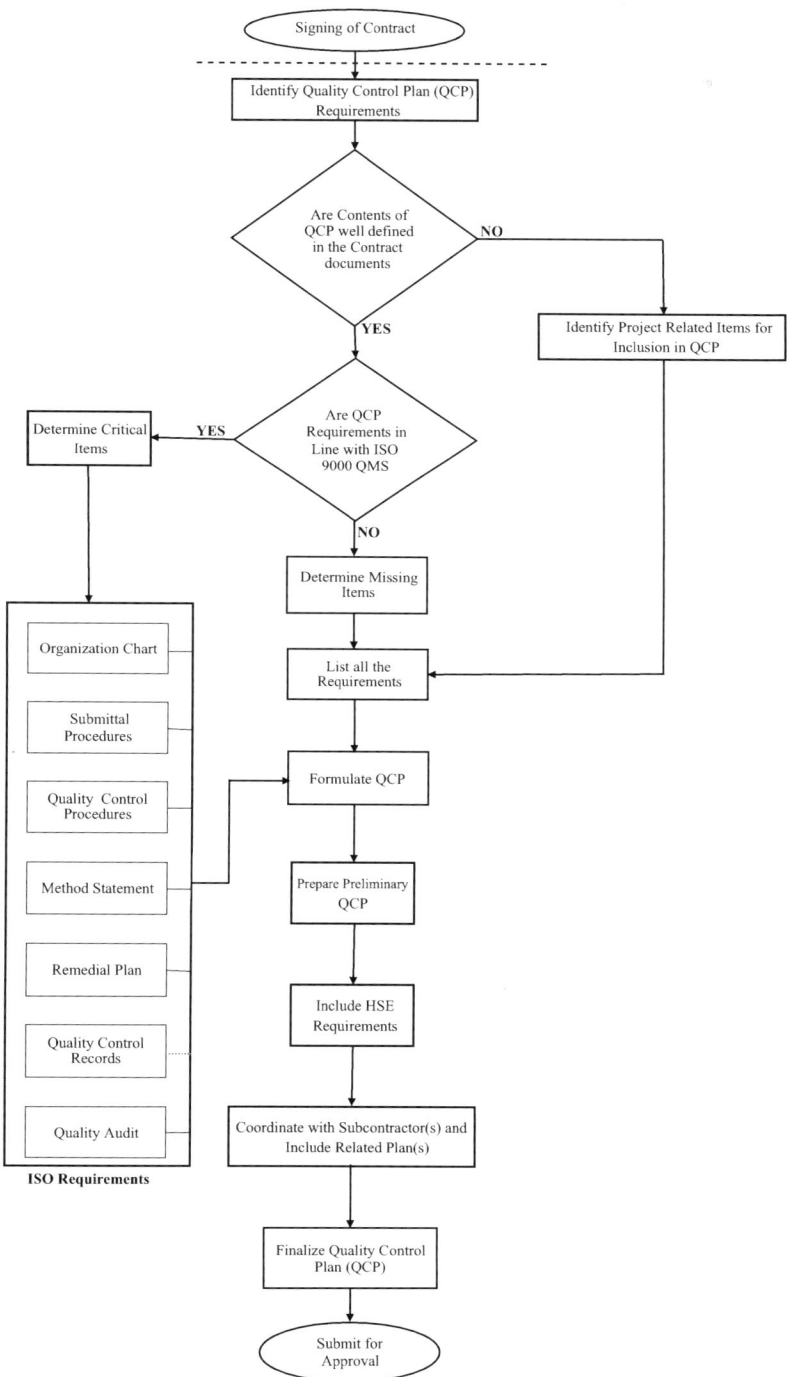

FIGURE 1.9 Logic flow diagram for development of contractor's quality control plan.

TABLE 1.8
Contents of Contractor's Quality Control Plan

Section	Topic
1	Introduction
2	Description of project
3	Quality control organization
4	Qualification of quality control staff
5	Responsibilities of quality control personnel
6	Procedure for submittals
7	Quality control procedure
7.1	Procurement
7.2	Inspection of site activities (checklists)
7.3	Inspection and testing procedure for systems
7.4	Off-site manufacturing, inspection, and testing
7.5	Procedure for laboratory testing of material
7.6	Inspection of material received at site
7.7	Protection of works
8	Method statement for various installation activities
9	Project-specific procedures
10	Quality control records
11	Company's quality manual and procedure
12	Periodical testing
13	Quality updating program
14	Quality auditing program
15	Testing, commissioning, and handover
16	Health, safety, and environment

Source: Abdul Razzak Rumane. (2016). *Handbook of Construction Management: Scope, Schedule, and Cost Control.* Reprinted with permission from Taylor & Francis Group.

1.5.3 Quality Control

Quality control in construction projects is performed at every stage through the use of various control charts, diagrams, check lists, and so on and can be defined as follows:

- Checking and review of project design
- Checking and review of bidding/tendering documents
- Analysis of contractor's bids
- Checking of executed/installed works to confirm that works have been performed/executed as specified, using specified/approved materials, installation methods and specified references, codes, and standards to meet intended use
- Controlling budget
- Planning, monitoring, and controlling project schedule

TABLE 1.9
Responsibilities for Site Quality Control

Sr. No.	Description	Owner/Project Manager Construction Manager	Consultant/Designer	Contractor Manager	Quality In-charge	Quality Engineers	Site Engineers	Safety Officer	Head Office
		Owner	Consultant	Contractor					
1	Specify quality standards	□	■					□	□
2	Prepare quality control plan			□	■	□			
3	Control distribution of plans and specifications			□	□	□			
4	Submittals			■			■		■
5	Prepare procurement documents			□			■		
6	Prepare construction method procedures			□			■		
7	Inspect work in progress		■				■		
8	Accept work in progress		■		□				
9	Stop work in progress	■	■						
10	Inspect materials upon receipt		■	□	■		■		
11	Monitor and evaluate quality of works		■	□	■	■	■		
12	Maintain quality records		□	■					
13	Determine disposition of nonconforming items	□	□	■					
14	Investigate failures	□	□	■	■	□	■		
15	Site safety			□				■	
16	Testing and commissioning	□	■	□			■		
17	Acceptance of completed works	■	□	□					

■, primary responsibility; □, advise/assist

Source: Abdul Razzak Rumane (2017). *Quality Management in Construction Projects*, Second Edition. Reprinted with permission from Taylor & Francis Group.

The construction project quality control process is a part of contract documents, which provide details about specific quality practices, resources, and activities relevant to the project. The purpose of quality control during construction is to ensure that the work is accomplished in accordance with the requirements specified in the contract. Inspection of construction works is carried out throughout the construction period either by the construction supervision team (consultant) or appointed inspection agency. Quality is an important aspect of construction project. The quality of construction project must meet the requirements specified in the contact documents. Normally, contractor provides on-site inspection and testing facilities at construction site. On a construction site, inspection and testing is carried out at three stages during the construction period to ensure quality compliance.

1. During construction process. This is carried with the checklist request submitted by the contractor for testing of ongoing works before proceeding to next step.
2. Receipt of subcontractor or purchased material or services. This is performed by a material inspection request submitted by the contractor to the consultant upon receipt of material.
3. Before final delivery or commissioning and handover.

1.6 INTEGRATED QUALITY MANAGEMENT SYSTEM

The integrated quality management system (IQMS) is the integration and proper coordination of functional elements of quality to achieve efficiency and effectiveness in implementation and maintaining an organization's quality management system to meet customer requirements and satisfaction. IQMS consists of any element or activity that has an effect on quality. Customer satisfaction is the goal of quality objectives.

During the past three decades, many programs have been implemented for organizational improvements. In the 1980s, programs such as statistical process control, various quality tools, and total quality management were implemented. In the 1990s, the ISO 9000 came into being, which resulted in improved productivity, cost reduction, improved time, improved quality, and customer satisfaction.

With globalization and competition, it became necessary for organizations to continuously improve to achieve the highest performance and a competitive advantage.

In the 1980s, the major challenge facing most organizations was to improve quality. In the 1990s, it was to improve faster by restructuring and reengineering all operations.

In today's global competitive environment, organizations are facing many challenges due to an increase in customer demand for higher performance requirements at a competitive cost. They are finding that their survival in the competitive market is increasingly in doubt. To achieve a competitive advantage, effective quality improvement is critical.

Processes and systems are essential for the performance and expansion of any organization. The ISO 9000 is an excellent tool to develop a strong foundation for good processes and systems. The ISO 9000 quality management system is accepted worldwide, and the ISO 9000 certification has global recognition.

An IQMS is developed by merging recommendations and specifications from ISO 9000 (quality management system), ISO 14000 (environmental management system),

Overview of Quality 43

and OHSAS 18000 (occupational health and safety management) together with other contract documents. If an organization has a certified quality management system (ISO 9000), it can build an IQMS system by adding environmental, health, safety, and other requirements of management system standards.

The benefits of implementing an IQMS are as follows:

- Reduced duplication and, therefore, cost
- Improved resource allocation
- Standardized process
- Elimination of conflicting responsibilities and relationship
- Consistency
- Improved communication
- Reduced risk and increase profitability
- Facilitated training development
- Simplified document maintenance
- Reduced record keeping
- Ease of managing legal and other requirements

Construction projects are unique and nonrepetitive in nature and have their own quality requirements that can be developed by integration of project specifications and an organization's quality management system. Normally, quality management system manuals consist of procedures to develop project quality control plans, taking into consideration contract specifications. This plan is called the CQCP. Certain projects specify which value engineering studies should be undertaken during the construction phase. The contractor is required to include these while developing the CQCP. This plan can be termed an IQMS for construction projects. The contractor has to implement a quality system to ensure that the construction is carried out in accordance with the specification details and approved CQCP. Figure 1.10 illustrates the logic flowchart for development of IQMS for construction projects.

1.7 RISK MANAGEMENT

Risk management is the process of identifying, assessing, prioritizing different kinds of risks, planning risk mitigation, implementing mitigation plan, and controlling the risks. It is a process of thinking systematically about the possible risks, problems, or disasters before they happen and setting up the procedure that will avoid the risk, or minimize the impact, or cope with the impact. The objectives of project risk management are to increase the probability and impacts of positive events and decrease the probability and impacts of events adverse to the project objectives. Risk is the probability that the occurrence of an event may turn into undesirable outcome (loss, disaster). It is virtually anything that threatens or limits the ability of an organization to achieve its objectives. It can be unexpected and unpredictable events that have the potential to damage the functioning of organization in terms of money, or in worst scenario, it may cause the business to close.

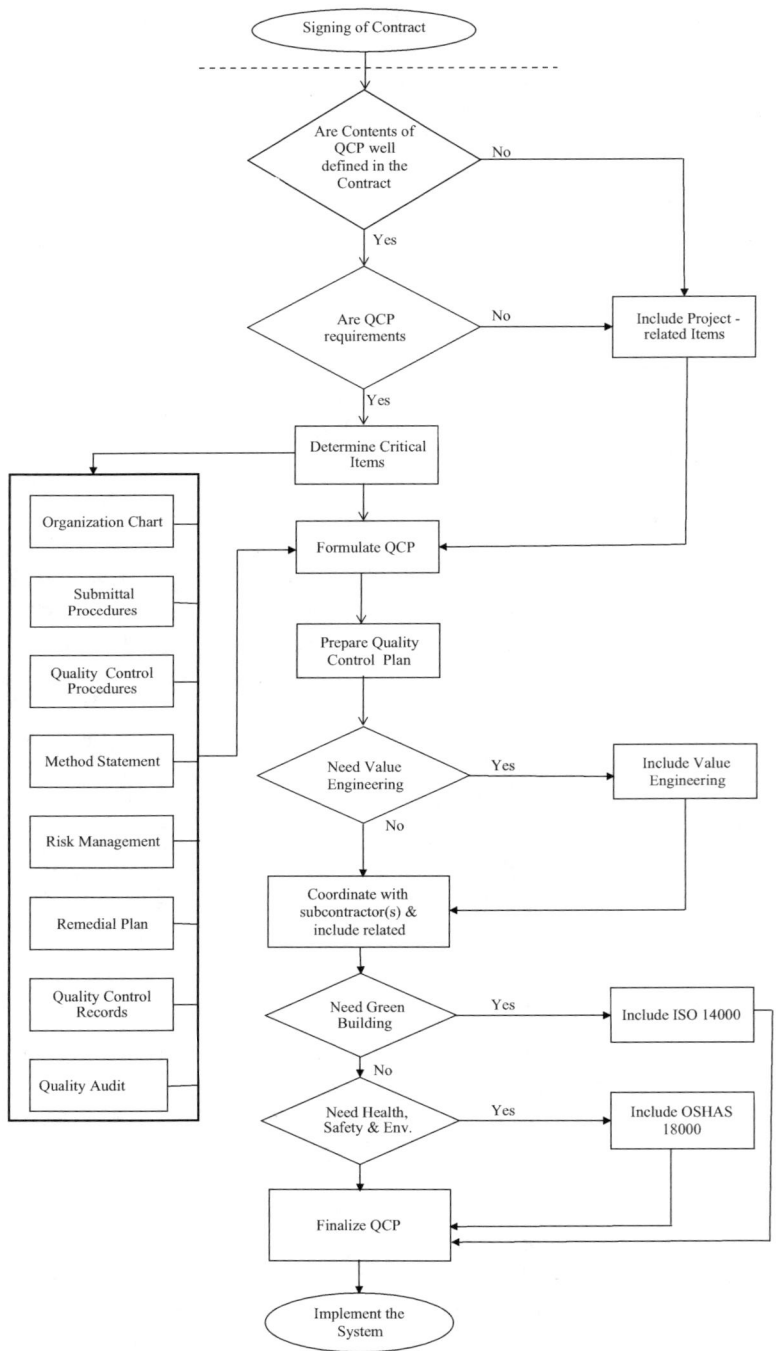

FIGURE 1.10 Logic flowchart for development of IQMS. IQMS, integrated quality management system.

Overview of Quality

1.7.1 Risk Management Process

Risk management process is designed to reduce or eliminate the risk of certain kinds of events happening (occurring) or having an impact on the project. The risk process consists of the following steps:

1. Identify the potential sources of risk on the project.
2. Analyze their impact on the project.
   a. Qualitative
   b. Quantitative
3. Select those with a significant impact on the project.
   a. Prioritization
4. Determine how the impact of risk can be reduced.
   a. Avoidance
   b. Transfer
   c. Reduction
   d. Retention (acceptance)
5. Select the best alternative.
6. Develop and implement mitigation plan.
7. Monitor and control the risks by implementing risk response plan, tracking identified risks, identifying new risks, and evaluating the risk impact.

Figure 1.11 illustrates risk management cycle (process).

1.7.1.1 Identify Risk

Risk identification involves determining the source and type of risk, which may affect the project.

The following tools and techniques are used to identify risks:

- Benchmarking
- Brainstorming
- Delphi technique
- Interviews

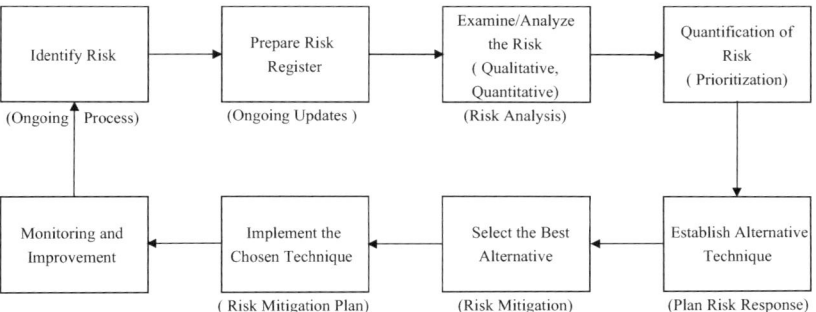

FIGURE 1.11 Risk management cycle. (Abdul Razzak Rumane. (2013). *Quality Tools for Managing Construction Projects*. Reprinted with permission from Taylor & Francis Group.)

- Past database and historical data from similar projects
- Questionnaires
- Risk breakdown structure
- Workshops

The identified risks are classified into the following:

1. Internal
2. External

These are further divided into the following main categories:

- Management
- Project (contract)
- Technical
- Construction
- Physical
- Logistic
- Health, safety, and environmental (HSE)
- Statutory/regulatory
- Financial
- Commercial
- Economical
- Political
- Legal
- Natural

Each identified risk is documented in a risk register.

1.7.1.1.1 Risk Register

Risk register is a document that records details of all the identified risks at the beginning of the project and during the life cycle of the project in a format that consists of comprehensive lists of significant risks along with the actions and cost estimated with the identified risks. Risk register is updated every time a new risk is identified or relevant actions are taken. Table 1.10 illustrates an example risk register.

1.7.1.2 Analyze/Assess Risk

Risk analysis is the process to analyze the listed risks. There are two methods of analyzing risks. These are as follows:

1. Qualitative analysis
2. Quantitative analysis

1.7.1.2.1 Qualitative Analysis

Qualitative analysis is a process to assess the probability of occurrence (likelihood) of the risk and its impact (consequence).

Overview of Quality

TABLE 1.10
Risk Register

Serial Number	Risk Identification Number (Risk ID)	Description of Risk	Owner of Risk	Estimated Likelihood of Risk	Impact	Estimated Severity	Prioritization	List of Activities Influenced	Leading Indicators for Risk	Risk Mitigation Plan	Risk Mitigation Plan on Leading Indicator	Timeline for Mitigation Action	Tracking of Leading Indicators	Date of Review/Update	Forecasting Risk Happenings	Action to be Taken in Future

Source: Abdul Razzak Rumane. (2016). *Handbook of Construction Management: Scope, Schedule, and Cost Control.* Reprinted with permission from Taylor & Francis Group.

The following tools and techniques are used for qualitative analysis:

- Failure mode and effects analysis
- Group discussion (workshop)
- Pareto diagram
- Probability and impact assessment
- Probability levels
- Risk categorization

1.7.1.2.2 Quantitative Analysis

Quantitative analysis is a process to quantify the probability of risk and its impact based on numerical estimation.

The following tools and techniques are used for quantitative analysis:

- Event tree analysis
- Probability analysis
- Sensitivity analysis
- Simulation techniques (Monte Carlo simulation)

Table 1.11 illustrates risk probability levels normally assumed for probability and impact matrix. The % probability of occurrence shown in the table is indicative. The organization can determine the probability level as per the nature of business.

Similarly, the risk impact is analyzed for each of the identified risk that may be classified into different levels such as follows:

- Very high
- High
- Substantial
- Possible
- Slight

TABLE 1.11
Risk Probability Levels

Serial Number	Value	Definition	Meaning	% Probability of Occurrence
1	Level 5	Very high (frequent)	Almost certain that the risk will occur. Frequency of occurrence is very high.	40–80
2	Level 4	High (likely)	It is likely to happen. Frequency of occurrence is less.	20–40
3	Level 3	Moderate (occasional)	Its occurrence is occasional.	10–20
4	Level 2	Low (unlikely)	It is unlikely to happen.	5–10
5	Level 1	Very low (rare)	The probability to occur is rare.	0–5

Source: Abdul Razzak Rumane. (2016). *Handbook of Construction Management: Scope, Schedule, and Cost Control.* Reprinted with permission from Taylor & Francis Group.

Overview of Quality

TABLE 1.12
Risk Probability and Impact Matrix

			Probability			
			Level 4	Level 3	Level 2	Level 1
			High (Likely)	Moderate (Occasional)	Low (Unlikely)	Very Low (Rare)
Probability		Severity				
Impact/severity	Level 5	Very high				
	Level 4	High				
	Level 3	Substantial				
	Level 2	Possible				
	Level 1	Slight				

Risk assessment matrix for; scope/schedule/cost/quality

Based on the impact level, probability and impact matrix is prepared. Table 1.12 is a sample risk probability and impact matrix.

1.7.1.3 Prioritization

It is the process to prioritized list of quantified risks. The results of risk assessment are used to prioritize risks to establish very high to very low ranking. Prioritization of risks depends on following factors:

1. Probability (occurrence)
2. Impact (consequences)
3. Urgency
4. Proximity
5. Manageability
6. Controllability
7. Responsiveness
8. Variability
9. Ownership ambiguity

The prioritization list helps the project manager to plan actions and assign the resources to mitigate the realization of high-value probability.

1.7.1.4 Plan Risk Response

Plan risk response is a process that determines what action (if any) will be taken to address the identified and assessed risks, which are listed under risk register on prioritization basis. Risk response process is used for developing options and actions to enhance opportunities and reduce the threats to the identified risk activities in the project.

For each identified risk, a response must be identified. The risk owner and project team have to select the risk response for each of the identified risk.

The probability of the risk event occurring and the impacts (threats) is the basis for evaluating the degree to which the response action is to be evolved. Based on the risk probability and impact matrix, risk response strategy for scope, schedule, cost, quality, and HSE can be listed as per Table 1.13.

Generally, risk response strategies for impact (consequences) on the project fall into one of the following categories:

1. Avoidance
2. Transfer
3. Mitigation (reduction)
4. Acceptance (retention)

1.7.1.4.1 Avoidance

Avoidance is changing the project scope, objectives, or plan to eliminate the risk or to protect the project objectives from the impact (threat).

1.7.1.4.2 Transfer

Transfer is transferring the risk to someone else who will be responsible to manage the risk. Transferring the threat does eliminate the threat, however it still exists, but it is owned and managed by other party and to be resolved by that party.

1.7.1.4.3 Mitigation

Mitigation is reduction in the probability and/or impact to an acceptable threshold. It is done by taking series of control actions.

TABLE 1.13
Risk Response Strategy

Potential Risk		Probability Level	Impact/Severity Level	Risk Response Strategy
Scope	1			
	2			
Schedule	1			
	2			
Cost	1			
	2			
Quality	1			
	2			
HSE	1			
	2			

HSE, health, safety, and environmental.

Overview of Quality

1.7.1.4.4 Acceptance

It is acceptance of consequences after response actions understanding the risk impact should it occur.

1.7.1.5 Reduce Risk

It is identifying various steps to reduce the probability and/or impact of the risk. Taking early steps to reduce the probability of risk is more effective and less costly than repairing the damage after the occurrence of the risk.

1.7.1.6 Monitor and Control Risk

It is a systematic process of tracking identified risks, monitoring residual risks, identifying new risks, execution of risk response plan, and evaluating the effectiveness of implementation of actions against established levels of risk in the area of scope, time, cost, and quality throughout the life cycle of the project. It involves timely implementation of risk response to identified risk to ensure the best outcome for a risk to a project.

1.7.2 RISKS IN CONSTRUCTION PROJECTS

Construction projects have many varying risks. Risk management throughout the life cycle of the project is important and essential to prevent unwanted consequences and effects on the project. Construction projects have involvement of many stakeholders such as project owners, developers, design firms (consultants), contractors, banks, and financial institutions funding the project who are affected by the risk. Each of these parties has involvement with certain portion of overall construction project risk; however, the owner has a greater share of risks as the owner is involved from the inception until completion of project and beyond. The owner must take initiatives to develop risk consciousness and awareness among all the parties emphasizing upon the importance of explicit consideration of risk at each stage of the project as the owner is ultimately responsible for overall project const. Traditionally,

1. Owner/client is responsible for the investment/finance risk.
2. Designer (consultant) is responsible for the design risk.
3. Contractors and subcontractors are responsible for the construction risk.

Construction projects are characterized as very complex projects, where uncertainty comes from various sources. Construction projects involve a cross section of many different participants. They have varying project expectations. Those both influence and depend on each other in addition to the "other players" involved in the construction process. The relationships and the contractual groupings of those who are involved are also more complex and contractually varied. Construction projects often require large amount of materials and physical tools to move or modify these materials. Most items used in construction projects are normally produced by other construction-related industries/manufacturers. Therefore, risk in construction projects is multifaceted. Construction projects inherently contain a high degree of risk in their projection of cost and time as each is unique. No construction project is without any risk. Risk management in construction projects is mainly focused on delivering the project with the following:

1. What was originally accepted (as per **defined scope**)
2. Agreed-upon time (as per **schedule** without any delay)
3. Agreed-upon budget (no overruns to accepted **cost**)

Risk management is an ongoing process. In order to reduce the overall risk in construction projects, the risk assessment (identification, analysis, and evaluation) process must start as early as possible to maximize project benefits. There are a number of risks that can be identified at each stage of the project. Early risk identification can lead to better estimation of the cost in the project budget, whether through contingencies, contractual, or insurance. Risk identification is the most important function in construction projects.

Risk factors in construction projects can be categorized into a number of ways according to level of details or selected viewpoints. These are categorized based on various risk factors and source of risk. Contractor has to identify related risks affecting the construction, analyze these risks, evaluate the effects on the contract, and evolve the strategy to counter these risks, before bidding for a construction contract. Construction project risks mainly relate to the following:

- Scope and change management
- Schedule/time management
- Budget/cost management
- Quality management
- Resources and manpower management
- Communication management
- Procurement/contract management
- HSE management

Table 1.14 illustrates typical categories of risks in construction projects.

Quality of construction projects is defined as follows: Construction project quality is fulfillment of owner's needs as per defined scope of works within a budget and specified schedule to satisfy owner's/user's requirements. The phenomenon of these three components can be called as "construction project trilogy". Risk management in all these elements is required to maintain the quality of construction projects. Table 1.15 lists probable risks that occur during construction phase and their effects on scope, schedule, and cost.

1.8 SUPPLY CHAIN MANAGEMENT

Supply chain management is a new concept of doing business, which has been recognized and practiced by the manufacturing industry for the last two decades. It has also emerged as new practice in construction industry.

Because of emerging technological advances and to maintain competitive advantage in business environment, it has become essential to plan, organize, and control

TABLE 1.14
Typical Categories of Risks in Oil and Gas Projects

Sr. No.	Category	Types
1	Management	Selection of project delivery and contracting system
		Selection of project management consultant
2	Contract (project)	Scope/design changes
		Schedule
		Cost
		Conflict resolution
		Delay in changer order negotiations
3	Statutory	Statutory/regulatory delay
4	Technical	Incomplete design
		Incomplete scope of work
		Design changes
		Design mistakes
		Errors and omissions in contract documents
		Incomplete specifications
		Ambiguity in contract documents
		Inconsistency in contract documents
		Inappropriate schedule/plan
		Inappropriate construction method
		Conflict with different trades
		Improper coordination with regulatory authorities
		Inadequate site investigation data
5	Technology	New technology
6	Construction	Delay in mobilization
		Delay in transfer of site
		Different site conditions to the information provided
		Changes in scope of work
		Resource (labor) low productivity
		Equipment/plant productivity
		Insufficient skilled workforce
		Union and labor unrest
		Failure/delay of machinery and equipment
		Quality of material
		Failure/delay of material delivery
		Delay in approval of submittals
		Extensive subcontracting
		Subcontractor's subcontractor
		Failure of project team members to perform as expected
		Information flow breaks

(*Continued*)

TABLE 1.14 (*Continued*)
Typical Categories of Risks in Oil and Gas Projects

Sr. No.	Category	Types
7	Physical	Damage to equipment
		Structure collapse
		Damage to stored material
		Leakage of hazardous material
		Theft at site
		Fire at site
8	Logistic	Resources availability
		Spare parts availability
		Consistent fuel supply
		Transportation facility
		Access to worksite
		Unfamiliarity with local conditions
9	Health, safety, and environment	Injuries
		Health and safety rules
		Environmental protection rules
		Pollution rules
		Disposal of waste
10	Financial	Inflation
		Recession
		Fluctuations in exchange rate
		Availability of foreign exchange (certain countries)
		Availability of funds
		Delays in payment
		Local taxes
11	Economical	Variation of construction material price
		Sanctions
12	Commercial	Import restrictions
		Custom duties
13	Legal	Permits and licenses
		Professional liability
		Litigation
14	Political	Change in laws and regulations
		Constraints on employment of expatriate workforce
		Use of local agent and firms
		Civil unrest
		War
15	Natural/environmental	Flood
		Earthquake
		Cyclone
		Sandstorm
		Landslide
		Heavy rains
		High humidity
		Fire

(*Continued*)

TABLE 1.14 (*Continued*)
Typical Categories of Risks in Oil and Gas Projects

Sr. No.	Category	Types
16	Geological	Estimation of accessible reserves
17	Supply and demand	Financial crises and macroeconomic factors
18	Operational	Availability of skilled workers

EPC, engineering, procurement, and construction.
Source: Abdul Razzak Rumane. (2013). *Quality Tools for Managing Construction Projects.* Reprinted with permission from Taylor & Francis Group.

TABLE 1.15
Potential Risks on Scope, Schedule, and Cost during Construction Phase and Its Effects and Mitigation Action

Sr. No.	Potential Risk	Probable Effects	Control Measures/Mitigation Action
		1.0 Scope	
1.1	Scope/design changes	• Project schedule • Project cost • Claim	• Compress schedule. • Resolve change order issues in order not to delay the project.
1.2	Different site conditions to the information provided	• Change in scope of work • Delay in project	• Contractor has to investigate site conditions prior to starting the relevant activity.
1.3	Inadequate site investigation data	• Additional work • Scope change	• Contractor has to investigate site conditions prior to starting the relevant activity.
1.4	Conflict in contract documents	• Project delay	• Amicably resolve the issue.
1.5	Incomplete design	• Project scope • Project schedule • Project cost	• Raise request for information (RFI). • Resolve issue in accordance with ontract documents
1.6	Incomplete scope of work	• Project scope • Project schedule • Project cost	• Raise RFI. • Resolve issue in accordance with contract documents.
1.7	Design changes	• Project scope • Project schedule • Project cost	• Follow contract documents for change order.
1.8	Design mistakes	• Project scope • Project schedule • Project cost	• Raise RFI. • Resolve issue in accordance with contract documents.
1.9	Errors and omissions in contract documents	• Project scope • Project schedule • Project cost	• Raise RFI. • Resolve issue in accordance with contract documents.

(*Continued*)

TABLE 1.15 (*Continued*)
Potential Risks on Scope, Schedule, and Cost during Construction Phase and Its Effects and Mitigation Action

Sr. No.	Potential Risk	Probable Effects	Control Measures/Mitigation Action
1.10	Incomplete specifications	• Project scope • Project schedule • Project cost	• Raise RFI. • Resolve issue in accordance with contract documents.
1.11	Conflict with different trades	• Project delay	• Coordinate with all trades while preparing coordination and composite drawings.
1.12	Inappropriate construction method	• Project delay • Claim	• Raise RFI and correct the method statement.
1.13	Quality of material	• Project delay	• Locate suppliers having proven record of supplying quality product.
2.0 Schedule			
2.1	Incompetent subcontractor	• Project delay • Project quality	• Contractor has to monitor the workmanship and work progress.
2.2	Delay in transfer of site	• Project delay	• Contractor has to adjust the construction schedule.
2.3	Delay in mobilization	• Project delay	• Adjust construction schedule accordingly.
2.4	Project schedule	• Project completion	• Compress duration of activities.
2.5	Inappropriate schedule/plan	• Project delay	• Contractor has to prepare schedule taking into consideration site conditions all the required parameters.
2.6	Delay in change order negotiations	• Project schedule	• Request owner/supervisor/project manager to expedite the negotiations and resolve the issue.
2.7	Resource availability (material)	• Project delay	• Contractor has to make extensive search.
2.8	Resource (labor) low productivity	• Project quality • Project delay	• Contractor has to engage competent and skilled labors.
2.9	Equipment/plant productivity	• Project delay	• Contractor has to hire/purchase equipment to meet project productivity requirements.
2.10	Insufficient skilled workforce	• Project duration	• Contractor has to arrange workforce from alternate sources.
1.11	Failure/delay of machinery and equipment	• Project delay	• Contractor has to plan procurement well in advance.
1.12	Failure/delay of material delivery	• Project delay	• Contractor has to plan procurement well in advance.
2.13	Delay in approval of submittals	• Project delay	• Notify owner/project manager.
2.14	Delays in payment	• Project delay • Claim	• Contractor has to have contingency plans. • Owner has to pay as per contract.

(*Continued*)

Overview of Quality

TABLE 1.15 (*Continued*)
Potential Risks on Scope, Schedule, and Cost during Construction Phase and Its Effects and Mitigation Action

Sr. No.	Potential Risk	Probable Effects	Control Measures/Mitigation Action
2.15	Statutory/regulatory delay	• Project delay	• Regular follow-up by the contractor and owner with the regulatory agency.
		3.0 Cost	
3.1	Low bid project cost	• Project quality	• Contractor has to try competitive material and improve method statement and higher production rate from its manpower.
3.2	Variation in construction material price	• Project quality • Project cost	• Contractor has to negotiate with supplier/manufacturer for best price. Contractor has to request for change order if applicable as per contract.
3.3	Damage to equipment	• Schedule	• Regularly maintain the equipment. Take immediate action to repair damage equipment.
3.4	Damage to stored material	• Project delay • Material quality	• Contractor has to follow proper storage system.
3.5	Structure collapse	• Injuries • Project delays	• Contractor has to ensure that formwork and scaffolding is properly installed.
3.6	Leakage of hazardous material	• Safety Hazards	• Contractor has to take necessary protect to avoid leakage. Store in safe area.
		4.0 General	
4.1	Failure of team members not performing as expected	• Project quality • Project delay	• Select competent candidate. Provide training.
4.2	Change in laws and regulations	• Scope/specification changes • Variation order	• Contractor has to inform owner/consultant and raise RFI.
4.3	Access to worksite	• Extra/additional time to access site	• Access road to be planned in coordination with adjacent area and local authority.
4.4	Theft at site	• Project delay	• Contractor has to monitor access to site. Record entry/exit to the site. Provide fencing around project site.
4.5	Fire at site	• Project delay	• Contractor has to install temporary firefighting system. Inflammable material should be stored in safe and secured place with necessary safety measures.
4.6	Injuries	• Project delay	• Contractor has to keep first aid provision at site. Take immediate action to provide medical aid.

(*Continued*)

TABLE 1.15 (*Continued*)
Potential Risks on Scope, Schedule, and Cost during Construction Phase and Its Effects and Mitigation Action

Sr. No.	Potential Risk	Probable Effects	Control Measures/Mitigation Action
4.7	New technology	• Scope change • Schedule • Cost	• Owner/contractor has to mutually agree for changes in the contract for better performance of project.t

Source: Abdul Razzak Rumane. (2016). *Handbook of Construction Management: Scope, Schedule, and Cost Control.* Reprinted with permission from Taylor & Francis Group.

all supply chain–related activities at all the levels of product/project life cycle to ensure competitiveness of end product. It is a process to ensure that the customer gets right product or service having right quality and quantity on time.

Supply chain management is very complex process which has involvement of many participants both within and outside the organization. Each of these participants has influence over the activities to be performed to ensure availability of the product for further processing.

1.8.1 SUPPLY CHAIN MANAGEMENT PROCESS

Supply chain management is a process to manage and optimize the flow of products, materials, services, and information starting from the creation/inception of need/demand until its installation/usage for specified purpose to achieve high level of performance for organizational success. It is a process for reducing cost, enhancing quality, and reducing operation time.

Supply chain management process starts from the inception of need and covers all the stages and systems till the installation/fixing/implementation of the product, material, and services.

Figure 1.12 illustrates supply chain management process.

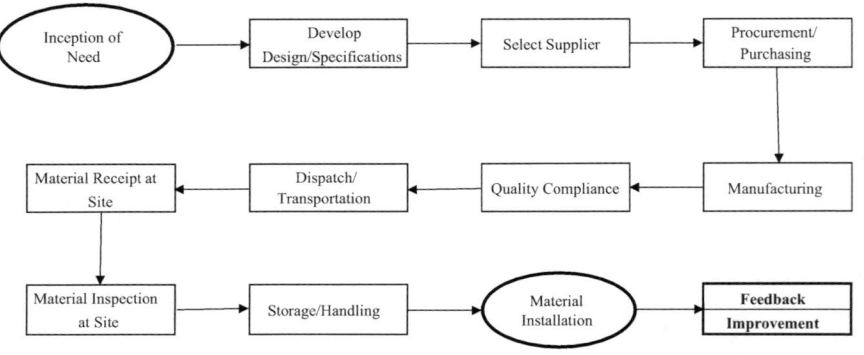

FIGURE 1.12 Supply chain management process.

Overview of Quality

1.8.2 Supply Chain Stakeholder

Supply chain management has involvement of many participants. Following are the main stakeholders and participants involved in supply chain management process:

1. Owner
2. Designer/specification developer
3. Purchasing/procurement department
4. Finance department
5. Manufacturer/supplier
6. Quality department
7. Dispatch section
8. Transportation
9. Receiving party
10. Incoming inspection
11. Material handling
12. Storage
13. Installation/usage

Table 1.16 illustrates matrix of stakeholders' responsibilities in supply chain management.

TABLE 1.16
Stakeholder's Responsibilities in Supply Chain Management

Serial Number	Stakeholder	Responsibilities
1	Owner	Inception of need
2	Designer	Develop design specifications
3	Procurement manager	Select competitive manufacturer/supplier
4	Finance manager	Arrange finance for procurement
5	Manufacturer/supplier	Produce/supply product as per specifications
6	Quality manager	Ensure the product meets specifications/product quality
7	Dispatch department	Ensure timely dispatch of material properly identified
8	Transportation department	Ensure economical and short duration process
9	Receiving department	Receive material in good condition
10	Incoming inspection department	Ensure incoming material conforms with specifications
11	Material handling	Proper handling of material
12	Store keeper	Proper storage of material
13	Installer/user	Installed as specified/recommended Give feed about the quality of the product

Source: Abdul Razzak Rumane. (2017). *Quality Management in Construction Projects*, Second Edition. Reprinted with permission from Taylor & Francis Group.

1.8.3 Supply Chain Management Processes

The following are the major processes that need to be implemented and monitored in supply chain process to achieve optimal result:

1. Product specification
2. Supplier selection
3. Purchasing
4. Product quality
5. Product handling
6. Improvement

1.8.3.1 Product Specification

In order to achieve efficient and smooth supply management process, it is essential to develop product design and specifications based on the customers' need/product requirements, goals, and objectives. The following are the major points that the designer has to consider while developing the design and product specifications:

- Suitability to the purpose and objectives
- Performance parameters and compatibility
- Sustainability (environmental, social, and economical)
- Availability
- Reliability
- Safety and security

1.8.3.2 Supplier Selection

The selection of supplier mainly depends on the following:

- Experience in supply of similar type of product/services
- Quality management certification
- Technical capability
- Current work load and capacity to supply future requirements
- Resources such as manpower and material
- Financial capability and stability
- Track record in timely execution of the past orders
- Product/supply rejection record
- Litigation record

1.8.3.3 Purchasing

Purchasing process starts with identification of potential suppliers who can deliver the needed product/material on time at competitive price conforming to the product/material quality.

Figure 1.13 illustrates purchasing process.

Overview of Quality

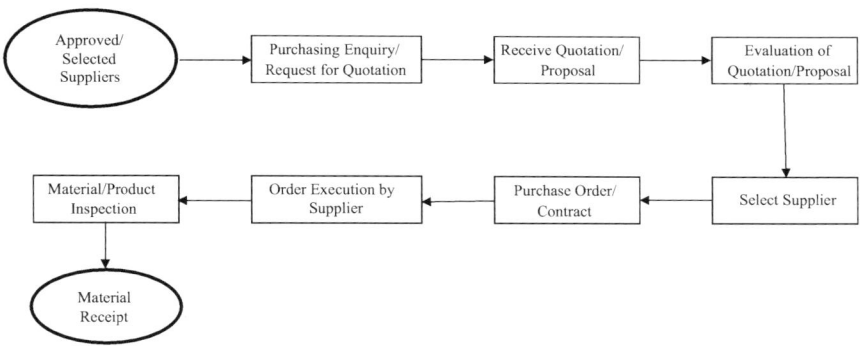

FIGURE 1.13 Purchasing process.

1.8.3.4   Product Quality

It is essential that both suppliers and customers are involved to ensure quality compliance of the product/material. The customer must discuss with the supplier about the acceptable level of product/material quality/performance. A quality metrics and method of inspection and quality control should be prepared for acceptance level. Quality management process to meet the customer requirements of the specific product/material should be placed in order to ensure smooth supply. Both parties should establish inspection and testing procedure to be mutually agreed and to be followed. Supplier and customer should work together to achieve zero defects policy.

1.8.3.5   Product Handling

Product handling is important to ensure the customer gets right quantity in right time. The following points are considered for proper handling of the product:

- Proper packing
- Product traceability and verification
- Proper shipping documentation
- Safe and risk free handling of the product
- Safe loading/unloading
- Shortest and reliable mode of transport to ensure product delivery in time
- Appropriate mode of shipment/transportation
- Maximum/minimum batch size to meet the ordered quantity
- Avoid breakage/damage during handling
- Safety during handling

1.8.3.6   Improvement

For competitive advantage, continual improvement is essential. The supplier has to ensure performance improvement. Customer feedback is necessary to improve the product performance. Supplier has to maintain proper communication with the customer to know the problems or potential problems related to the supplied product and take necessary steps to improve the quality and customer satisfaction.

1.8.4 Supply Chain in Construction Projects

Supply chain management in construction project is managing and optimizing the flow of construction materials, systems, equipment, and resources to ensure timely availability of all the construction resources without affecting the progress of works at the site. Figure 1.14 illustrates supply chain management process in construction projects.

In construction projects, the supply chain management starts from the inception of project. The designer has to consider the following points while specifying the products (materials, systems, equipment) for use/installation in the project:

- Quality management system followed by the manufacturer/supplier
- Quality of product
- Reliability of product
- Reliability of manufacturer/supplier
- Durability of product
- Availability of product for entire project requirement
- Price economy/cost efficient
- Sustainability
- Conformance to applicable codes and standards
- Manufacturing time
- Location of the manufacturer/supplier from the project site
- Interchangeability
- Avoid monopolistic product

Product specifications are documented in the construction documents (particular specifications). In certain projects, the document lists the names of recommended

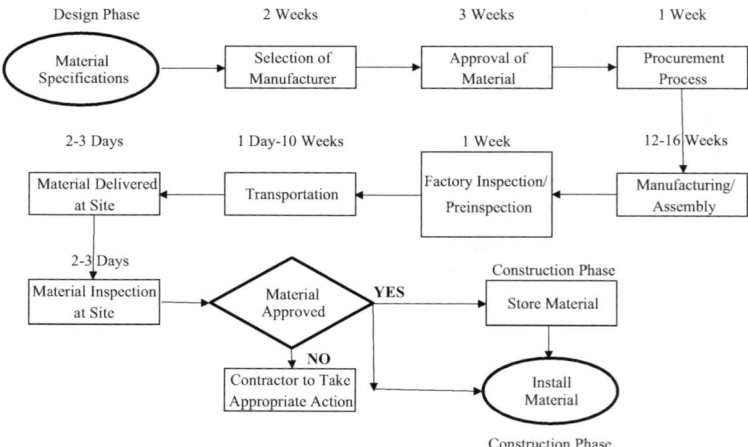

FIGURE 1.14 Supply chain process in construction project.

Overview of Quality

manufacturers/suppliers. However, in order of continuous and uninterrupted supply of specified product, the contractor has to consider the following:

- Quality management system followed by the manufacturer/supplier
- Historical rejection/acceptance record
- Reliability of the manufacturer/supplier
- Product certification
- Financial stability
- Proximity to the project site
- Manufacturing/lead time
- Availability of product as per the activity installation/execution schedule
- Manufacturing capacity
- Availability of quantity to meet the project requirements
- Timeliness of delivery
- Location of manufacturer/supplier
- Product cost
- Transportation cost
- Product certification
- Risks in delivery of product
- Responsiveness
- Cooperative and collaborative nature to resolve problem

In major construction projects, the following items are required in bulk quantities:

- Concrete
- Concrete block
- Conduit
- Utility pipes
- Light fixtures
- Electrical devices
- Plumbing fixtures

The contractor can follow just-in-time method to procure these items and avoid large inventory at project site. The contractor has to select reliable manufacturer/supplier for these products and to ensure that the manufacturer/supplier is capable of maintaining continuous flow of these items at a short notice. The contractor can sign agreement with the manufacture/supplier for entire project quantity with agreed-upon delivery schedule as per the requirement at site.

In order to ensure supply chain payments are made promptly, the cash flow system should be projected accordingly as the supply chain may affect due to interruption in payments toward supply of products.

2 Quality Management System

2.1 QUALITY STANDARDS

A quality system is a framework for quality management. It embraces the organizational structure, procedure, and processes needed to implement quality management system (QMS). The quality system should address everything in the organization related to quality of product, services, processes, projects, and operations. The adequacy of the quality system and the quality of products, services, processes, projects, operations, and customer satisfaction are judged by their compliance to specified/ relevant standards. Standards have important economic and social repercussions. They are useful to industrial and business organizations of all types, to government, to other regulatory bodies, to conformity assessment professionals, to suppliers and customers of products, projects, and services in both the public and private sectors, and to people, in general, in their role as customers and users. Standards provide governments with a technical base for health, safety, and environmental legislation.

A standard is simply a definition of how something should be. According to Pyzdek (1999):

> Standards are documents used to define acceptable conditions or behaviors and to provide a base line for assuring that conditions or behaviors meet the acceptable criteria. In most cases standards define minimum criteria; world class quality is, by definition, beyond the standard level of performance. Standards can be written or unwritten, voluntary or mandatory. Unwritten quality standards are generally not acceptable. (p. 2)

Pyzdek (1999) further states that quality standards serve the following purpose:

> Standards educate—They set forth ideals or goals for the guidance of manufacturers and users alike. They are invaluable to the manufacturer who wishes to enter a new field and to the naïve purchaser who wants to buy a new product.
>
> Standards simplify—They reduce the number of sizes, the variety of process, the amount of stock, and the paperwork that largely accounts for the overhead costs of making and selling.
>
> Standards conserve—By making possible large-scale production of standard designs, they encourage better tooling, more careful design, and more precise controls, and thereby reduce the production of defective and surplus pieces. Standards also benefit the user through lower costs.
>
> Standards provide a base upon which to certify—They serve as hallmarks of quality which are of inestimable value to the advertiser who points to proven values, and to the buyer who sees the accredited trademark, nameplate, or label. (p. 3)

Chung (1999) defines standards as a reference base that is required to judge the adequacy of a quality system. He further states that a "quality system has to cover all the activities leading to the finished product. Depending on the scope of operation of the organization, these activities include planning, design, development, purchasing, production, inspection, storage, delivery, and after-sales service". A reference base is required against which the adequacy of a quality system can be judged. Such a reference is called "quality system standard" (p. 14).

2.1.1 Importance of Standards

The International Organization for Standardization (ISO) has given the importance of standards as follows:

> Standards make an enormous contribution to most aspects of our lives. Standards ensure desirable characteristics of products and services such as quality, environmental fitness, safety, reliability, efficiency, and interchangeability and at an economical cost.
>
> When products and services meet our expectations, we tend to take this for granted. And be unaware of the role of standards. However, when standards are absent, we soon notice. We soon care when products turn out to be poor quality, do not fit, are incompatible with equipment that we already have, are unreliable or dangerous. When products, systems, machinery and devices work well and safely, it is often because they meet standards.

Standard setting is one of the first issues in developing a quality assurance system, and increasingly, organizations are relying on readily available standards rather than developing their own. Each standard should be

- clearly written in simple language that is unambiguous,
- convenient in understanding,
- specific in setting out precisely what is expected,
- measurable so that the organization can know whether it is being met,
- achievable, that is, the organization must have the resources available to meet the standard, and
- constructible.

Standards are published documents that establish specifications, guidelines, and procedures designed to ensure reliability of materials, products, systems, processes, and services that people use every day to satisfy their requirement. In the construction sector, the compliance with minimum standards in implementation of workmanship, materials, products, systems, and process provides a competitive advantage in terms of customer satisfaction and achieve qualitative, competitive, and economical project. Standards have important economic, social, and environmental repercussions.

Standards are used to ensure that a product, system, or service measures up to its specifications and is safe for use. Standards are the key to any conformity assessment activity.

2.2 STANDARDS ORGANIZATIONS

There are many organizations that produce standards; some of the best known organizations in the quality field of oil and gas industry are as follows:

- American Concrete Institute (ACI)
- American Institute of Steel Construction (AISC)
- American National Standards Institute (ANSI)
- American Petroleum Institute (API)
- American Society of Mechanical Engineers (ASME)
- American Society for Heating, Refrigerating, and Air-Conditioning Engineers (ASHRAE)
- American Society for Quality (ASQ)
- American Society for Testing and Materials (ASTM)
- American Welding Society (AWS)
- British Standards Institution (BSI)
- European Committee for Electrotechnical Standardization (CENELEC)
- International Association of Plumbing and Mechanical Officials (IAPMO)
- International Electrotechnical Commission (IEC)
- International Organization for Standardization (ISO)
- Engineering Equipment and Material Users Association (EEMUA)
- Expansion Joint Manufacturers Association (EJMA)
- National Association of Corrosion Engineers (NACE)
- National Electrical Code (NEC)
- National Fire Protection Association (NFPA)
- Tubular Exchange Manufacturers Association (TEMA)

Standards produced by these organizations/institutes are recognized worldwide. These standards are referred in the contract documents by the designers to specify products or systems or services to be used in a project. They are also used to specify the installation method to be followed or the fabrication works to be performed during the construction process. Table 2.1 lists some of the most common standards used in oil and gas projects.

Apart from these, there have been many other national and international quality system standards. These various standards have commonalities and historical linkage. However, in order to facilitate international trade, delegates from 25 countries met in London in 1946 to create a new international organization. The objective of this organization was to facilitate international coordination and unification of industrial standards. The new organization, ISO, officially began operation on February 23, 1947.

TABLE 2.1
Most Common Standards Used in Oil and Gas Projects

Sr. No.	Related Documents	Reference Items	Codes/Standards
1	Quality Management System	Quality management system manual	ISO 9001:215
2	Environment	Environmental system manual	ISO 14001
3	HSE	Health and safety	ISO 18000 (OHSAS)
4	HSE	Health and safety	ISO 45000
5	Information Security	Information security management system	ISO 27000
6	Supply Chain Security	Supply chain security management system	ISO 28000
7	Oil and Gas Sector Specifications	Oil and gas sector specification quality management system	ISO/TS 29001
8	Project Management	Project management system	ISO 21500
9	Civil	Specifications for Portland cement	ASTM.C150-02a
		Specifications for aggregate	ASTM.C 33-02a
		Reinforcement concrete	BS 4449, ASTM A615/A615M
		Cement concrete	ASTM C150, ASTM C295, ASTM C33
		Concrete	ASTM C94, ACI 301, ACI 117
		Masonry units concrete	ASTM C140, ASTM C1314, ASTM C270, ASTM C1019
		Comprehensive strength of cylindrical concrete specimens	ASTM.C39
		Standard method of testing piles	ASTM.D1143
		Design of blast resistant buildings in petrochemical facilities	ASCE 40265
		Minimum design loads for buildings and other structures	ASCE 7-98
		Standard method of testing piles under static axial compressive load	ASTM D 1143
		Building code requirements for structural concrete	ACI-318
		Recommendation for design, manufacture, and installation of concrete piles	ACI-543R-00

(Continued)

TABLE 2.1 (Continued)
Most Common Standards Used in Oil and Gas Projects

Sr. No.	Related Documents	Reference Items	Codes/Standards
		Hot weather concreting	ACI-305R-99
		Requirement for environmental engineering concrete structure	ACI-350-01 and 350R-01
		Recommended practice for concrete pavement	ACI-325.12R-02
		Standard method for density of soil in place by sand-cone method	ASTM.D1556-00
		Specifications for PVC (polyvinyl chloride), plastic drain, waste, and vent pipe and fitting	ASTM.2665-02a
10	Thermal and Moisture Protection	Liquid waterproofing, bituminous waterproofing, membrane waterproofing	ASTM C836, D412, D570, D903, ASTM D1187, ASTM D312, ASTM D1227, ASTM D4479
		Building insulation	ASTM C558, ASTM C578, ASTM C612, ASTM 665
11	Finishes	Gypsum plaster	ASTM C11, ASTM E119, ASTM E90, ASTM E413
		Cement plaster	ASTM C150, ASTM C260, ASTM C897, ASTM C926, ASTM C932, ASTM C1063, ASTM E488, ASTM A641M, ASTM A653M, ASTM C847
		Tiling	ANSI A137.1
		Acoustic ceiling	ASTM E1264, ASTM E795, ASTM E1264, ASTM C635, ASTM E488, ASTM B633, ASTM A641
		Metallic ceiling	ASTM B209, ASTM 591, ASTM E1264, ASTM E795, ASTM C635, ASTM C636, ASTM A641, ASTM A653
12	Doors and Windows	Steel doors	SDI 105, SDI 108, SDI 111
		Aluminum doors	AAMA 101, AAMA 603
		Wooden doors	NWWDA-ISI-A
		Curtain wall	AAMA 101, ASTM E283

(*Continued*)

TABLE 2.1 (*Continued*)
Most Common Standards Used in Oil and Gas Projects

Sr. No.	Related Documents	Reference Items	Codes/Standards
13	Metals	Glazing	AAMATIR A7, ANSIZ 971
		Material fabrication	ASTM A6/A6M, ASTM A36, ASTM 490, AISC 89
14	Coating	Preparation of steel substrates before application of paint and related product—visual assessment of surface cleanliness	ISO 8501-1
		Preparation of steel substrates before application of paint and related product—surface roughness characteristics of blast cleaned steel substrates	ISO 8503-1
		Preparation of steel substrates before application of paint and related product—abrasive blast cleaning	ISO 8504-2
		Preparation of steel substrates before application of paint and related product—hand and power tool cleaning	ISO 8504-3
15	Electrical/Instrumentation	Safety colors and safety signs—Part 1: Design principles for safety signs in workplaces and public areas	ISO 3864-1
		Specification for residual current-operated circuit breakers	BS 4293:1983
		Specification for electrical apparatus for explosive Atmospheres	BS 4683 Part 1: 1971
		Fire tests on building materials and structures	BS 476 P31.1:1983/BS 476 P33: 1993
		Instrumentation cables—Part 1. Specification for polyethylene insulated cables	BS 5308 P1: 1986
		Instrumentation Cables—Part 2. Specification for polyethylene insulated cables	BS 5308 P2: 1986
		Electrical apparatus for explosive gas atmospheres—Part 14: Electrical	BS EN-60079-14:1997
		Electrical apparatus for potentially explosive gas atmosphere—General requirements	EN 50014 (BS)1998

(*Continued*)

Quality Management System

TABLE 2.1 (Continued)
Most Common Standards Used in Oil and Gas Projects

Sr. No.	Related Documents	Reference Items	Codes/Standards
		Electrical apparatus for potentially explosive gas atmosphere	EN 50015 (BS) 1998
		Electrical apparatus for explosive gas atmospheres—Part 0: General requirements	IEC 60079-0
		Low-voltage control gear	IEC 60947-4-1
		Instrument transformers—Part 1: Current transformers	IEC 44-1
		Dry-type power transformers	BS 726
		Specification for industrial plugs, socket outlets, and couplers for industrial purposes—Part 1: General requirements	IEC 60309-1
		Low-voltage switchgear and control gear—Part 5-1: Control circuit devices and switching elements—Electromechanical control circuit devices	IEC 60947-5-1
		High-voltage motor starters—for voltages above 1 kV a.c. and 1.2 kV d.c.—Part 1: Direct online (full voltage) a.c. starters	BS 5856 P1
		Functional safety of electrical/electronic/programmable electronic safety-related systems	IEC 61508
		Fire detection and fire alarm systems for buildings—Part 1: Code of practice for system design, installation, commissioning, and maintenance	BS 5839-1:2002
		Common test methods for insulating and sheathing materials of electric cables and optical cables—Part 1-1	IEC 60811-1-1
		Measurement of flow by pressure differential devices—Part 1: Orifice plates, nozzles, and venture tubes inserted in circular cross-section conduits running full	ISO 5167-1
		Industrial valves/pressure testing on valves	ISO 5208

(Continued)

TABLE 2.1 (Continued)
Most Common Standards Used in Oil and Gas Projects

Sr. No.	Related Documents	Reference Items	Codes/Standards
		Industrial valves: Multiturn valve actuator attachments	ISO 5210
		Industrial valves: Part-turn valve actuator attachments	ISO 5211
		Refrigerate light hydrocarbon liquid/measurement of temperature in tanks containing liquefied gases—resistance thermometers and thermocouples	ISO 8310
		Sizing, selection, and installation of pressure-relieving devices in refineries: Part 1—Sizing and selection	API/RP 520 Part-1
		Sizing, selection, and installation of pressure-relieving devices in refineries: Part II—Installation	API/RP 520 Part-II
		Guide for pressure-relating and depressurizing system	API/RP 521
		Process measurement instrumentation	API/RP 551
		Gas meters, specification for rotating displace turbine meters for gas pressure up to 100 bar	BS 4161 Part 6: 1987
		Specification for 600/1000V and 1900/3300V armored electric cables having thermosetting insulation	BS 5467: 1997
		Specification for 600/1000V and 1900/3300V armored electric cables having PVC insulation	BS 6346 : 1997
		Specification for cables with extruded cross-linked polyethylene or ethylene propylene rubber insulation for rated voltages from 3.8/6.6 kV up to 19/33 kV	BS 6622: 1999
		Specification for electrical apparatus for explosive atmospheres with type of protection N	BS 6941: 1988
		Code of practice for earthing	BS 7430: 1998
16	Mechanical	Petroleum and liquid petroleum products—calibration of vertical cylindrical tanks, Part 1: strapping method	ISO 7507-1

(Continued)

TABLE 2.1 (*Continued*)
Most Common Standards Used in Oil and Gas Projects

Sr. No.	Related Documents	Reference Items	Codes/Standards
		Venting atmospheric and low-pressure storage tanks: nonrefrigerated and refrigerated	API STD 2000
		Design and construction of LPG installations, 8th edition	API STD 2510
		Pipeline valves (gate, ball, and check valves)	API Spec.6D
		Flanged steel pressure safety valves	API STD 526
		Check valves: wafer, wafer-lug, and double flanged type	API STD 594
		Valves inspection and testing	API STD 598
		Butterfly valves: double flanged, lug, and wafer type	API STD 609
		Valve-flanged, threaded, and welding end	ASME B 16.34
		Ball valves with flanged or butt-welding ends for general service	MSS SP 72
		Bronze gate, globe, angle and check valves	MSS SP 80
		Centrifugal pumps for general process services	API STD 610
		Specification for horizontal, end suction centrifugal pumps for chemical process	ASME B 73.1
		General purpose steam turbines for petroleum, chemical, and gas industry services	API STD 611
		Axial and centrifugal compressors and expander compressors for petroleum, chemical, and gas industry services—for general refinery service	API STD 617
		Rotary-type positive displacement compressors for petroleum, chemical, and gas industry services	API STD 619
		Air cooled heat exchanger for general refinery services	API STD 661

(*Continued*)

TABLE 2.1 (Continued)
Most Common Standards Used in Oil and Gas Projects

Sr. No.	Related Documents	Reference Items	Codes/Standards
		Plate heat exchangers for general refinery services	API STD 662
		Packaged, integrally geared, centrifugal air compressors for petroleum, chemical, and gas industry services	API STD 672
		Welded steel tanks for oil storage	API STD 650
		Specification for welding rods, electrodes, and filler metals	ASME S0002C
		American society of heating, refrigerating and air conditioning engineers handbooks: fundamentals, equipment, system, application	ASHRAE
17	Other	Installation of sprinkler systems	NFPA
		National electrical code	NFPA
		National fire alarm code	NFPA
		Standard on water-cooling towers	NFPA
		Standard for water spray fixed systems for fire protection	NFPA
		Utility LP-gas plant code	NFPA

2.3 INTERNATIONAL ORGANIZATION FOR STANDARDIZATION

The ISO is an independent nongovernmental organization with membership of 164 (as of April 2020) national standards bodies/institutes, formed on the basis of one member per country, with a Central Secretariat in Geneva, Switzerland, that coordinates the system.

The ISO is the world's largest developer and publisher of international standards. It is a nongovernmental organization that forms a bridge between the public and private sectors. The work of preparing international standards is normally carried out by technical committees. The ISO has more than 21,000 international standards. Of all the standards produced by ISO, the ones that are most widely known are the ISO 9000 and ISO 14000 series. ISO 9000 has become an international reference for quality requirements in business-to-business dealings, and ISO 14000 looks to achieve at least as much, if not more, in helping organizations to meet their environmental changes. ISO 9000 and ISO 14000 families are known as "generic management system standards".

The ISO 9000 is a series, or family of standards, primarily concerned with "quality management". This means what the organization does to fulfill the following:

- The customer's quality requirements
- Applicable regulatory requirements, while aiming to enhance customer satisfaction
- Continual improvement of its performance in pursuit of the objectives

The ISO 9000 family addresses various aspects of quality management and contains some of the ISO's best known standards. The standards provide guidance and tools for companies and organizations who want to ensure that their products and services consistently meet customer's requirements and that quality is consistently improved. ISO 9000 outlines the ways to achieve, as well as benchmark, consistent performance and service. Its application makes the business more competitive and credible. ISO 9000 QMS help to continually monitor and manage quality across all the operations from inception through completion and enhance customer satisfaction.

The ISO 14000 family is primarily concerned with "environmental management". This includes

- minimize harmful effect on the environment caused by its activities and
- achieve continual improvement with its environmental performance.

ISO standards are updated periodically since they were originally published in 1987. ISO 9000 actually comprises several standards.

ISO 9000:2000 specifies requirements for a QMS for any organization that needs to demonstrate its ability to consistently provide product that meets customer and applicable regulatory requirements and to enhance customer satisfaction.

In keeping with the process of updating the standards, certain clauses of ISO 9001:2000 of the QMS were amended during 2008 in order to improve the QMS, and accordingly, the amended standard is known as ISO 9001:2008. ISO 9001:2008 includes Annex B, which outlines the text changes that have been made to specific clauses.

ISO 9001:2008 was revised by various committees, societies, and institutes, and ISO 9001:2015 was published in September 2015. It has the following clauses:

1. Context of organization/QMS (Clause 4)
2. Leadership (Clause 5)
3. Planning for QMS (Clause 6)
4. Support (Clause 7)
5. Operation (Clause 8)
6. Performance evaluation (Clause 9)
7. Improvement (Clause 10)

Changes in ISO 9001:2015 are an opportunity to revisit organizational areas that yet need to be improved. An awareness of the upcoming changes in ISO 9001:2015 will enable quality professionals to better prepare for the future. The change is to incorporate risk-based thinking into the management system by considering the context of the organization. In other words, all processes are not equal for all organizations with some being more critical than others, resulting in different levels of risk.

The tremendous impact of ISO 9001 and ISO 14001 on organizational practices and on trade stimulated the development of other ISO standards and deliverables that adapt the generic management system to specific sectors or aspects. These are as follows:

1. Food safety management systems—ISO 22000
2. Information security management Systems—ISO 27001
3. Supply chain security management systems—ISO 28000

ISO 22000:2005, published on September 1, 2005, is related to the safe food supply management system to ensure that food is safe at the time of human consumption. ISO 27001:2005 is related to information security system. ISO 28000:2005 is related to supply management system to help combat threats to safe and smooth flow of international trade.

2.4 ISO 9000 QUALITY MANAGEMENT SYSTEM

ISO 9000 quality system standards are a tested framework for taking a systematic approach to managing the business process so that organizations turn out products or services conforming to customer's satisfaction. The typical ISO QMS is structured on four levels, usually portrayed as a pyramid. Figure 2.1 illustrates this.

On top of the pyramid is the quality policy, which sets out what management requires its staff to do in order to ensure QMS. Underneath the policy is the quality manual, which details the work to be done. Beneath the quality manual are work instructions or procedures. The number of manuals containing work instructions or procedures is determined by the size and complexity of the organization. The procedures mainly discuss the following:

- What is to be done?
- How is it done?

Quality Management System

FIGURE 2.1 QMS pyramid. QMS, quality management system.

- How does one know that it has been done properly (for example, by inspecting, testing, or measuring)?
- What is to be done if there are problems (for example, failure)?

The bottom level of hierarchy contains forms and records that are used to capture the history of routine events and activities.

The ISO 9000 QMS requires documentation that includes a quality manual and quality procedures, as well as work instructions and quality records. All documentation (including quality records) must be controlled according to a document control procedure. The structure of the QMS depends largely on the management structure in the organization.

ISO 9001:2000 identifies certain minimum requirements that all QMSs must meet to ensure customer satisfaction. ISO 9001:2000 specifies requirements for QMSs when an organization

- needs to demonstrate its ability to consistently provide product that meets customer and applicable regulatory requirements and
- aims to enhance customer satisfaction through the effective application of the system, including processes for continual improvement of the system and the assurance of conformity to customer and applicable regulatory requirements.

2.4.1 QUALITY SYSTEM DOCUMENTATION

A quality system has to cover all the activities leading to the final product or service. The quality system depends entirely on the scope of operation of the organization and particular circumstances such as number of employees, type of organization, and physical size of the premises of the organization. The quality manual is the document that identifies and describes the QMS.

The adoption of a QMS should be a strategic decision of an organization. The design and implementation of an organization's QMS (documentation) is influenced by

1. its organizational environment, changes in that environment, and the risks associated with that environment,
2. its varying needs,
3. its particular objectives,
4. the products it provides,
5. the processes it employs, and
6. its size and organizational structure.

The QMS requirements specified in international standard are complementary to requirements for products.

This international standard can be used by internal and external parties, including certification bodies, to assess the organization's ability to meet customer, statutory, and regulatory requirements applicable to the product and the organization's own requirements.

The QMS is based on the guidelines for performance improvement per ISO 9004:2000 and the quality management requirements. The quality management principles stated in ISO 9000 and ISO 9004 have been taken into consideration for development of this international standard. ISO 9000:2000 outlines the necessary steps to implement the QMS. These are as follows:

1. Identify the process (activities and necessary elements) needed for QMS.
2. Determine the sequence and interaction of these processes and how they fit together to accomplish quality goals.
3. Determine how these processes are effectively operated and controlled.
4. Measure, monitor, and analyze these processes and implement action necessary to correct the process and achieve continual requirements.
5. Ensure that all information is available to support the operation and monitoring of the process.
6. Display the most options, thus helping make the right management system.

ISO 9001:2000 requirements fall into the following sections:

1. QMS
2. Management responsibility
3. Resource management

Quality Management System

4. Product realization
5. Measurement analysis and improvement

These documents are regularly updated and revised to meet the changing industry requirements due to the following:

- Globalization
- Higher levels of expectations by customers
- Higher-performance requirements
- Complex nature of environment and works
- Specialized services requirements

ISO 9000:2008 was revised in 2015 and is called as 9000:2015. The revised QMS standard has ten clauses against eight in the previous version.

In general, QMSs begin with the writing of a quality manual. The quality manual serves as a roadmap for the QMS. It is more practical to build the "road" before preparing the "map". The manual is generic at first, and as the QMS develops, the manual is updated.

In the construction industry, a contractor may be working at any time on a number of projects of varied natures. These projects have their own contract documents to implement project quality, which require a contractor to submit a contractor's quality control plan to ensure that specific requirements of the project are considered to meet client's requirements. Therefore, while preparing a QMS at a corporate level, the organization has to take into account tailor-made requirements for the projects, and accordingly, the manual should be prepared.

2.4.2 Quality Management System Manual

A QMS is a set of coordinated documentation that includes a quality manual, quality procedures, processes as well as work instructions (details of works to be done), quality forms, and records that are developed to meet customer and regulatory requirements taking into consideration organization's quality policies and business objectives and improve the effectiveness and efficiency on continuous basis.

ISO 9001 is an international standard that is most recognized and used as international reference for development of an effective QMS.

ISO 9001 contain eight quality management principles (CLIPSCFM). These are as follows:

Principle 1—Customer focus
Principle 2—Leadership
Principle 3—Involvement of people
Principle 4—Process approach
Principle 5—System approach to management
Principle 6—Continual improvement

Principle 7—Factual approach to design making
Principle 8—Mutual beneficial supplier relationship

2.4.2.1 Development of Quality Management System

ISO 9000 family is primarily concerned with "quality management". This means what the organization does to fulfill:

- The customers quality requirements
- Applicable regulatory requirements, while aiming to enhance customer satisfaction
- Achieve continual improvement of its performance in pursuit of the objectives

Although any ISO 9000 QMS should be created to address organization's business objectives, needs, and customer satisfaction, there are some general elements that are common in all ISO compliant QMS. These elements have importance as per the levels known as QMS documentation pyramid that includes quality policy and objectives, a quality manual, quality procedures as well as work instructions, and quality records. The manual is developed taking into consideration the following:

1. Eight principles (CLIPSCFM) of QMS as defined by ISO Technical Committee, TC 176.
2. All the related and applicable documents produced taking into consideration ten sections/clauses listed under ISO 9001:2015 to ensure that the manual is in compliant with ISO 9001:2015. Table 2.2 illustrates these sections/clauses.

Figure 2.2 illustrates relationship between QMS principles and ISO sections.

QMS is made up of elements (components) having functional relationship to achieve common objective (customer satisfaction). These elements need to be coordinated taking a systematic approach to improve and sustain the overall performance of the products and services. Figure 2.3 illustrates simple behavioral approach to system and is generally known as black box.

Systems engineering or process-based approach can be used to develop QMS. Figure 2.4 illustrates that the concept of system phenomenon can be applied to develop QMS.

The application of the process approach in a QMS enhances the following:

- Overall performance of the organization by effectively controlling the interrelationship and interdependencies among the QMS processes
- Consistency in meeting the customer requirements
- Consideration of processes in terms of added value
- Achievement of effective process performance
- Improvement of processes based on evaluation of data and information

The process-based QMS approach incorporates the plan–do–check–act (PDCA) cycle and risk-based thinking. The PDCA cycle can be applied to all processes and to the QMS as a whole. Figure 2.5 illustrates organizational structure of ISO 9001:2015 sections (clauses) 4–10, and Figure 2.6 describes in brief the related activities in PDCA cycle.

TABLE 2.2
ISO 9001:2015 Sections (Clauses)

Section (Clause) No.	Relevant Clause in 9001:2015	Description
1	Scope	
2	Normative references	
3	Terms and references	
4	Context of the organization	
	4.1	Understanding the organization and its context
	4.2	Understanding the needs and expectations of interested parties
	4.3	Determining the scope of quality management system
	4.4	Quality management system and its processes
5	Leadership	
	5.1	Leadership and commitment
	5.2	Policy
	5.3	Organizational roles, responsibilities, and authorities
6	Planning	
	6.1	Actions to address risks and opportunities
	6.2	Quality objectives and planning to achieve them
	6.3	Planning of changes
7	Support	
	7.1	Resources
	7.2	Competence
	7.3	Awareness
	7.4	Communication
	7.5	Documented information
8	Operation	
	8.1	Operational planning and control
	8.2	Requirements for product and services
	8.3	Design and development of products and services
	8.4	Control of externally provided processes, products, and services
	8.5	Production and service provision
	8.6	Release of products and services
	8.7	Control of nonconforming outputs
9	Performance evaluation	
	9.1	Monitoring, measurement, analysis, and evaluation
	9.2	Internal audit
	9.3	Management review
10	Improvement	
	10.1	General
	10.2	Nonconformity and corrective action
	10.3	Continual improvement

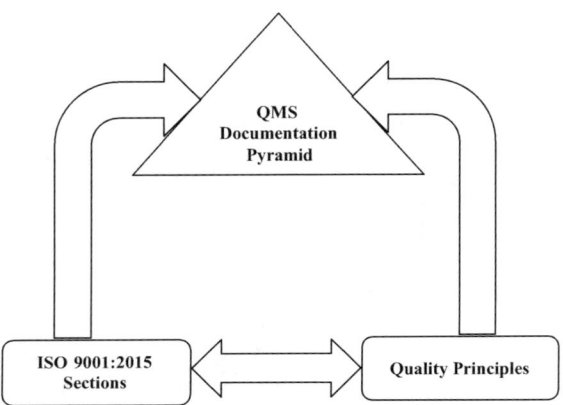

FIGURE 2.2 QMS documentation model. QMS, quality management system.

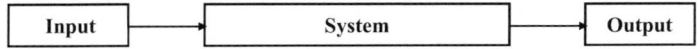

FIGURE 2.3 Black box.

FIGURE 2.4 Process-based QMS development procedure. QMS, quality management system.

The PDCA cycle can be applied to all the processes. Figure 2.7 illustrates how clauses 4–10 can be grouped in relation to PDCA cycle.

Development and implementation of an effective QMS is a strategic decision of the organization that helps to improve its overall performance and provide a sound basis for sustainable development initiatives.

Traditional oil and gas construction projects have involvement of three main groups. These are as follows:

1. Owner (client, project owner)
2. Designer (consultant)
3. EPC contractor (constructor)

Each of these groups should have their own QMS to meet their business objectives. While developing the QMS, the organizations have to include those documents that are required to perform relevant processes and specific requirements that the organizations have business interest.

Quality Management System

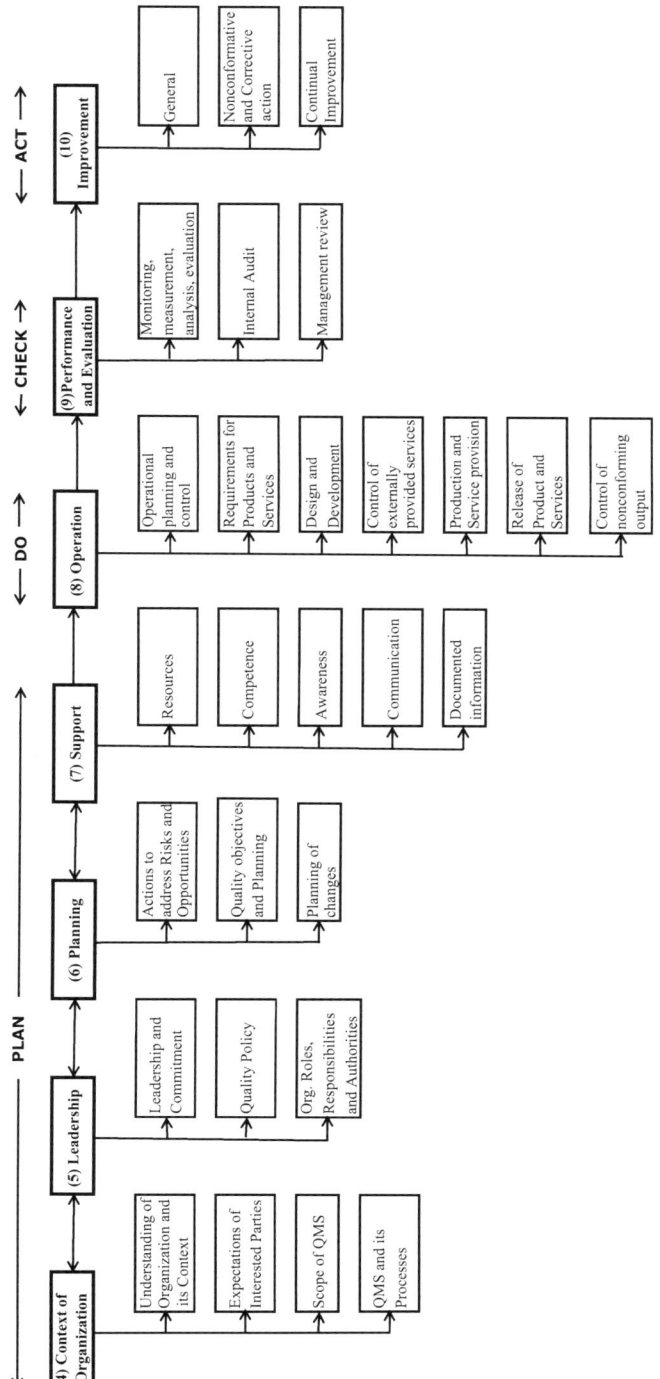

FIGURE 2.5 Organizational structure of ISO 9001:2015 sections. QMS, quality management system.

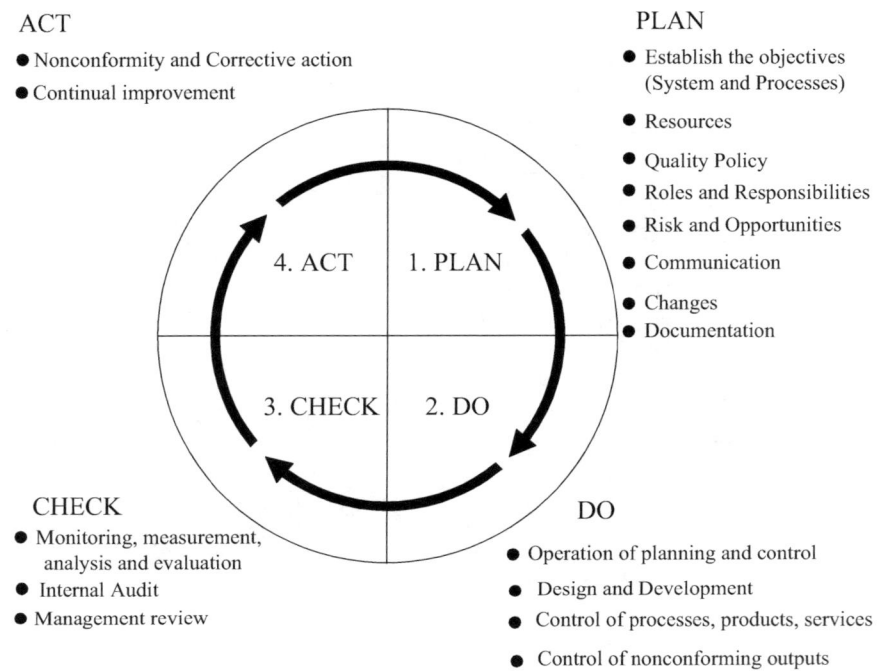

FIGURE 2.6 PDCA cycle for QMS activities. PDCA, plan–do–check–act; QMS, quality management system.

Normally, QMS consists of documents produced taking into consideration relevant sections of ISO 9001:2015 and business-related activities.

Table 2.3 lists example contents of QMS manual for owner (client, project owner).
Table 2.4 lists example contents of QMS manual for designer (designer, consultant).
Table 2.5 lists example contents of QMS manual for EPC contractor.
Table 2.6 lists example contents of QMS manual for project manual consultant.
Figure 2.8 illustrates ISO QMS design and implementation model

2.5 ISO CERTIFICATION

Following are the details of ISO certification, registration, and accreditations, as mentioned on the ISO website (2008):

> In the context of ISO 9000 or ISO 14000, "certification" refers to the issuing of written assurance (the certificate) by an independent, external body that has audited an organization's management system and verified that it conforms to the requirements specified in the standard.
>
> "Registration" means that the auditing body then records the certification in its client register so the organization's management system has therefore been both certified and registered. Therefore, in the ISO 9000 and ISO 14000 contexts, the difference between the two terms is not significant and both are acceptable for general use.

Quality Management System

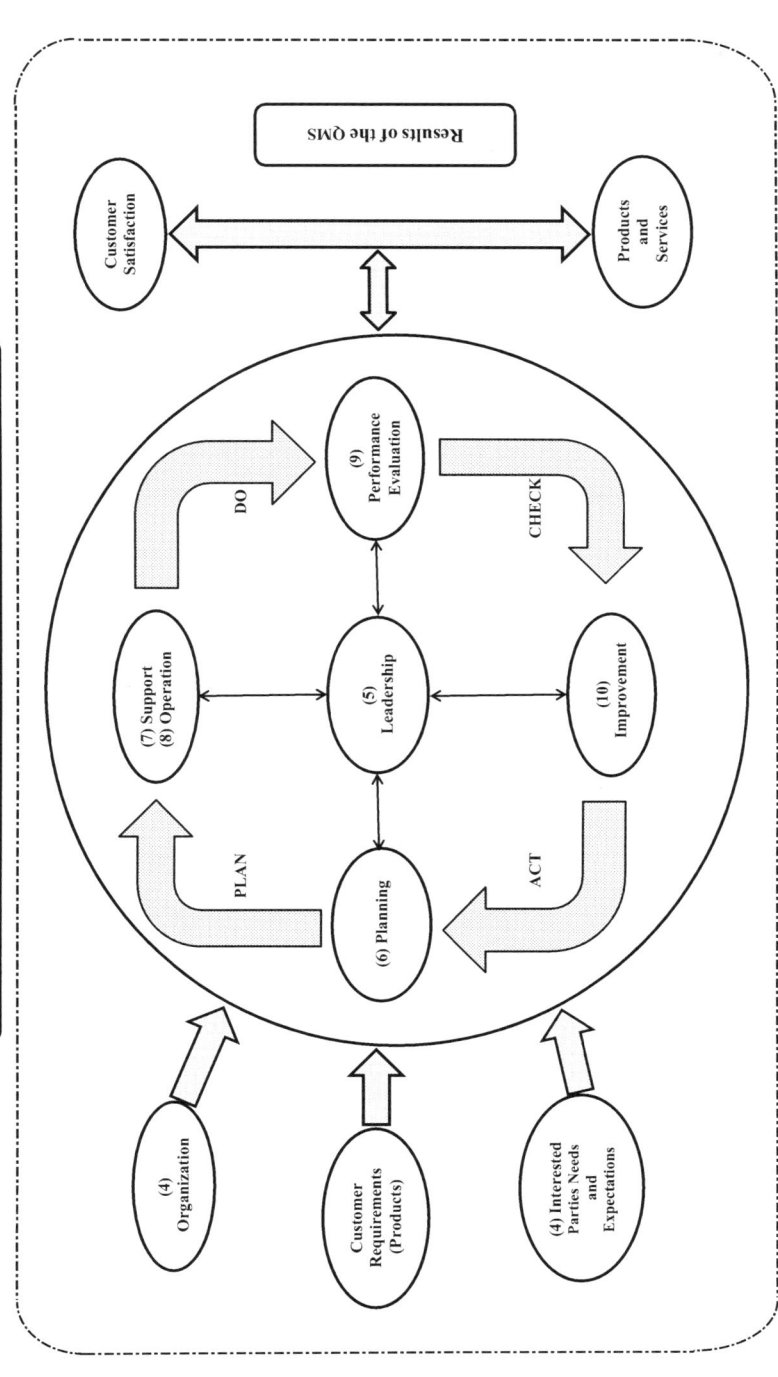

FIGURE 2.7 Quality management process model.

TABLE 2.3
List of Quality Manual Documents—Owner (Client, Project Owner)

Document No.	Document Title	Relevant Clause in 9001:2015	Version/ Revision Date
QC-0	Circulation list		
QC-00	Records of revision		
QC-01	Scope		
QC-02	Normative references		
QC-03	Terms and references		
QC-1.1	Understanding the organization and its context	4.1	
QC-1.2	Monitoring and review of internal and external issues	4.1	
QC-2.1	Relevant requirements of stakeholders	4.2	
QC-2.2	Monitoring and review of stakeholder's information	4.2	
QC-3	Scope of quality management system	4.3/4.4	
QC-4	Project quality management system	4.4	
QC-5	Management responsibilities	5.1	
QC-6	Customer focus	5.1.2	
QC-7.1	Quality policy (organization)	5.2	
QC-7.2	Quality policy (project)	5.2.2	
QC-8	Organizational roles, responsibilities, and authorities (organization chart)	5.3	
QC-9	Preparation and control of project quality plan	6.0	
QC-10.1	Project risk (study stage)	6.1	
QC-10.2	Project risk (during design)	6.1	
QC-10.3	Project risk (during bidding and tendering)	6.1	
QC-10.4	Project risk (during construction)	6.1	
QC-10.5	Project risk (selection of designer and contractor)	6.1	
QC-11	Project quality objective	6.2	
QC-12.1	Change management (during design phase)	6.3	
QC-12.2	Change management (during construction phase)	6.3	
QC-12.3	Change management (owner initiated)	6.3	
QC-13.1	Office resources (office equipment)	7.1	
QC-13-2	Human resources (project management and construction management)	7.1.2	
QC-13.3	Human resources (office staff)	7.1.2	
QC-14.1	Infrastructure	7.1.3	
QC-14.2	Work environment	7.1.4	
QC-15	Monitoring and measuring office equipment	7.1.5	
QC-16	Organizational knowledge	7.1.6	
QC-17.1	Competence	7.2	

(*Continued*)

TABLE 2.3 (*Continued*)
List of Quality Manual Documents—Owner (Client, Project Owner)

Document No.	Document Title	Relevant Clause in 9001:2015	Version/ Revision Date
QC-17.2	Training in quality system	7.2	
QC-17.3	Training in quality auditing	7.2	
QC-17.4	Training in supervision/technical skills	7.2	
QC-17.5	Records of training, skills, experience, and qualifications	7.2	
QC-18	Communication internal and external	7.4	
QC-19.1	Control of documents for general application	7.5.2/3	
QC-19.2	Control of documents for specific projects	7.5.2/3	
QC-20	Records updates	7.5.2	
QC-21	Control of quality records	7.5.3	
QC-22.1	Planning of project/facility and quality plan	8.1/8.3/8.3.1/2/3	
QC-22.2	Project study and feasibility	8.1/8.3/8.3.1/2/3	
QC-22.3	Project brief	8.1/8.3/8.3.1/2/3	
QC-22.4	Records of feasibility inputs	8.3.3	
QC-22.5	Records of terms of reference	8.3.3	
QC-22.6	Record of changes	8.3.6	
QC-23.1	Selection and evaluation of designer, project manager, construction manager, contractor	8.4	
QC-23.2	Control of design and contracting services	8.4	
QC-23.3	Communication with designer and contractor	8.4.3	
QC-24.1	Designer selection procedure	8.5	
QC-24.2	Contractor selection procedure	8.5	
QC-25.1	Design review procedure	8.5	
QC-25.2	Construction supervision procedure	8.5	
QC-25.3	Project management procedure	8.5	
QC-25.4	Construction management procedure	8.5	
QC-26.1	Release of project documents for tender	8.6	
QC-26.2	Release of project documents for construction	8.6	
QC-27	Control of nonconforming work (project)	8.7	
QC-28.1	Project document review (during design)	9.1	
QC-28.2	Project document review (during construction)	9.1	
QC-28.3	Project manager/construction manager comments	9.1.2	
QC-29	Internal quality audits	9.2	
QC-30	Management review	9.3	
QC-31	Nonconformity and corrective action	10.2.2	
QC-32	Control of client complaints	10.2.2	
QC-33	Continual improvement	10.3	

TABLE 2.4
List of Quality Manual Documents—Consultant (Design, Supervision)

Document No.	Document Title	Relevant Clause in 9001:2015	Version/ Revision Date
QC-0	Circulation list		
QC-00	Records of revision		
QC-01	Scope		
QC-02	Normative references		
QC-03	Terms and references		
QC-1.1	Understanding the organization and its context	4.1	
QC-1.2	Monitoring and review of internal and external issues	4.1	
QC-2.1	Relevant requirements of stakeholders	4.2	
QC-2.2	Monitoring and review of stakeholder's information	4.2	
QC-3	Scope of quality management system	4.3/4.4	
QC-4	Project quality management system	4.4	
QC-5	Management responsibilities	5.1	
QC-6	Customer focus	5.1.2	
QC-7.1	Quality policy (organization)	5.2	
QC-7.2	Quality policy (project)	5.2.2	
QC-8	Organizational roles, responsibilities, and authorities (organization chart)	5.3	
QC-9	Preparation and control of project quality plan	6.0	
QC-10.1	Project risk (design proposal)	6.1	
QC-10.2	Project risk (supervision proposal)	6.1	
QC-10.3	Project risk (during design)	6.1	
QC-10.4	Project risk (during construction supervision)	6.1	
QC-11	Project quality objective	6.2	
QC-12.1	Change management (during design phase)	6.3	
QC-12.2	Change management (during construction phase)	6.3	
QC-13.1	Office resources (office equipment, design software)	7.1	
QC-13-2	Human resources (design team, supervision team)	7.1.2	
QC-13.3	Human resources (office staff)	7.1.2	
QC-14.1	Infrastructure	7.1.3	
QC-14.2	Work environment	7.1.4	
QC-15	Monitoring and measuring office equipment	7.1.5	
QC-16	Organizational knowledge	7.1.6	
QC-17.1	Competence	7.2	
QC-17.2	Training in quality system	7.2	
QC-17.3	Training in quality auditing	7.2	
QC-17.4	Training in operational/technical skills	7.2	

(*Continued*)

TABLE 2.4 (*Continued*)
List of Quality Manual Documents—Consultant (Design, Supervision)

Document No.	Document Title	Relevant Clause in 9001:2015	Version/ Revision Date
QC-17.5	Record of training, skills, experience, and qualifications	7.2	
QC-18	Communication internal and external	7.4	
QC-19.1	Control of documents for general application	7.5.2/3	
QC-19.2	Control of documents for specific projects	7.5.2/3	
QC-20	Records updates	7.5.2	
QC-21	Control of quality records	7.5.3	
QC-22.1	Proposal (design, supervision) documents	8.2	
QC-22.2	Proposal review	8.2	
QC-22.3	Records of proposal changes	8.2	
QC-23.1	Planning of engineering design and quality plan	8.1/8.3/8.3.1/2/3	
QC-23.2	Design development (design–bid–build)	8.1/8.3/8.3.1/2/3	
QC-23.3	Design development (design–build)	8.1/8.3/8.3.1/2/3	
QC-23.4	Design development (oil and gas projects)	8.1/8.3/8.3.1/2/3	
QC-23.5	Record of design development inputs	8.3.3	
QC-23.6	Records of design development control	8.3.4	
QC-23.7	Records of design development outputs	8.3.5	
QC-23.8	Records of design development changes	8.3.6	
QC-24.1	Selection and evaluation of subconsultant	8.4	
QC-24.2	Control of subconsultant services	8.4	
QC-24.3	Communication with subconsultant	8.4.3	
QC-24.4	Evaluation and selection of equipment, provisions	8.4	
QC-25	Engineering design procedure	8.5	
QC-25.1	Design review procedure	8.5	
QC-26.1	Construction supervision procedure	8.5	
QC-26.2	Project management procedure	8.5	
QC-26.3	Construction management procedure	8.5	
QC-27	Release of project documents	8.6	
QC-28	Control of nonconforming work (design errors)	8.7	
QC-29.1	Project document review (management and control)	9.1	
QC-29.2	Client/owner comments	9.1.2	
QC-30	Internal quality audits	9.2	
QC-31	Management review	9.3	
QC-32.1	Nonconformity and corrective action	10.2.2	
QC-32.2	Preventive action	10.2.2	
QC-33	Control of client complaints	10.2.2	
QC-34	Continual improvement	10.3	

TABLE 2.5
List of Quality Manual Documents–EPC Contractor

Document No.	Document Title	Relevant Clause in 9001:2015	Version/ Revision Date
QC-0	Circulation list		
QC-00	Records of revision		
QC-01	Scope		
QC-02	Normative references		
QC-03	Terms and references		
QC-1.1	Understanding the organization and its context	4.1	
QC-1.2	Monitoring and review of internal and external issues	4.1	
QC-2.1	Relevant requirements of stakeholders	4.2	
QC-2.2	Monitoring and review of stakeholder's information	4.2	
QC-3	Scope of quality management system	4.3/4.4	
QC-4	Project quality management system	4.4	
QC-5	Management responsibilities	5.1	
QC-6	Customer focus	5.1.2	
QC-7.1	Quality policy (organization)	5.2	
QC-7.2	Quality policy (project)	5.2.2	
QC-8	Organizational roles, responsibilities, and authorities (organization chart)	5.3	
QC-9	Preparation and control of project quality plan	6.0	
QC-10.1	Project risk management (tendering)	6.1	
QC-10.2	Project risk management (construction)	6.1	
QC-10.3	Project risk management (HSE)	6.1	
QC-10.4	Project risk management (selection of subcontractor, supplier)	6.1	
QC-10.5	Project risk management (tendering)	6.1	
QC-10.6	Project risk management (detailed engineering)	6.1	
QC-10.7	Project risk management (construction)	6.1	
QC-10.8	Project risk management (resources, skilled manpower)	6.1	
QC-11	Project quality objectives	6.2	
QC-12.1	Change management (scope)	6.3	
QC-12.2	Change management (variation orders. Site work instructions)	6.3	
QC-12.3	Change management (design changes)	6.3	
QC-12.4	Change management (changes during construction)	6.3	
QC-13.1	Office resources (office staff, office equipment)	7.1	
QC-13.2	Construction resources (human resources, equipment and machinery)	7.1	
QC-13.3	Design resources (design team, design software)	7.1	
QC-14.1	Infrastructure	7.1.3	
QC-14.2	Work environment	7.1.4	

(*Continued*)

TABLE 2.5 (*Continued*)
List of Quality Manual Documents–EPC Contractor

Document No.	Document Title	Relevant Clause in 9001:2015	Version/ Revision Date
QC-15	Control of construction, material, measuring, and test equipment	7.1.5	
QC-16	Control of human resources	7.1.6	
QC-17	Organizational knowledge	7.1.6	
QC-18.1	Competence	7.2	
QC-18.2	Training and development in quality system	7.2	
QC-18.3	Training in quality auditing	7.2	
QC-18.4	Training in operational/technical skills	7.2	
QC-18.5	Records of training, skills, experience, and qualifications	7.2	
QC-19	Communication internal and external	7.4	
QC-20.1	Control of documents for general application	7.5.2/3	
QC-20.2	Control of documents for specific projects	7.5.2/3	
QC-21	Records updates	7.5.2	
QC-22	Control of quality records	7.5.3	
QC-23	Documents control (logs)	7.5.3.2	
QC-24	Project planning and control	8.1	
QC-25	Project specific requirements	8.2	
QC-25.1	Planning of engineering design	8.1/8.3	
QC-25.2	Engineering design development	8.1/8.3	
QC-25.3	Records of design development inputs	8.3.3	
QC-25.4	Planning of procurement	8.1	
QC-25.5	Planning of quality management	8.1	
QC-25.6	Planning of HSE management	8.1	
QC-26	Project-specific quality control plan (contractor's quality control plan)	8.2	
QC-27-1	Tender documents	8.2	
QC-27.2	Tender review	8.2	
QC-27.3	Contract review	8.2.1	
QC-28	Construction processes	8.2.2	
QC-29	Variation review	8.2.3	
QC-30.1	Engineering and shop drawings	8.3	
QC-30.2	Records of engineering and shop drawing input	8.3.3	
QC-31	Design developments for EPC projects	8.3	
QC-32	Selection and evaluation of subcontractors	8.4	
QC-32.1	Selection and evaluation of suppliers and vendors	8.4.1	
QC-33	Communication with subcontractors, material suppliers, and vendors	8.4.3	
QC-34.1	Inspection of subcontracted work	8.4.2	
QC-34.2	Incoming material inspection and testing	8.4.3	
QC-35	Installation procedures	8.5	

(*Continued*)

TABLE 2.5 (Continued)
List of Quality Manual Documents–EPC Contractor

Document No.	Document Title	Relevant Clause in 9001:2015	Version/ Revision Date
QC-36	Product identification and traceability	8.5.2	
QC-37	Identification of inspection and test status	8.5.2	
QC-38	Control of owner supplied items	8.5.3	
QC-39	Handling and storage	8.5.4	
QC-40	Construction inspection, testing, and commissioning	8.6	
QC-41	Project handover	8.6	
QC-41.1	Release of project design/detailed engineering design	8.6	
QC-42.1	Control of nonconforming work	8.7	
QC-42.2	Control of nonconforming subcontractor	8.7	
QC-42.3	Control of nonconforming work (design errors)	8.7	
QC-43.1	Project performance review	9.1.1	
QC-43.2	Project quality assessment and measurement	9.1.2	
QC-44	Client/owner/PMC/CM comments	9.1	
QC-45	Internal quality audits	9.2	
QC-46	Management review	9.3	
QC-47	New technology in construction	10.1	
QC-48.1	Nonconformity and corrective action	10.2.2	
QC-48.2	Preventive action	10.3	
QC-49	Control of client complaints	10.2.2	
QC-50	Continual improvement	10.3	

PMC, project manual consultant.

TABLE 2.6
List of Quality Manual Documents—Project Management Consultant

Document No.	Document Title	Relevant Clause in 9001:2015	Version/ Revision Date
QC-0	Circulation list		
QC-00	Records of revision		
QC-01	Scope		
QC-02	Normative references		
QC-03	Terms and references		
QC-1.1	Understanding the organization and its context	4.1	
QC-1.2	Monitoring and review of internal and external issues	4.1	
QC-2.1	Relevant requirements of stakeholders	4.2	
QC-2.2	Monitoring and review of stakeholder's information	4.2	
QC-3	Scope of quality management system	4.3/4.4	
QC-4	Project quality management system	4.4	
QC-5	Management responsibilities	5.1	

(Continued)

TABLE 2.6 (*Continued*)
List of Quality Manual Documents—Project Management Consultant

Document No.	Document Title	Relevant Clause in 9001:2015	Version/ Revision Date
QC-6	Customer focus	5.1.2	
QC-7.1	Quality policy (organization)	5.2	
QC-7.2	Quality policy (project)	5.2.2	
QC-8	Organizational roles, responsibilities, and authorities (organization chart)	5.3	
QC-9	Preparation and control of project quality plan	6.0	
QC-10.1	Project risk (project management proposal)	6.1	
QC-10.2	Project risk (project team selection)	6.1	
QC-10.3	Project risk (during design management)	6.1	
QC-10.4	Project risk (during construction management)	6.1	
QC-11	Project quality objective	6.2	
QC-12.1	Change management (during design phase)	6.3	
QC-12.2	Change management (during construction phase)	6.3	
QC-13.1	Office resources (office equipment, design software, project management software)	7.1	
QC-13-2	Human resources (design team, project management team)	7.1.2	
QC-13.3	Human resources (office staff)	7.1.2	
QC-14.1	Infrastructure	7.1.3	
QC-14.2	Work environment	7.1.4	
QC-15	Monitoring and measuring office equipment	7.1.5	
QC-16	Organizational knowledge	7.1.6	
QC-17.1	Competence	7.2	
QC-17.2	Training in quality system	7.2	
QC-17.3	Training in quality auditing	7.2	
QC-17.4	Training in operational/technical skills	7.2	
QC-17.5	Record of training, skills, experience, and qualifications	7.2	
QC-18	Communication internal and external	7.4	
QC-19.1	Control of documents for general application	7.5.2/3	
QC-19.2	Control of documents for specific projects	7.5.2/3	
QC-20	Records updates	7.5.2	
QC-21	Control of quality records	7.5.3	
QC-22.1	Proposal documents	8.2	
QC-22.2	Proposal review	8.2	
QC-22.3	Records of proposal changes	8.2	
QC-23.1	Planning of engineering design and quality plan	8.1/8.3/8.3.1/2/3	
QC-23.2	Design development (EPC contract)	8.1/8.3/8.3.1/2/3	
QC-23.3	Record of project design development inputs	8.3.3	
QC-23.4	Records of project design development control	8.3.4	
QC-23.5	Records of project design development outputs	8.3.5	

(*Continued*)

TABLE 2.6 (Continued)
List of Quality Manual Documents—Project Management Consultant

Document No.	Document Title	Relevant Clause in 9001:2015	Version/ Revision Date
QC-23.6	Records of project design development changes	8.3.6	
QC-24.1	Evaluation and selection of equipment, provisions	8.4	
QC-25	Engineering design review procedure	8.5	
QC-26.1	Project management procedure	8.5	
QC-27	Release of project documents	8.6	
QC-28	Control of nonconforming work (design errors)	8.7	
QC-29.1	Project document review (management and control)	9.1	
QC-29.2	Client/owner comments	9.1.2	
QC-30	Internal quality audits	9.2	
QC-31	Management review	9.3	
QC-32.1	Nonconformity and corrective action	10.2.2	
QC-32.2	Preventive action	10.2.2	
QC-33	Control of client complaints	10.2.2	
QC-34	Continual improvement	10.3	

"Certification" seems to be the term most widely used worldwide, although "registration" is often preferred in North America, and the two are also used interchangeably.

On the contrary, using "accreditation" as an interchangeable alternative for "certification" or "registration" is a mistake, because it means something different.

In the ISO 9000 or ISO 14000 context, accreditation refers to the formal recognition by a specialized body—an accreditation body—that a certification body is competent to carry out ISO 9000 or ISO 14000 certification in specified business sectors.

In simple terms, accreditation is like certification of the certification body. Certificates issued by accredited certification bodies may be perceived on the market as having increased credibility.

Thus, it should be understood that the certification body is a third-party company registered with an established national accreditation board and is authorized to issue a certificate of conformance after evaluating the conformance of an organization's management system to the requirements of appropriate standard.

With the advent of globalization and competitive market, it has become essential to implement a QMS in an organization and to get it certified from a third party to enhance business opportunities in the international market. The ISO 9000 QMS is accepted worldwide, and international customers prefer to do business with organizations having ISO certification. An ISO QMS includes all activities and overall management functions that determine quality policy, objectives, and responsibilities and their implementation.

ISO certification is not compulsory; however, it is required for competitive advantage. Certification can be a useful tool to add credibility, by demonstrating that the product or services meet the expectations of the customers. For some industries, certification is a legal or contractual requirement. The ISO does not perform certification.

ISO certification is valuable to firms because it provides a framework so they can assess where they are, where they would like to be, and what is their standing in the international market. Implementation of an ISO management system in the organization brings in increased effectiveness and efficiency of operations and ensures that the product satisfies customer requirements. ISO 9000 and ISO 14000 concern the way an organization goes about its work and processes. ISO 9000 and ISO 14000 are not product standards.

There are three types of audits that can be done on ISO QMSs:

1. First-party audit—Audit your own organization (internal audit).
2. Second-party audit—Audit of supplier by the customer.
3. Third-party audit—Totally independent of the customer–supplier relationship. The best certification of a firm is through third party.

ISO 9000 certification audit is done by a certification body that has been accredited or has been officially approved as competent to carry out certification in a specified business sector by a national accreditation body.

Figure 2.8 discussed earlier under Section 2.5 diagrammatically summarizes the ISO certification process.

With the certification of ISO standards, organizations obtain the following advantages:

- Customer satisfaction and confidence in the organization's products/services
- Increase in revenues
- Increased market share
- Continuous improvement in organizational process
- Consistency in products/services quality
- Improvement in staff performance
- Effectiveness in the utilization of staff
- Efficient utilization of time, money, and other resources
- Environmental benefits

2.6 ISO/TS 29001

ISO/TS 29001 is petroleum, petrochemical, and natural gas industries sector–specific QMSs—requirements for product and service supply organizations. It is an advanced version of ISO 9001 standard with additional requirements/features imposed by taking into accounts the requirements of petroleum, petrochemical, and natural gas industries (oil and gas sector) projects. Implementation of this standard ensures that all the requirements of owner/client are properly met and the requirements are taken care. It specifies the QMS requirements for the layout, establishment, production, and implementation of products and services for the petroleum, petrochemical, and natural gas industries.

The requirements have been developed separately to ensure that they are clear and audible. They also provide global consistency and improved assurance in the supply quality of goods and services from providers. This is particularly important

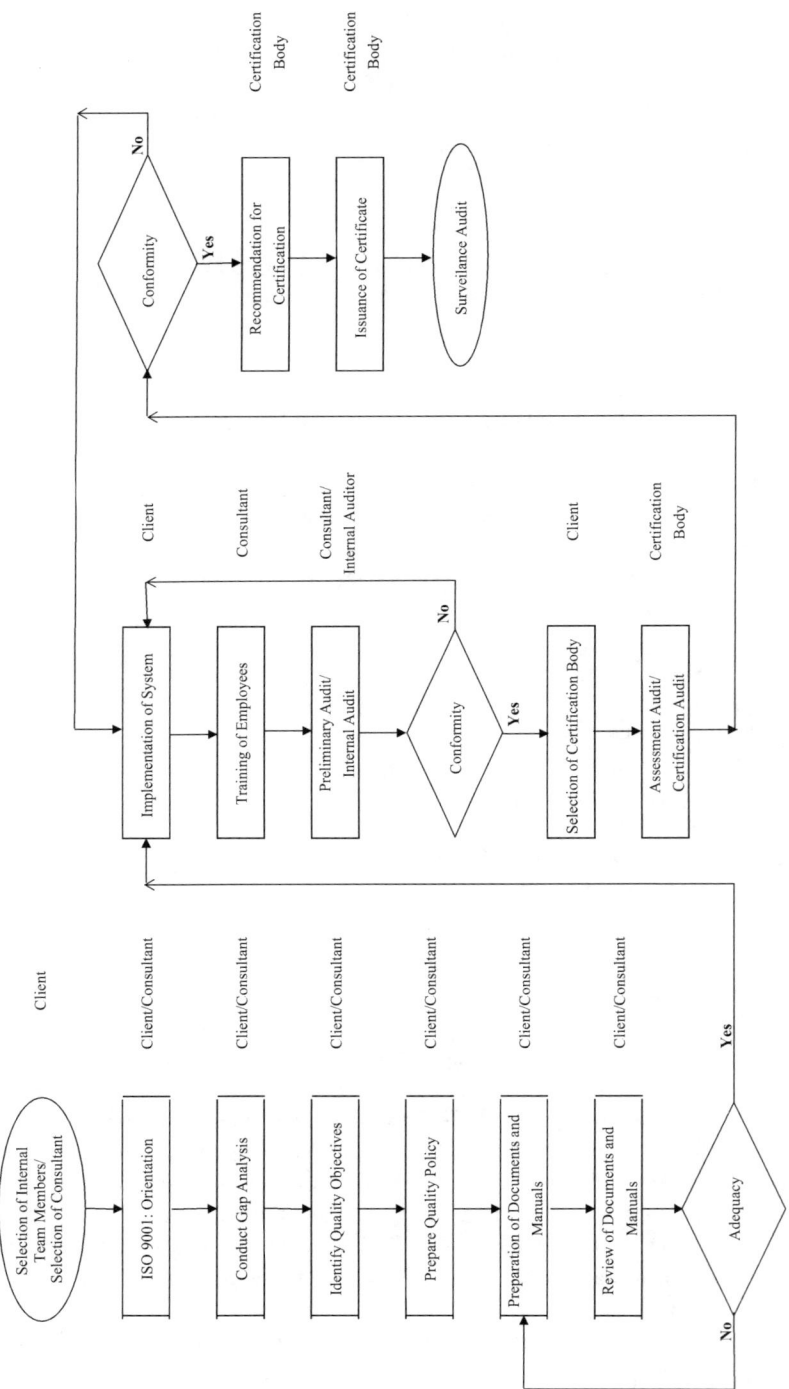

FIGURE 2.8 ISO design and implementation process flow diagram. (Abdul Razzak Rumane. (2019). *Quality Auditing in Construction Projects*. Reprinted with permission from Taylor & Francis Group.) ISO, International Organization for Standardization.

when the failure of goods or services have severe ramification for the companies and industries involved.

ISO/TS 29001 is also based on ISO 29001, which contains requirements on error prevention, reduction of variation, and waste management from the service provider. The requirements have been written separately in order to ensure clarity and perceptibility. This standard is suitable for all companies within the oil and gas industry as it was developed to ensure quality improvement within this particular sector.

Certification of ISO/TS 29001 ensures standardization and improvement within the sector.

Benefits of ISO/TS 29001
- Gain competitive advantage
- Enhance the organization's performance
- Increase efficiency
- Improve marketing capability
- Contribute to cost reduction
- Ensure quality and safety
- Reduce waste
- Improve risk management
- Ensure customer reliability
- Continual improvement

2.7 QUALITY AUDIT

A quality audit is formal or methodical examining, reviewing, and investigating existing system to determine whether agreed-upon requirements are being met. An audit is a systematic independent and documented process to verify or evaluate and report the degree of compliance to the agreed-upon quality criteria, or the specification or contract requirements of the product, services, or project. There are mainly three types of audits. These are as follows:

1. Product audit
2. Process audit
3. System audit
4. Compliance audit
5. Adequacy audit

To achieve competitive advantage, effective quality improvement is critical for an organization's growth. Quality audit serves the purpose of examining the effectiveness of the management-directed control programs. The audit of existing quality system, which is being implemented, provides the factual information to developing long-term organizational strategies for quality.

2.7.1 Categories of Auditing

Audits are mainly classifieds as follows:

First party—Audit your own organization (internal audit)
Second party—Customer audits the supplier (external audit)
Third party—Audits performed by independent audit organization (independent audit)

Third-party audits may result in independent certification of a product, process, or system such as ISO9000 QMS certification. Third-party certification enhances organization's image in the business circle. Quality audit provides feedback to the management on the adequacy implementation and effectiveness of quality system.

2.7.2 Quality Auditing Process

Figure 2.9 illustrates quality auditing process.

Quality Management System

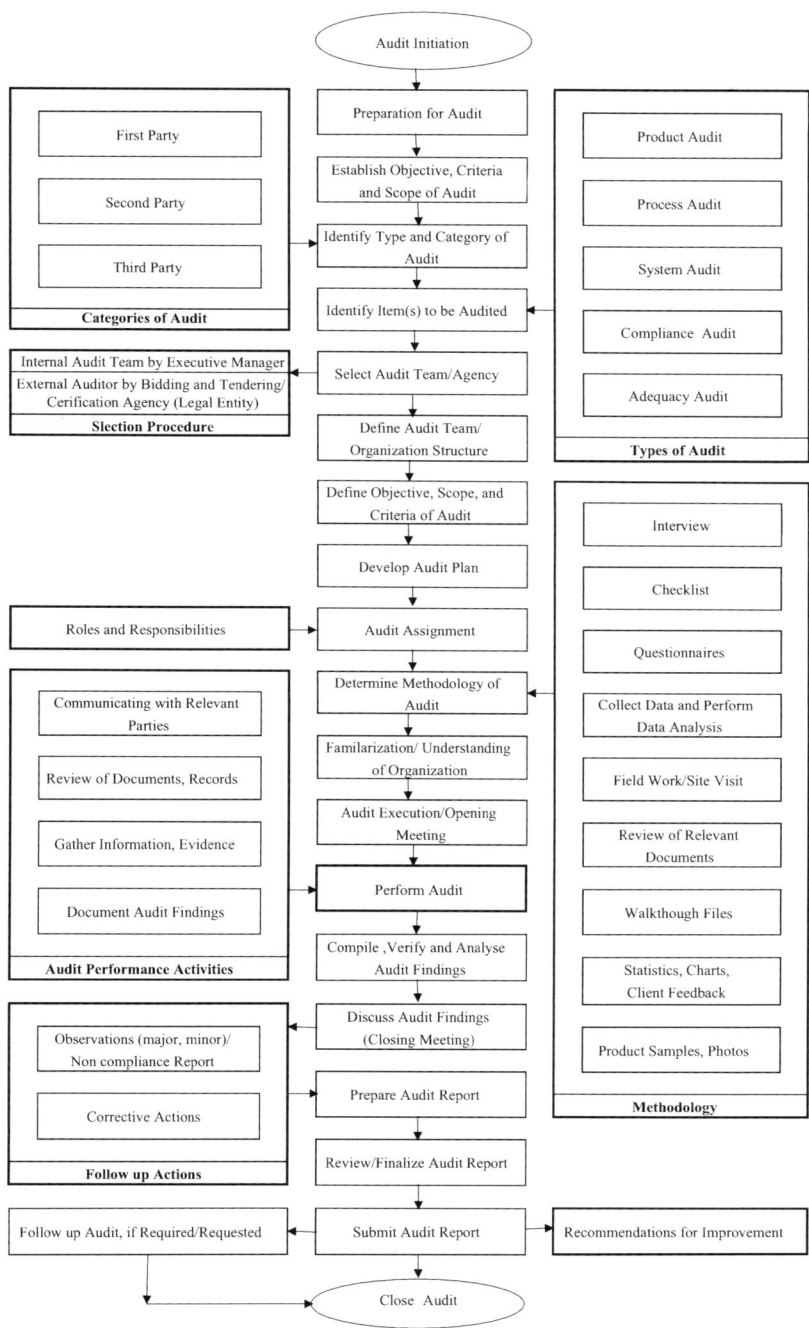

FIGURE 2.9 Typical quality auditing process for construction projects. (Abdul Razzak Rumane. (2019). *Quality Auditing in Construction Projects.* Reprinted with permission from Taylor & Francis Group.)

3 Overview of Oil and Gas Projects

3.1 OIL AND GAS PROJECTS

A project is a temporary endeavor undertaken to create a unique product or service. "Temporary" means that every project has a definite beginning and a definite end. "Unique" means that the product or service is different in some distinguishing way from all similar product or services. Projects are often critical components of the performing organization business strategy. Examples of projects include the following:

- Developing a new product or service
- Effecting a change in structure, staffing, or style of an organization
- Designing a new transportation vehicle/aircraft
- Implementing a new business procedure or process
- Constructing a building or facilities
- Constructing process projects
- Constructing non process projects
- Constructing manufacturing plants

The duration of a project is finite, projects are not ongoing efforts, and the project ceases when its declared objectives have been attained. Some of the characteristics of projects, for example, are as follows:

1. Performed by people
2. Constrained by limited resources
3. Planned, executed, and controlled

Based on various definitions, the project can be defined as:

> A project is a plan or program performed by the people with assigned resources to achieve an objective within a finite duration.

Construction is the translation of owner's goals and objectives into a facility built by the contractor/builder as stipulated in the contract documents, plans, and specifications on schedule and within budget. Construction project is a custom rather than a routine, repetitive business and differs from manufacturing. Construction projects work against defined scope, schedule, and budget to achieve the specified result.

Construction has history of several thousand years. The first shelters were built from stone or mud and the materials collected from the forests to provide protection against cold, wind, rain, and snow. These buildings were primarily for residential purpose, although some may have commercial function.

The scientific revolution of the 17th and 18th centuries gave birth to the great Industrial Revolution of the 18th century. After some delay, construction followed these developments in the 19th century.

The first half of the 20th century witnessed that the construction industry became an important sector throughout the world, employing many workers. During this period, skyscrapers, long-span dams, shells, and bridges were developed to satisfy new requirements and marked the continuing progress of construction techniques. The provision of services such as heating, air conditioning, electrical lighting, mains water, sanitation, and elevators to building became common. The 20th century has seen the transformation of construction and building industry into a major economic sector. During the second half of the 20th century, construction industry began to industrialize, introducing mechanization, prefabrication, and system building. The design of building services systems changed considerably in the past 20 years of the 20th century. Construction projects are constantly increasing in technological complexity. In addition, the requirements of construction clients are on the increase, and as a result, construction projects must meet the varied performance standards. Therefore, to ensure the adequacy of client brief, which addresses the numerous complex client/user needs, it has become the responsibility of designer to evaluate the requirements in terms of activities and their relationship and follow health, safety, and environmental regulation while designing any building.

There are several types of construction projects. Figure 3.1 illustrates types of construction projects.

Construction projects comprised of a cross section of many different participants. These participants are influenced by and depend on each other in addition to "other players" involved in the construction process. Figure 3.2 illustrates concept of traditional construction project organization.

Traditional construction projects have involvement of three main groups. These are as follows:

1. Owner—A person or an organization that initiates and sanctions a project. They request the need of the facility and responsible to arrange the financial resources for creation of the facility/project.
2. Designer (A/E)—This consists of architects or engineers or consultant. They are the owner's appointed entity accountable to convert owner's conception and need into specific facility/project with detailed directions through drawings and specifications within the economic objectives. They are responsible for the design of the project and in certain cases supervision of construction process.
3. Contractor—A construction firm engaged by the owner to complete the specific facility/project by providing the necessary staff, work force, materials, equipment, tool, and other accessories to the satisfaction of the owner/end user in compliance with the contract documents. The contractor is responsible for implementing the project activities and to achieve owner's objectives.

Construction projects are executed based on predetermined set of goals and objectives. Under traditional construction projects, the owner heads the team, designating

Overview of Oil and Gas Projects

1	**Process Type Projects**			
1.1	Liquid chemical Plants			
1.2	Liquid/solid Plants			
1.3	Solid process plants			
1.4	Petrochemical Plants			
1.5	Petroleum refineries			
2	**NonProcess Type Projects**			
2.1	Power plants			
2.2	Manufacturing plants			
2.3	Support facilities			
2.4	Miscellaneous (R&D) projects			
2.5	Civil construction projects	Categories of Civil construction projects and Commercial A/E projects	Residential construction	Family homes, Multiunit town houses, Garden, Apartments, Condominiums, High-rise apartments, Villas.
2.6	Commercial A/E projects		Building construction (institutional and commercial)	Schools, Universities, Hospitals, Commercial office complexes, Shopping malls, Banks, Theaters, Stadiums, Government buildings, Warehouse, Recreation centers, Amusement parks, Holiday resorts, Neighborhood centers.
			Industrial construction	Petroleum refineries, Petroleum plants, Power plants, Heavy manufacturing plants, Steel mills, Chemical processing plants.
			Heavy engineering	Dams, Tunnels, Bridges, Highways, Railways, Airports, Urban rapid transit system, Ports, Harbors, Power lines and Communication network.
			Environmental	Water treatment and clean water distribution, Sanitary and sewage system, Waste management.

FIGURE 3.1 Types of construction projects.

as a project manager. The project manager is a person/member of the owner's staff or independently hired person/firm who has overall or principal responsibility for the management of the project as a whole.

Complex and major construction projects have many challenges such as delays, changes, disputes, and accidents at site, and therefore, the projects need to be efficiently managed from the beginning to the end to meet the intended use and owner's expectations. The owner/client may not have necessary staff/resources in-house to design and manage planning, monitoring, and construction of the project to achieve the desired results. Therefore, in such cases, owners engage professional firms for design, supervision, and construction/project management who has expertise in the management of construction processes, to assist in designing, developing bid documents, overseeing, and coordinating project for the owner. The main area of construction management covers planning, organizing, executing, and controlling to ensure that the project is built as per defined scope, maintaining the completion schedule

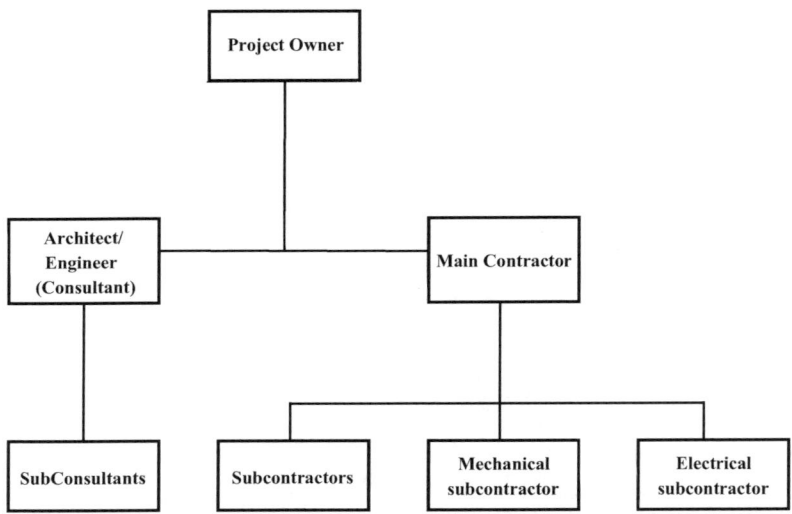

FIGURE 3.2 Design–bid–build (traditional contracting system).

and within agreed-upon budget. These firms bring knowledge and experience that contribute to decision at every stage of project for successful completion of project.

Construction projects are mainly capital investment projects. They are customized and nonrepetitive in nature. Construction projects have become more complex and technical, and the relationships and the contractual grouping of those who are involved are also more complex and contractually varied. The products used in construction projects are expensive, complex, immovable, and long lived. Generally, a construction project is composed of building materials (civil), electromechanical items, finishing items, and equipment. For oil and gas projects, there are other items such as processing equipment/machines, pumps, piping material, values, storage tanks, and instrumentation. These are normally produced by other construction-related industries/manufacturers. These industries produce products as per their own quality management practices complying with certain quality standards or against specific requirements for a particular project. The owner of the construction project or his representative has no direct control over these companies unless he/his representative/appointed contractor commit to buy their product for use in their facility. These organizations may have their own quality management program. In manufacturing or service industries, the quality management of all in-house manufactured products is performed by the manufacturer's own team or under the control of same organization having jurisdiction over their manufacturing plants at different locations. Quality management of vendor-supplied items/products is carried out as stipulated in the purchasing contract as per the quality control specifications of the buyer.

Oil and gas sector projects are considered as biggest sector in terms of investment as well as in employment generation. Oil and gas projects are categorized as process type of projects. These types of projects are undertaken by large contraction

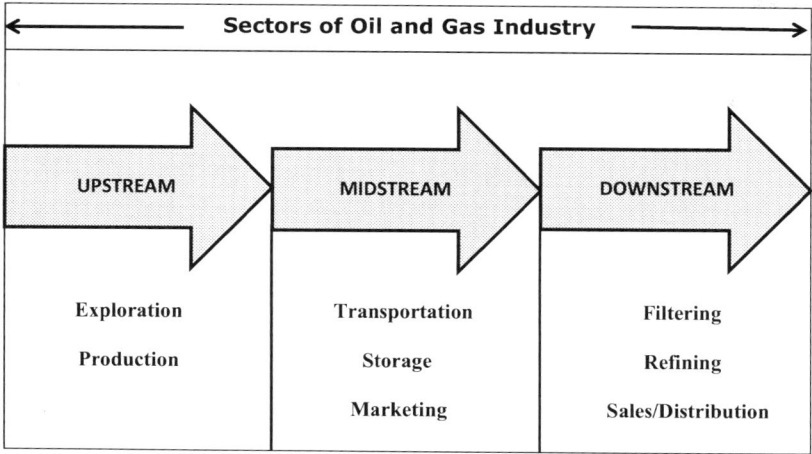

FIGURE 3.3 Sectors of oil and gas industry.

companies on behalf of large corporations/governmental authorities. These companies are required to have a wide array of expertise and knowledge in such construction. While safety is a prime concern in construction projects, the health, safety, and environmental (HSE) requirements in process industry projects are to be taken care right from the design and generally need more planning and teamwork coordination. The oil and gas industry has mainly three sectors. These are as follows:

1. Upstream
   - Upstream is exploration and production of crude oil and gas
   - Production and stabilization of oil and gas
2. Midstream
   - Transportation, storage, and processing of oil and gas
   - Gas treatment, LNG production, and regasification plants.
   - Oil and gas pipeline system
3. Downstream
   - Filtering, refining of crude oil, and purifying natural gas
   - Processing of oil and condensates into marketable products with defined specifications

Figure 3.3 illustrates sectors of oil and gas industry

3.1.1 Oil and Gas Project Life Cycle

Most construction projects are custom oriented having a specific need and a customized design. It is always the owner's desire that their project should be unique and better. Furthermore, it is the owner's goal and objective that the facility is completed on time. The expected time schedule is important both from financial viewpoint and for the acquisition of the facility by the owner/end user.

Oil and gas downstream projects are process type of projects that produce different types of products to satisfy and meet the consumers' and end users' requirements. The processing of raw materials is very important, and necessary precautions are required to construct the project focusing mainly on sustainability to ensure that the output products meet the required specifications.

The system life cycle is fundamental to the application of systems engineering. A systems engineering approach to construction projects helps to understand the entire process of project management and to manage and control its activities at different levels of various phases to ensure timely completion of the project with economical use of resources to make the construction project the most qualitative, competitive, and economical.

Systems engineering starts from the complexity of the large-scale problem as a whole and moves toward structural analysis and the partitioning process until the questions of interest are answered. This process of decomposition is called a work breakdown structure (WBS). The WBS is a hierarchical representation of system levels. Being a family tree, the WBS consists of a number of levels, starting with the complete system at level 1 at the top and progressing downward through as many levels as necessary, to obtain elements that can be conveniently managed.

Benefits of systems engineering applications are as follows:

- Reduction in cost of system design and development, production/construction, system operation and support, system retirement, and material disposal
- Reduction in system acquisition time
- More visibility and reduction in the risks associated with the design decision-making process

However, it is difficult to generalize project life cycle to system life cycle. However, considering that there are innumerable processes that make up the construction process, the technologies and processes, as applied to systems engineering, can also be applied to construction projects. The number of phases shall depend on the complexity of the project. Duration of each phase may vary from project to project. The methodology applied in this book is based on dividing life cycle of oil and gas downstream project into six phases. These phases are as follows:

1. Feasibility study
2. Concept design
3. Front-end engineering design (FEED)
4. Bidding and tendering
5. Engineering, procurement, and construction (EPC)
   - Detailed engineering
   - Procurement
   - Construction
6. Testing, commissioning, and handover

The life cycle division into six phases will help to conveniently manage and control the project at each phase/stage. Each phase is further subdivided on WBS principle to reach a level of complexity where each element/activity can be treated as a single unit, which can be conveniently managed. WBS represents a systematic and logical breakdown of the project phase into its components (activities). It is constructed by dividing the project into major elements with each of these being divided into subelements. This is done till a breakdown is done in terms of manageable units of work for which responsibility can be defined. WBS involves envisioning the project as a hierarchy of goal, objectives, activities, subactivities, and work packages. The hierarchical decomposition of activities continues until the entire project is displayed as a network of separately identified and nonoverlapping activities. Each activity is single purposed, of a specific time duration, and manageable; its time and cost estimates are easily derived; deliverables are clearly understood; and responsibility for its completion is clearly assigned. The WBS helps in the following:

- Effective planning by dividing the work into manageable elements, which can be planned budgeted and controlled
- Assignment of responsibility for work elements to project personnel and outside agencies
- Development of control and information system

WBS facilitates the planning, scheduling, budgeting, managing, and controlling activities for the project manager and its team. By application of WBS phenomenon, the construction phases are further divided into various activities. Division of these phases will improve the control and planning of the construction project at every stage before a new phase starts. The components/activities of construction project life cycle phases divided on WBS principle are listed in the following:

1. **Feasibility study phase**
    i. Project initiation
    ii. Identify need
    iii. Feasibility study
    iv. Establish project goals and objectives
    v. Identify alternatives/options
    vi. Analyze and evaluate alternatives/options
    vii. Select preferred alternative/option
    viii. Finalize project delivery and contracting system
    ix. Project charter
2. **Concept design phase**
    i. Select project team member
    ii. Identify project stakeholders
    iii. Establish concept design phase requirements
    iv. Identify concept design deliverables
    v. Develop concept design
    vi. Prepare concept design
    vii. Prepare preliminary schedule
    viii. Estimate conceptual cost
    ix. Establish project quality requirements

- x. Manage concept design quality
- xi. Estimate project resources
- xii. Identify project risks
- xiii. Identify project HSE issues and requirements
- xiv. Review concept design
- xv. Finalize concept design

3. FEED phase
- i. Identify design team members
- ii. Identify project stakeholders
- iii. Establish FEED requirements
- iv. Develop FEED (technical study)
- v. Prepare FEED
- vi. Prepare project schedule/plan
- vii. Evaluate and estimate project cost
- viii. Manage FEED quality
- ix. Estimate resources
- x. Develop communication plan
- xi. Manage risks
- xii. Manage HSE requirements
- xiii. Develop FEED documents
- xiv. Perform value engineering study
- xv. Review FEED
- xvi. Finalize FEED
- xvii. Develop tender documents

4. Bidding and tendering phase
- i. Organize tender documents
- ii. Identify project team
- iii. Identify tendering procedure
- iv. Identify bidders
- v. Manage tendering process
- vi. Distribute tender documents
- vii. Conduct pre-bid meeting
- viii. Submit/receive bids
- ix. Manage bidding and tendering quality
- x. Manage risk
- xi. Review bid documents
- xii. Evaluate bid documents
- xiii. Select contractor
- xiv. Award contract

5. EPC phase
- i. Develop project site facilities
- ii. Notice to proceed
- iii. Kick off meeting
- iv. Mobilization
- v. Identify stakeholders

    vi. Develop responsibility matrix
    vii. Establish construction phase requirements
    viii. Develop construction phase scope
    ix. Develop contractor's construction schedule
    x. Develop project S-curve
    xi. Develop contractor's quality control plan
    xii. Develop stakeholder management plan
    xiii. Develop resource management plan
    xiv. Develop communication management plan
    xv. Develop risk management plan
    xvi. Develop contract management plan
    xvii. Develop HSE management plan

5.1 EPC (detailed engineering)
    i. Identify detailed engineering deliverables
    ii. Develop detailed engineering
    iii. Perform interdisciplinary coordination
    iv. Develop project schedule
    v. Estimate project cost
    vi. Manage detailed engineering design quality
    vii. Estimate project resources
    viii. Manage project risk
    ix. Manage HSE plan
    x. Review detailed engineering
    xi. Finalize detailed engineering

5.2 EPC (procurement)
    i. Identify procurement deliverables
    ii. Develop procurement documents
    iii. Prepare material procurement list
    iv. Identify vendors
    v. Manage procurement quality
    vi. Finalize procurement

5.3 EPC (construction)
    i. Identify construction works deliverables
    ii. Manage execution/installation of works
    iii. Manage scope change
    iv. Manage construction quality
    v. Manage construction resources
    vi. Manage communication
    vii. Manage construction risks
    viii. Manage contract
    ix. Manage HSE requirements
    x. Manage project finances
    xi. Manage claims
    xii. Monitoring and control of project works
    xiii. Validate executed works

6. Testing, commissioning, and handover phase
   i. Identify testing, commissioning, and startup requirements
   ii. Identify stakeholders
   iii. Develop scope of work
   iv. Develop testing, commissioning, and handover plan
   v. Execute testing and commissioning works
   vi. Manage testing and commissioning quality
   vii. As-built drawings
   viii. Technical manuals and documents
   ix. Record books (project, construction, manufacturing)
   x. Train owner's/end user's personnel
   xi. Regulatory/authority approval
   xii. Handover the project to the owner/end user
   xiii. Substantial completion
   xiv. Settle payments
   xv. Settle claims

Table 3.1 illustrates construction project life cycle (downstream) phases with subdivided activities/components.

These activities may not be strictly sequential; however, the breakdown allows implementation of project management functions more effectively at different stages.

3.2 PROJECT DELIVERY SYSTEMS

A project delivery system is defined as the organizational arrangement among various participants comprising owner, designer, contractor, and many other professionals involved in the design and construction of a project/facility to translate/transform the owner's needs/goals/objectives into a finished facility/project to satisfy the owner/s/end user's requirements.

The project delivery system

- establishes scope and responsibility for how the project is delivered to the owner,
- includes project design and construction,
- defines responsibility/obligations that each of the participants is expected to perform, such as scheduling, cost control, quality management, safety management, and risk management, during various phases of construction project life cycle.
- is the approach by which the project is delivered to the owner, but is separate and distinct from the contractual arrangements for financial compensation, and
- establishes procedures, actions, and sequence of events to be carried out.

3.2.1 Types of Project Delivery Systems

There are several types of contract delivery systems. Table 3.2 illustrates most common project delivery systems followed in construction projects.

TABLE 3.1
Project Life Cycle (Downstream) Phases

	Feasibility Study			Construction Project Life Cycle (Downstream) Phases			
	Feasibility Study	Concept Design	Front End Engineering Design (FEED)	Bidding and Tendering	EPC (Engineering, Procurement, Construction) — Engineering / Procurement / Construction		Testing, Commissioning, and Handover
• Project initiation	• Select design team members	• Identify design team members	• Organize tender documents	• Notice to proceed • Kick off meeting	• Develop project site facilities	• Mobilization • Temporary facilities	• Identify testing, commissioning, and startup requirements
• Identification of need	• Identify project stakeholders	• Identify project stakeholders	• Identify project team/ stakeholders	• Identify stakeholders • Responsibility matrix	• Contractor's design team • Contractor's core staff • Subcontractors/ vendors	• Project management consultant • Supervision consultant	• Identify stakeholders
• Feasibility study	• Establish concept design requirements	• Establish feed requirements	• Identify tendering procedure	• Construction phase requirements	• Construction phase scope	• Contractor's construction schedule	• Develop scope of work

(*Continued*)

TABLE 3.1 (*Continued*)
Project Life Cycle (Downstream) Phases

• Establish project goals and objectives	• Identify concept design deliverables	• Develop FEED (technical study to select process)	• Identify bidders	• Project S-curve contractor's quality control plan	• Stakeholder management plan • Resource management plan • Communication management plan	• Risk management plan • Contract management plan • HSE management plan	• Develop testing and commissioning plan
• Identification of alternatives/options	• Develop concept design	• Prepare FEED	• Manage tendering process	• Identify detailed engineering deliverables ⇔	• Identify procurement deliverables ⇔	• Identify construction works deliverables	• Execute testing and commissioning works
• Analyze and evaluate alternatives/options	• Identify/collect regulatory requirement	• Prepare project schedule/plan	• Distribute tender documents	• Develop detailed engineering	• Develop procurement documents	• Manage execution/installation of works	• Manage testing and commissioning quality procedure
• Select preferred alternative	• Prepare preliminary schedule	• Evaluate and estimate project cost	• Conduct prebid meetings	• Perform interdisciplinary coordination	• Prepare material procurement list	• Manage scope change	• As-built drawings

(*Continued*)

TABLE 3.1 (Continued)
Project Life Cycle (Downstream) Phases

• Finalize project delivery and contracting system	• Prepare preliminary schedule	• Manage FEED quality	• Submit/receive bids	• Develop project schedule	• Identify vendors	• Manage construction quality	• Technical manuals and documents
• Project charter	• Estimate conceptual project cost	• Estimate resources	• Manage bidding and tendering quality	• Estimate project cost	• Manage procurement quality	• Manage construction resources	• Record books (PRB, CRB, MRB)
	• Establish project quality requirements	• Develop communication plan	• Manage risks	• Manage detailed engineering quality	• Finalize procurement	• Manage communication	• Train owner's/end user's personnel
	• Manage concept design quality	• Manage project risks	• Review bid documents	• Estimate project resources		• Manage construction risk	• Regulatory/authority approval
	• Estimate project resources	• Manage HSE requirements	• Evaluate bids	• Manage project risk		• Manage contract	• Handover of project to the owner/end user
	• Identify project risks	• Develop FEED documents	• Select contractor	• Manage project HSE requirements		• Manage HSE requirements	• Issue substantial completion

(*Continued*)

TABLE 3.1 (Continued)
Project Life Cycle (Downstream) Phases

• Identify project HSE issues and requirements • Review concept design • Finalize concept design	• Perform value engineering study • Review FEED • Finalize FEED • Develop tender documents	• Award contract	• Review detailed engineering • Finalize detailed engineering	• Manage project finances • Manage claims • Monitor and control project works • Validate executed works	• Settle payment • Settle claims

FEED, front-end engineering design; HSE, health, safety, and environmental.

TABLE 3.2
Categories of Project Delivery Systems

Serial Number	Category	Classification	Subclassification
1	Traditional system (separated and cooperative)	Design–bid–build	Design–bid–build
		Variant of traditional system	Sequential method
			Accelerated method
2	Integrated system	Design–build	Design–build
		Design–build	Joint venture (architect and contractor)
		Variant of design–build system	Package deal
		Variant of design–build system	Turnkey method (engineering, procurement, construction)
		Variant of design-build system (turnkey)	Build–operate–transfer (BOT)
			Build–own–operate–transfer
			Build–transfer–operate
			Design–build–operate–maintain
		Variant of design–build system (funding option)	Lease–develop–operate
			Wraparound (public–private partnership)
		Variant of design–build system	Build–own–operate
			Buy–build–operate
3	Management-oriented system	Management contracting	Project manager (program management)
		Construction management	Agency construction manager
			Construction manager at risk
4	Integrated project delivery system	Integrated form of contract	

Source: Abdul Razzak Rumane. (2013). *Quality Tools for Managing Construction Projects.* Reprinted with permission from Taylor & Francis Group.

However, the following types of project delivery systems are commonly used in oil and gas projects:

1. Design–build
2. Turnkey contract

3.2.1.1 Design–build

In design–build contract, owner contracts a firm singularly responsible to design and build the project. In this type of contracting system, the contractor is appointed based on an outline design or design brief to understand the owner's intent of the project. The owner has to clearly define his/her needs, performance requirements, and comprehensive scope of works prior to signing of the contract. It is a must that project definition is understood by the contractor to avoid any conflict in future as the contractor is responsible for detailed design and construction of the project. A design–build type of contract is often used to shorten the time required to complete a project. Since the contract with the design–build firm is awarded before starting any design or construction, a cost plus contract or reimbursable arrangement is normally used instead of lump-sum, fixed-cost arrangement. This type of contract requires extensive involvement of the owner during entire life cycle of project. They have to involve for taking decisions during the selection of design alternatives and the monitoring of costs and schedules during construction, and therefore, the owner has to maintain/hire a team of qualified professionals to perform these activities. Design–build contracts are used for relatively straightforward work, where no significant risk or change is anticipated and when the owner is able to specify precisely what is required. Table 3.3 illustrates main aspect, advantages, and disadvantages of design–build type of project delivery system, whereas Figure 3.4 illustrates contractual relationship.

In case of design–bid–build type of project delivery system, the contractor is contracted after completion of design based on successful bidding, whereas in design–build type of deliverable system, the contractor is contracted right from the early stage of the construction project and is responsible for design development of the project.

Figure 3.5 illustrates typical logic flowchart for design–build type of construction project contracting system.

3.2.1.2 The Turnkey Contract

As the name suggests, these are the types of contracts where, on completion, a key is turned in the door and everything is working to full operating standards. In this type of method, the owner employs a single firm to undertake design, procurement, construction, and commissioning of the entire works. The firm is also involved in management of project during entire process of the contract. The client is responsible for preparation of their statement requirements, which becomes strict responsibility of contractor to deliver. This type of contract is used mainly for the process type of projects and is sometimes called EPC.

In EPC contracting systems, the contractor has to execute the project based on FEED by the owner and has to deliver the project within an agreed time and cost, commonly known as lump sum turnkey contract.

Table 3.4 illustrates difference between design–build and EPC type of project delivery system, Figure 3.6 illustrates typical logic flowchart for EPC project, and Figure 3.7 illustrates EPC project system contractual relationship.

3.2.2 Types of Contracting/Pricing

Contract is a formalized means of agreement, enforceable by law, between two or more parties to perform certain works or provides certain services against agreed-upon financial incentives to complete the works/services. Regardless of the type of project delivery system selected, the contractual arrangement by which the parties are compensated must also be established. In the construction projects, determining how to procure the product is as important as determining what and when. In most procurement activities, there are several options available for purchase or subcontracts. The basis of compensation type relates to the financial arrangement among the

TABLE 3.3
Design-Build

Project Delivery System	Main Aspects	Advantages	Disadvantages
Design-build	In design–build contract, owner contracts a firm (contractor) singularly responsible for design and construction of the project. In this type of contracting system, the contractor is appointed based on an outline design or design brief to understand the owner's intent of the project. The owner has to prepare comprehensive scope of work and has to clearly define his/her needs and performance requirements/specifications prior to signing of the contract. It is a must that project definition is understood by the contractor to avoid any conflict in future as the contractor is responsible for detailed design and construction of the project. Owner has to involve for taking decisions during the selection of design alternatives and the monitoring of costs and schedules during construction, and therefore, the owner has to maintain/hire a team of qualified professionals to perform these activities.	• Reduce overall project time because construction begins before completion of design. • Singular responsibility, contractor takes care of the design, schedule, construction services, quality, methods, and technology. • For owner/client, the risk is transferred to design/build contractor. • Project cost defined in early stage and has certainty. • Early involvement of contractor assists constructability. • Suitable for straightforward projects where significant changes or risks are not anticipated and owner is able to precisely specify the objectives/requirements. • Risk management is better than design/bid/build	• Not suitable for complex projects. • Owner has reduced control over design quality. • Extensive involvement to ensure design deliverable meets project performance requirements. • Extensive involvement of owner during entire life cycle of project. • Real price for a contract cannot be estimated by the owner/client in the beginning. • Changes by owner can be expensive and may results in heavy cost penalties to the owner. • Poor identification of owner need and wrong understanding of project brief/concept can cause main problem during the project realization. • Project quality cannot be assured if it is not monitored properly by the owner. • For contractor, more risks in this type of contract. • Not suitable for renovation projects.

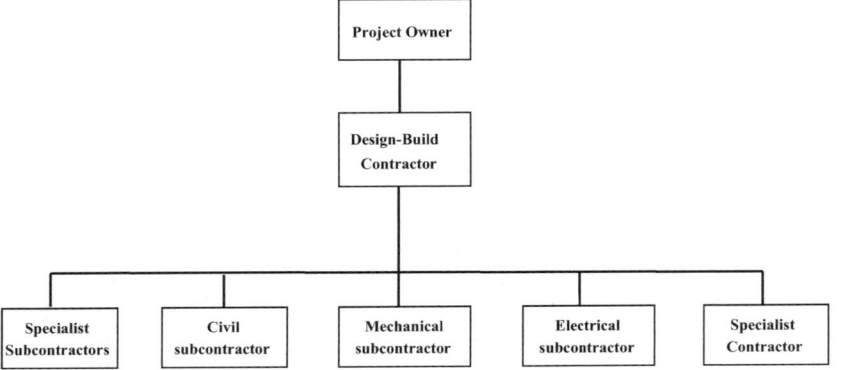

FIGURE 3.4 Design–build delivery system.

parties as to how the designer or contractor is to be compensated for their services. The following are the most common types of contract/compensation methods:

1. Firm fixed price or lump sum contract
   a. Firm fixed price
   b. Fixed price incentive fee
   c. Fixed price with economic adjustment price

Table 3.5 illustrates advantages and disadvantages of fixed price or lump sum price contracts.

The following are the common procurement methods for selection of project teams such as designer, consultant, project manager, construction manager, and contractor:

1. Low bid
   - Selection is based solely on the price.
2. Best value
   a. Total cost
      - Selection is based on total construction cost and other factors.
   b. Fees
      - Selection is based on weighted combination of fees and qualification.
3. Qualification-based selection
   - Selection is based solely on qualification.

3.3 MANAGEMENT OF OIL AND GAS PROJECT

Oil and gas projects are executed based on predetermined set of goals and objectives and are customized and nonrepetitive in nature. Oil and gas construction projects have involvement of many participants comprising owner, designer, contractor, and many other professionals from construction-related industries. Each of these participants is involved to implement quality in construction projects. These participants

Overview of Oil and Gas Projects

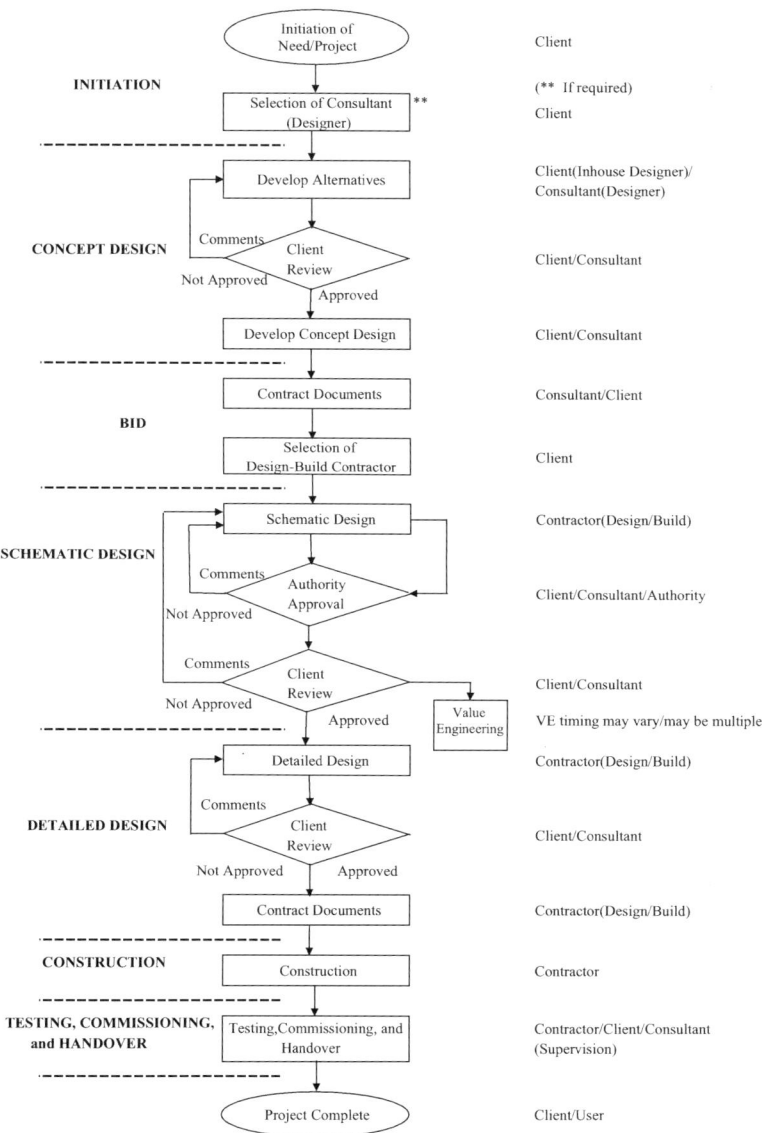

FIGURE 3.5 Logic flowchart for construction projects—design–build system. (Abdul Razzak Rumane. (2010). *Quality Management in Construction Projects.* Reprinted with permission from Taylor & Francis Group.)

are influenced by and depend on each other in addition to "others players" involved in the construction process. Therefore, the construction projects have become more complex, and technical and extensive efforts are required to reduce rework and costs associated with time, materials, and engineering. Oil and gas are more complex

TABLE 3.4
Difference between Design–Build and EPC

Design–Build	EPC
• Client/owner has an input into the outline design or design brief to understand owner's intent of the project • Client/owner contracts a designer/consultant (in case in-house facility not available) to decide design outline • Design–build contractor is selected after concept design phase • Schematic design and detailed design is carried out by the design–build contractor (singular responsibility) • Client/owner has to employ a team of professionals to perform supervision/management of detail design, monitor quality, schedule, and cost • Claims risk is higher • Mainly suitable for building projects	• Client has an input in the output (operating capacity) • Client/owner contracts a designer/consultant (in case in-house facility not available) for development of concept design and front-end engineering design • EPC contractor is selected after front-end engineering design phase • EPC is direct contract between client/owner to build complete project to meet agreed-upon output • All project activities are carried out by EPC contractor • Client/owner normally employs project management consultant • Client/owner is not involved with detailed design process, except in the event of variation and quality procedures • Claims risk is lower • Mainly used for projects such as power plants, process industry, and oil and gas sector

EPC, engineering, procurement, and construction

and technical, and the relationships and the contractual grouping of those who are involved are also more complex and contractually varied. The products used in construction projects are expensive, complex, immovable, and long lived. The products used in these projects are expensive, complex, immovable, and long lived.

Therefore, to ensure the adequacy of client need that addresses the numerous complex requirements, it is needed to evaluate the requirements in terms of activities and their interrelationship and manage all the construction activities and elements in an efficient and effective way.

Oil and gas construction projects have many challenges such as delays, changes, disputes, and accidents at site and therefore need efficient management of the project from the beginning to the end of construction of the facility/project to meet the intended use and owner's expectations. The main area of construction/project management covers planning, organizing, executing, and controlling to ensure that the project is built as per defined scope, maintaining the completion schedule and within agreed-upon budget. The owner/client may not have necessary staff/resources in-house to manage planning, design, and construction of the construction project to achieve the desired results. Therefore, in such cases, the owners engage a professional firm or a person, called construction/project manager, who is trained and has expertise in the management of construction processes, to assist in developing design, bid documents, overseeing, monitoring, controlling, and coordinating project for the owner.

Overview of Oil and Gas Projects

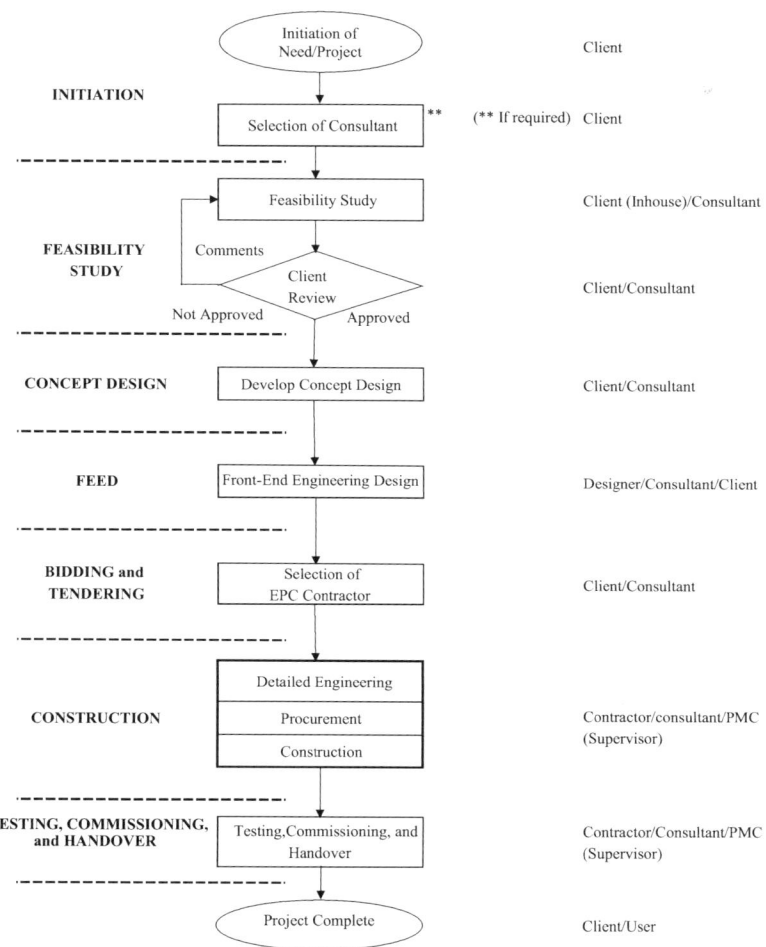

FIGURE 3.6 Logic flowchart for EPC projects. EPC, engineering, procurement, and construction.

3.3.1 Project Management Processes

As per *Project Management Body of Knowledge (PMBOK® Guide), Fifth Edition*, published by Project Management Institute (PMI), project management is the application of knowledge, skills, tools, and techniques to project activities to meet the project requirements. Project management totally deals with five process groups (initiating, planning, executing, monitoring and control, and closing) divided into 13 knowledge areas, which are spread over 47 locally grouped project management processes. Project management involves managing all types of project from start to finish.

Project management is a discipline and management system specially tailored to promote the successful execution of capital and complex projects.

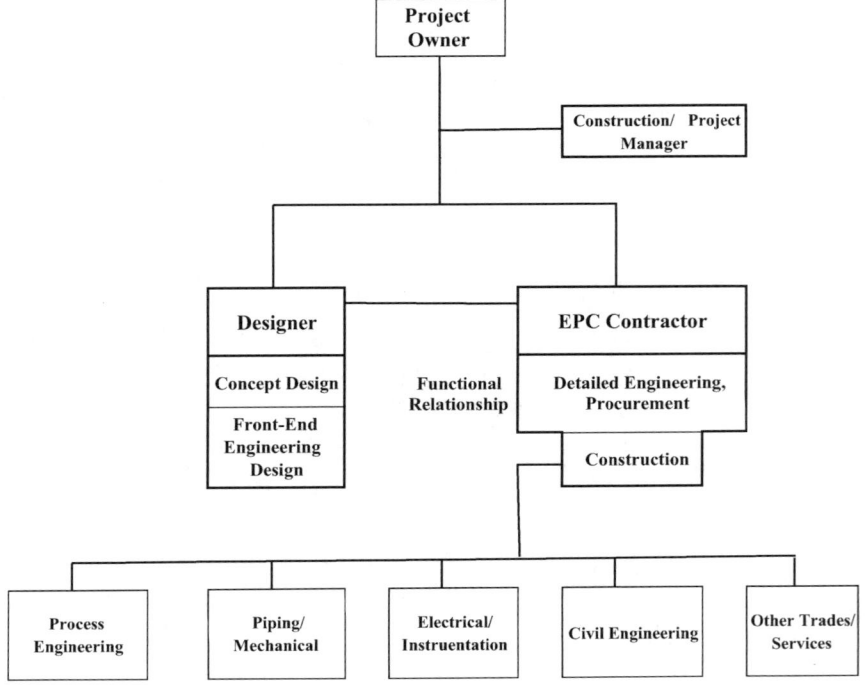

FIGURE 3.7 EPC contracting system contractual relationship. EPC, engineering, procurement, and construction.

Project management mainly involves construction-related management activities such as follows:

- Planning
- Scheduling
- Monitoring and control
- Quality control/quality assurance
- Human resources
- Material, equipment
- Safety and environmental protection

Project/construction management process is a systematic approach to manage construction project from project inception to the completion of construction and handover to the client/end user. Construction management is application of professional services, skills, and effective tools and techniques to manage project planning, design, and construction from project inception through to issuance of completion certificate. Some of these techniques are tailored to the specific requirements unique to construction projects.

The main objective of construction management is to ensure that the client/end user is satisfied with the quality of project delivery. In order to achieve project

TABLE 3.5
Fixed Price/Lump Sum Contracts

Serial Number	Project Contracting Type	Main Aspects	Advantages	Disadvantages
1	Fixed price or lump sum	With this type of contract, the contractor agrees to perform the specified work/services with a fixed-price (it is also called lump-sum) for the specified and contracted work. Any extra work is executed only upon receipt of instruction from the owner. Fixed price contracts are generally inappropriate for work involving major uncertainties, such as work involving new technologies.	• Low financial risk to the owner. • Total cost of project known before construction. • Project viability is known before a commitment is made. • Suitable for projects, which can be completely designed and whose quantities are definable. • Minimal owner supervision. • Contractor will usually assign the best personnel to the work. • Maximum financial motivation of contractor. • Contractor has to solve own problems and do so quickly. • Contractor selection by competitive bidding is fairly easy, apart from the deliberate low price.	• High financial risk to the contractor. • Owner bears the risk of poor quality from the contractor trying to maximize the profit within fixed cost. • Contractor's price may include high contingency. • Variation can be time-consuming and costly. • Variations (changes) are difficult and costly. • More time taken for bidding and for developing a good design basis. • Contractor will tend to choose the cheapest and quickest solutions, making technical monitoring and strict quality control by the owner essential.

performance goals and objectives, it is required to set performance measures that define what the contractor is to achieve under the contract. Therefore, to achieve the adequacy of client brief, which addresses the numerous complex needs of client/end user, it is necessary to evaluate the requirements in terms of activities and their functional relationships and establish construction management procedures and practices to be implemented and following toward all the work areas of the project to make the project successful to the satisfaction of the owner/end user and to meet owner needs.

Construction management involves project execution–related management works managed by a professional firm or a person called construction manager. For successful management of the project, it is essential to have professional knowledge of management functions, management processes, and project phases (technical processes). The construction project quality is the fulfillment of owner's needs per defined scope of works within a budget and specified schedule to satisfy the owner's/user's requirements.

There are mainly three key attributes in construction projects, which the construction/project manager has to manage effectively and efficiently to achieve successful project. These are as follows:

1. Scope
2. Time (schedule)
3. Cost (budget)

From the quality perspective, these three elements are known as "quality trilogy", whereas when considered with project/construction management perspective, these are known as "triple constraint".

From the project quality perspective, the phenomenon of these three components is called the "construction project quality trilogy" and is illustrated in Figure 3.8.

The construction project quality trilogy phenomenon when considered with project management/construction management perspective is known as triple constraint. Scope (defined scope), schedule (time), and cost (budget) are three sides of a triangle. Figure 3.9 illustrates principle of triple constraint.

Triple constraint is a framework for the construction manager or project manager to evaluate and balance these competing demands. It became a way to track and

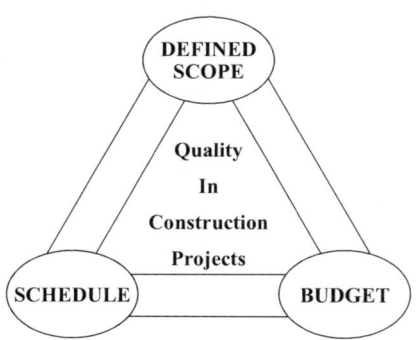

FIGURE 3.8 Construction project quality trilogy.

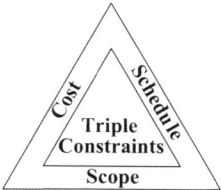

FIGURE 3.9 Triple constraint.

monitor projects. In pictorial form, triple constraint is a triangle in which one cannot adjust or alter one side of it without any effect or altering the other side(s).

- If scope is increased, the cost will increase or the time must be extended, or both.
- It time is reduced, then cost must increase or scope must decrease, or both.
- If cost is reduced, then the scope must be decreased, or the time must increase, or both.

In order to achieve successful project, the construction/project manager must handle these key attributes effectively and efficiently, and track the progress of the work from the inception to completion of construction and handover of the project for successful completion. These three attributes have functional relationship with many other processes, activities, and elements/subsystems of the project. To achieve a successful project to the satisfaction of owner/end user, the construction/project manager has to manage the project in a systematic manner at every stage of the project and balance these attributes in conjunction with all the other activities, which may affect the successful completion of the project. This can be done by implementing, amalgamating, and coordinating some or all the activities/elements of management functions, management processes, and project life cycle phases (technical processes). Thus, construction management process can be described as implementation and interaction of following functions and processes:

- Management functions
- Management processes
- Project phases (technical processes)

Figure 3.10 illustrates construction management process elements.

In practice, it is difficult to separate one element from others while executing a project. Interaction and/or combination among some or all of the elements/activities of these processes and their effective implementation and applications are essential throughout the life cycle of the project to conveniently manage the project. Figure 3.11 illustrates integration diagram of components/activities of three major elements of project/construction management process.

PMBOK® Guide published by PMI describes application of project management processes during the life cycle of projects to enhance the chances of success over a

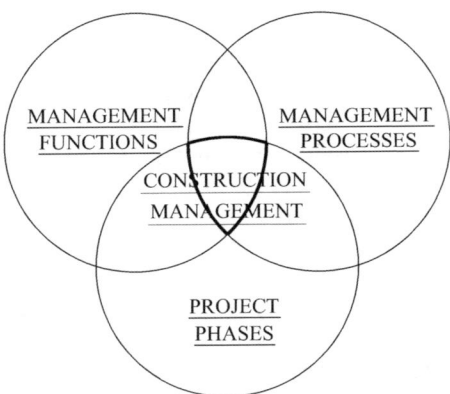

FIGURE 3.10 Construction management integration.

wide range of projects. PMBOK® Guide, Fifth Edition, identifies and describes five project management process groups required for successful completion of any project. These are as follows:

1. Initiating process group
2. Planning process group
3. Executing process group
4. Monitoring and controlling process group
5. Closing process group

Figure 3.12 illustrates overview of project management process groups.

In order to conveniently manage the project, each of these phases is treated as a project itself with all the five process groups operating as they do for overall project, and each phase is composed of activities, elements having functional relationship to achieve a common objective for useful purpose. Figure 3.13 illustrates construction project life cycle phases for EPC type of project delivery system.

Based on the principles of project management processes defined in PMBOK® Guide, the following are the management processes derived, based on knowledge areas, to manage and control various processes and activities to be performed in construction project management:

1. Integration management
2. Stakeholder management
3. Scope management
4. Schedule management
5. Cost management
6. Quality management
7. Resource management
8. Communication management
9. Risk management

10. Contract management
11. HSE management
12. Financial management
13. Claim management

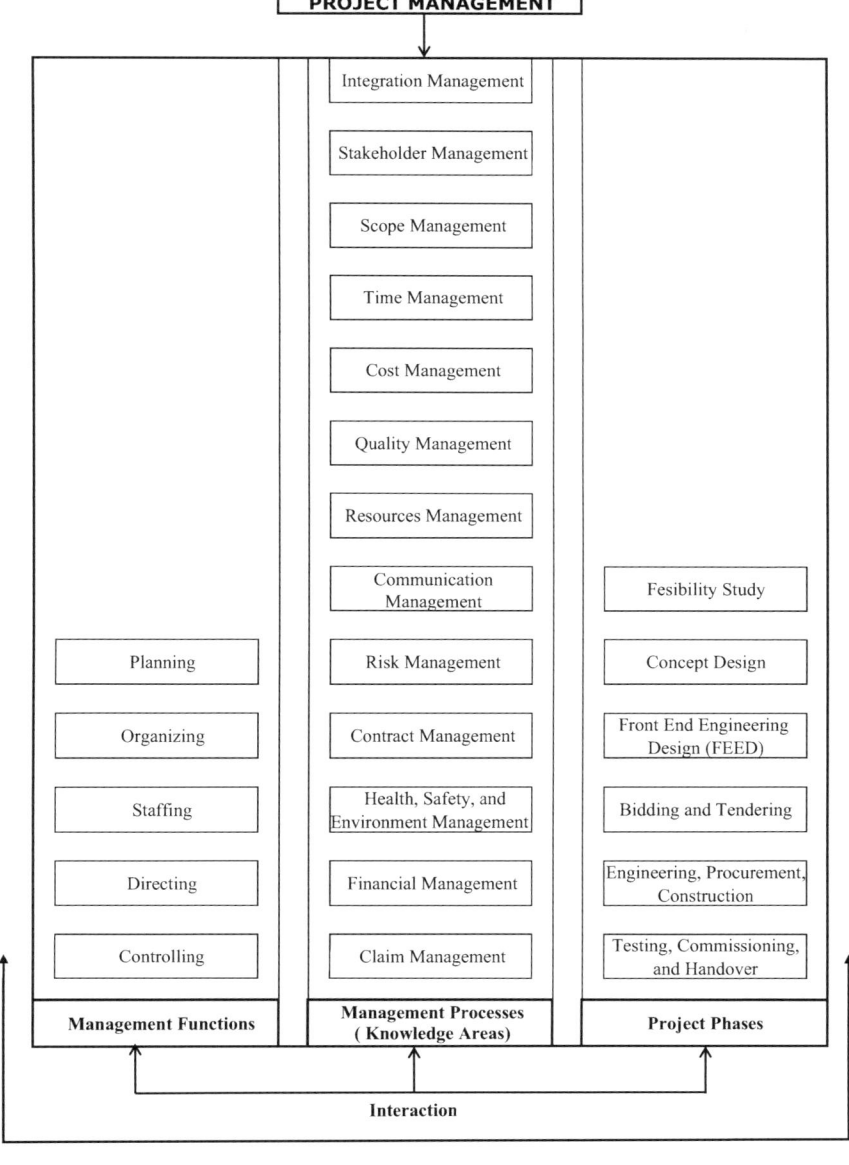

FIGURE 3.11 Project management process elements integration diagram. (Abdul Razzak Rumane. (2016). *Handbook of Construction Management: Scope, Schedule, and Cost Control*. Reprinted with permission from Taylor & Francis Group.)

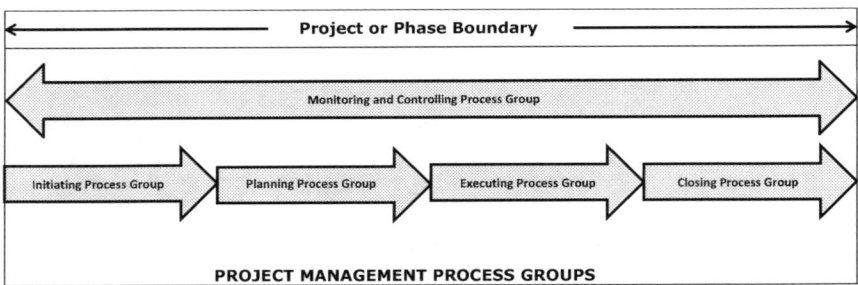

FIGURE 3.12 Overview of project management process groups.

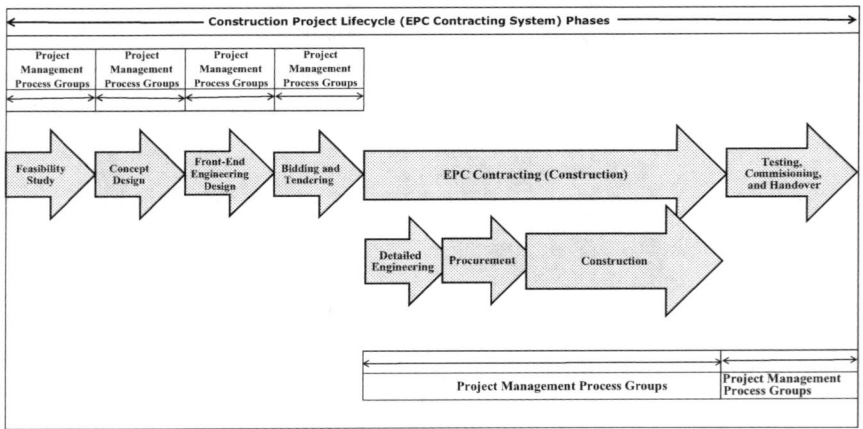

FIGURE 3.13 Construction project life cycle (EPC contracting system) phases. EPC, engineering, procurement, and construction.

The major activities in each phase of the oil and gas project life cycle are developed based on project management process groups methodology and processes in knowledge areas that help understand the entire process of project management and to manage and control its activities at different levels of various phases to ensure timely completion of the project with economical use of resources to make the construction project most qualitative, competitive, and economical.

4 Quality Tools for Oil and Gas Industry

4.1 INTRODUCTION

Quality tools are the charts, check sheets, diagrams, graphs, techniques, and methods that are used to create an idea, engender planning, analyze the cause, analyze the process, foster evaluation, and create a wide variety of situations for continuous quality improvement. Applications of tools enhance chances of success and help maintain consistency, accuracy, increase efficiency, and process improvement.

4.2 CATEGORIZATION OF TOOLS

There are several types of tools, techniques, and methods, in practice, which are used as quality improvement tools and have a variety of applications in manufacturing and process industry. However, all of these tools are not used in construction projects due to the nature of construction projects, which are customized and nonrepetitive. Some of these quality management tools that are most commonly used in construction industry are listed under the following broader categories:

1. Classic quality tools
2. Management and planning tools
3. Process analysis tools
4. Process improvement tools
5. Innovation and creative tools
6. Lean tools
7. Cost of quality
8. Quality function deployment (QFD)
9. Six Sigma
10. Triz

Brief description of these tools is given below.

4.2.1 Classic Quality Tools

Classic quality tools have a long history. These tools are listed in Table 4.1.

All of these tools have been in use for since World War II. Some of these tools date prior to 1920. The approach includes both quantitative and qualitative aspects, which, taken together, focus on companywide quality.

Brief definition of these quality tools is as follows (values shown in the figures and tables are indicative only).

TABLE 4.1
Classic Quality Tools

Serial Number	Name of Quality Tool	Usage
Tool 1	Cause-and-effect diagram	To identify possible cause and effect in processes.
Tool 2	Check sheet	To provide a record of quality. How often it occurs?
Tool 3	Control chart	A device in statistical process control to determine whether or not the process is stable.
Tool 4	Flowchart	Used for graphical representation of a process in sequential order.
Tool 5	Histogram	Graphs used to display frequency of various ranges of values of a quantity.
Tool 8	Run chart	Used to show measurement against time in a graphical manner with a reference line to show the average of the data.
Tool 9	Scatter diagram	To determine whether there is a correlation between two factors.
Tool 10	Stratification	Use to show the pattern of data collected from different sources.

Source: Abdul Razzak Rumane. (2013). *Quality Tools for Managing Construction Projects.* Reprinted with permission from Taylor & Francis Group.

4.2.1.1 Cause-and-Effect Diagram

Cause-and-effect diagram is also called as Ishikawa diagram, after its developer Kaoru Ishikawa, or fishbone diagram. It is used to identify possible causes and effects in process. It is used to explore all the potential or real causes that result in a single output. The causes are organized and displayed in graphical manner to their level of importance or details. It is a graphical display of multiple causes with a particular effect. The causes are organized and arranged mainly into six categories. These are as follows:

1. Machine
2. Manpower
3. Material
4. Method
5. Measurement
6. Environment

The effect or problem being investigated is shown at the end of horizontal arrow. Potential causes are shown as labeled arrows entering the main cause arrow. Each arrow may have a number of other arrows entering as the principal cause or factors are reduced to their subcauses. Figure 4.1 illustrates an example cause-and-effect diagram for rejection of concrete works for not complying with contract specifications

Quality Tools for Oil and Gas Industry

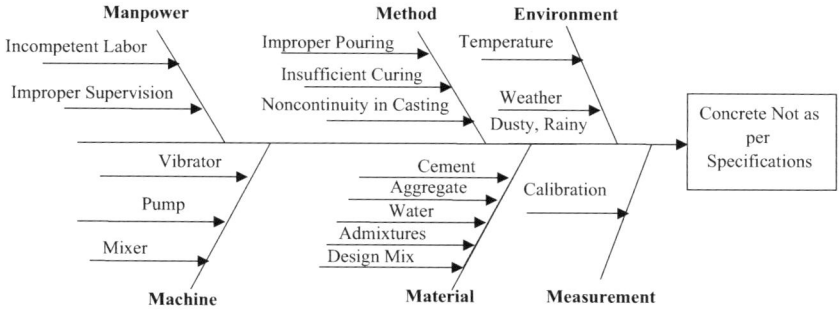

FIGURE 4.1 Cause-and-effect diagram for rejection of concrete work. (Abdul Razzak Rumane. (2013). *Quality Tools for Managing Construction Projects*. Reprinted with permission from Taylor & Francis Group.)

4.2.1.2 Check Sheet

Check sheet is a structured list, prepared from the collected data to indicate how often each item occurs. It is an organized way of collecting and structuring data. The purpose of check sheet is to collect the facts in the most efficient manner. Data are collected and ordered by adding tally or check marks against predetermined categories of items or measurements. Figure 4.2 illustrates check sheet for checklist approval record for control valves.

4.2.1.3 Control Chart

Control chart is the fundamental tool of statistical process control (SPC). The control chart is a graph used to analyze how a process behaves over time and to show whether the process is stable or is being affected by special cause of variation and creating an out-of-control condition. It is used to determine whether the process is stable or varies between the predictable limits. It can be employed to distinguish between the existence of a stable pattern of variation and the occurrence of an unstable pattern. With control charts, it is easy to see both special and common causes of variation in a process. There are many types of control charts. Each is designed for a specific

Approval Record for Control Valves

	Approved	Not-Approved	Total	% Not-Approved
25 mm dia	///// ///// //	///	50	4
50 mm dia	///// ///// ///// ///// //	///	85	6
75 mm dia	///// ///// ///// //	///	25	3
100 mm dia	///// //	///	15	2
150 mm dia	///// //	///	10	2

FIGURE 4.2 Check sheets.

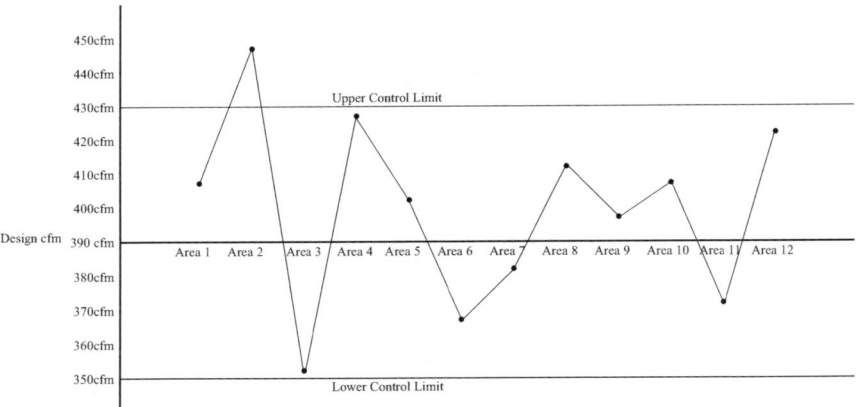

FIGURE 4.3 Control chart for HVAC air handling unit air distribution (cfm). HVAC, heating, ventilation, and air conditioning. (Abdul Razzak Rumane. (2013). *Quality Tools for Managing Construction Projects*. Reprinted with permission from Taylor & Francis Group.)

kind of process or data. Figure 4.3 illustrates control chart for distribution of HVAC (heating, ventilation, and air conditioning) air handling unit air.

4.2.1.4 Flowchart

Flowchart is a pictorial tool that is used for representation process in sequential order. Flowchart uses graphic symbols to depict the nature and flow of the steps in a process. It helps to see whether the steps of process are logical, uncover the problems or miscommunications, define the boundaries of a process, and develop a common base of knowledge about a process. The flow of steps is indicated with arrows connecting the symbols. Flowcharts can be applied at all stages of the project life cycle. Figure 4.4 illustrates flowchart for contractor's staff approval in construction projects.

4.2.1.5 Histogram

Histogram is a pictorial representation of frequency distribution of the data. It is created by grouping the data points into cell and displays how frequently different values occur in the data set. Figure 4.5 illustrates histogram for employee reporting time.

4.2.1.6 Pareto Chart

Pareto chart is named after Vilfredo Pareto, a 19th-century Italian economist who postulated that a large share (80%) of wealth is owned by a small (20%) percentage of population. Pareto chart is a graph bar having series of bars whose height reflects the frequency of occurrence. Pareto charts are used to display the Pareto Principle in action and arrange data so that the few vital factors that are causing most of the problems reveal themselves. The bars are arranged in descending order of height from left to right. Pareto charts are used to identify those factors that have greatest cumulative effect on the system, and thus, less significant factors can be screened out

Quality Tools for Oil and Gas Industry

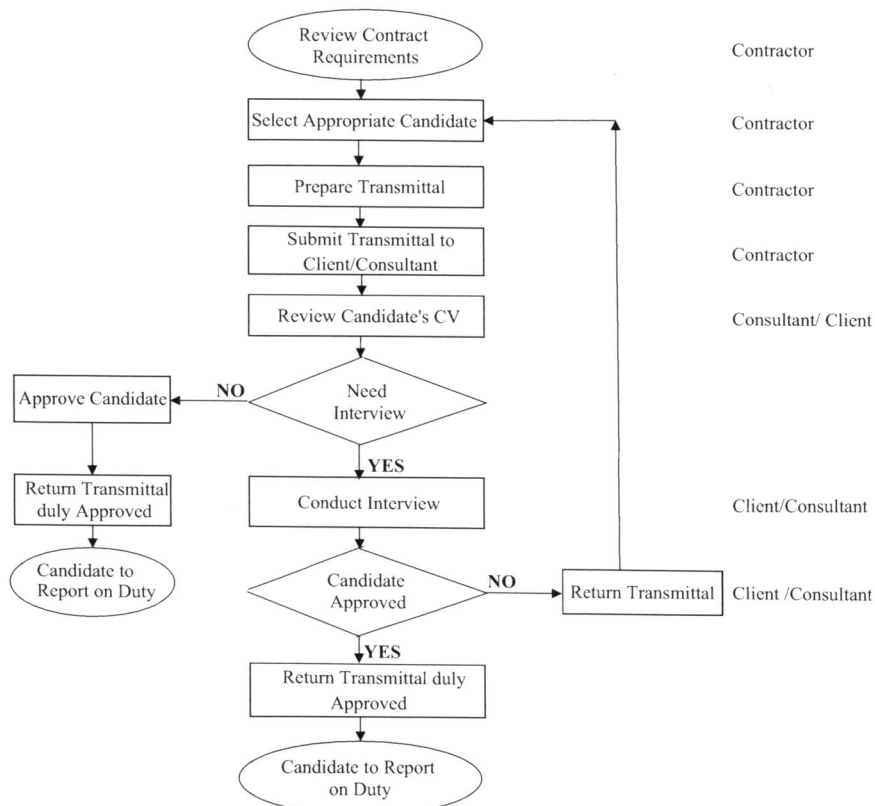

FIGURE 4.4 Flowchart for contractor's staff approval. (Abdul Razzak Rumane. (2013). *Quality Tools for Managing Construction Projects*. Reprinted with permission from Taylor & Francis Group.)

from the process. Pareto chart can be used at various stages in a quality improvement program to determine which step to take next. Figure 4.6 illustrates Pareto chart for division cost of construction project.

4.2.1.7 Pie Chart

Pie chart is a circle divided into wedges to depict proportion of data or information in order to understand how they make up the whole. The entire pie chart represents all the data, while each slice or wedge represents a different class or group within the whole. The portions of entire circle or pie sum to 100%. Figure 4.7 illustrates the contents of contractor's site staff at construction project site.

4.2.1.8 Run Chart

Run chart is a graph plotted by showing measurement (data) against time. Run charts are used to know the trends or changes in a process variation over time over the average and also to determine if the pattern can be attributed to common causes of

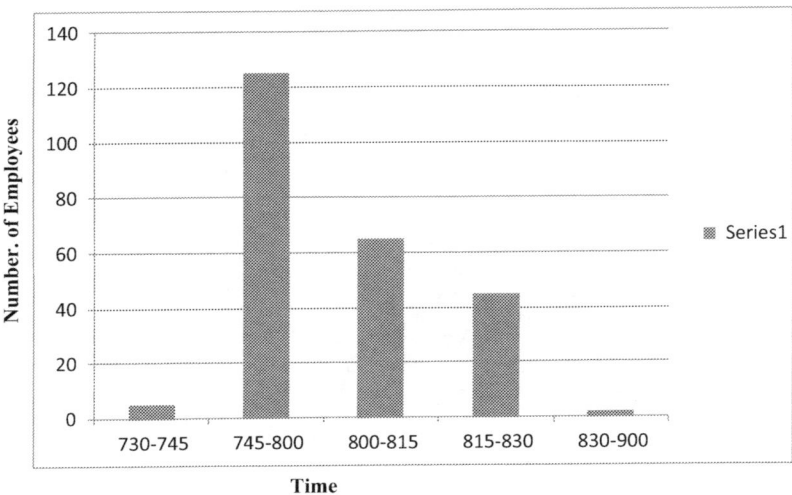

FIGURE 4.5 Employee reporting histogram. (Abdul Razzak Rumane. (2013). *Quality Tools for Managing Construction Projects*. Reprinted with permission from Taylor & Francis Group.)

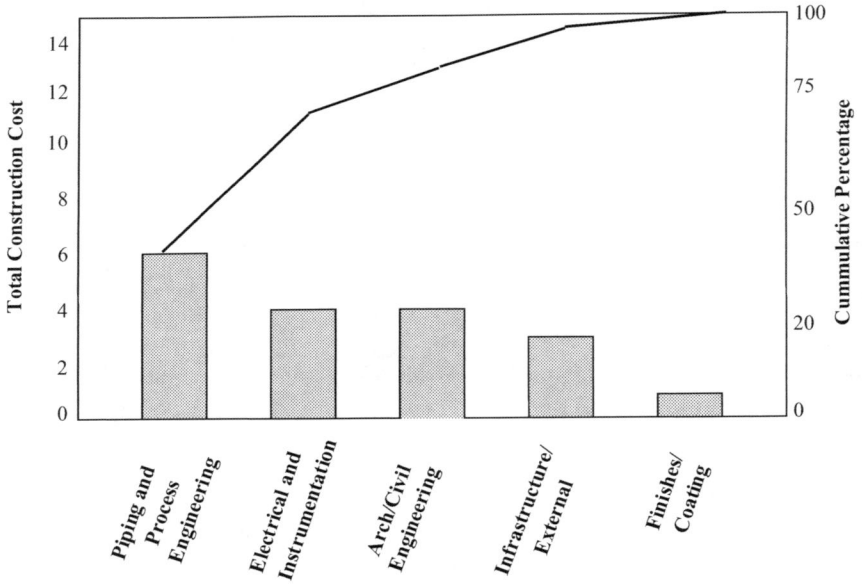

FIGURE 4.6 Pareto analysis for construction cost.

variation or if special causes of variation were present. A run chart is also used to monitor process performance. Run charts can be used to track improvements that have been put in place, checking to determine their success. Figure 4.8 illustrates run chart for weekly manpower of different trades of a project. It is similar to control chart but does not show control limits.

Site Staff	
Managers	2
Engineers	16
Technicians	9
Foremen	20
Skilled	230
Unskilled	570

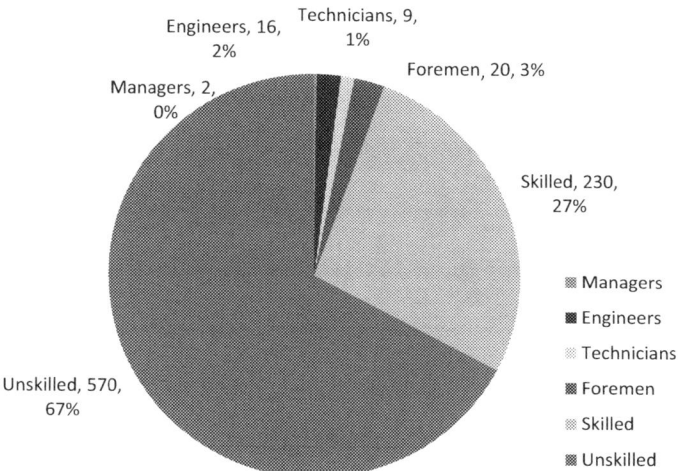

FIGURE 4.7 Pie chart for site staff. (Abdul Razzak Rumane. (2013). *Quality Tools for Managing Construction Projects.* Reprinted with permission from Taylor & Francis Group.)

4.2.1.9 Scatter Diagram

A scatter diagram is a plot of one variable versus another. It is used to investigate the possible relationship between two variables that both relate to the same event. It helps to know how one variable changes with respect to other. It can be used to identify potential root cause of problems and to evaluate cause-and-effect relationship. Figure 4.9 illustrates scatter diagram for pipe length quantity of various lengths.

4.2.1.10 Stratification

Stratification is a graphical representation of data collected from different data. Figure 4.10 illustrates stratification diagram for pipe bundles.

4.2.2 Management and Planning Tools

Seven management tools are most popular after classic quality tools. These tools are listed in Table 4.2.

These tools are focused on managing and planning quality improvement activities.

Brief definition of these quality tools is as follows (values shown in the figures are indicative only).

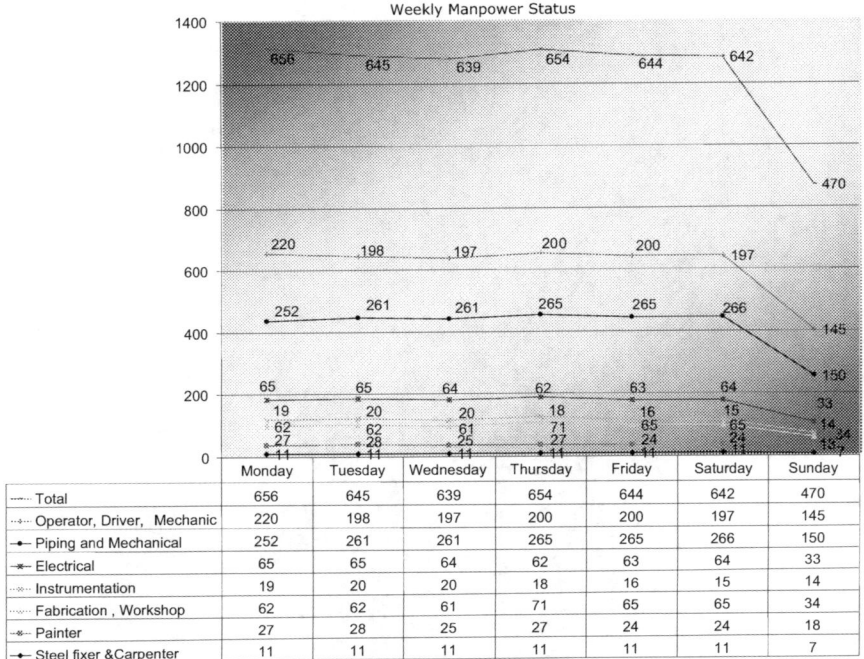

FIGURE 4.8 Run chart for manpower.

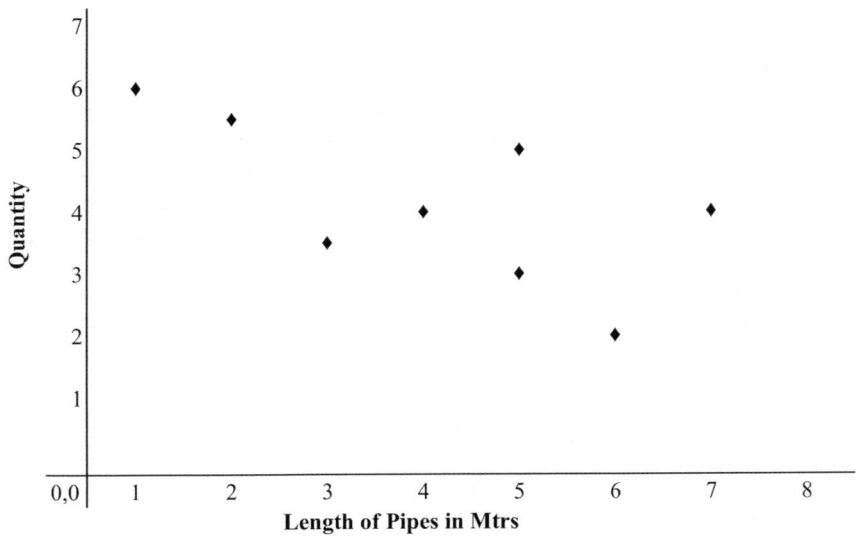

FIGURE 4.9 Scatter diagram.

Quality Tools for Oil and Gas Industry

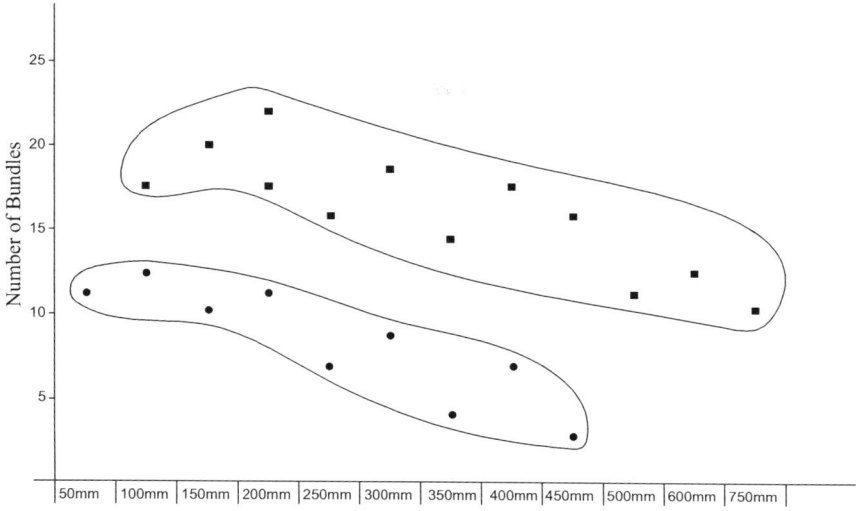

FIGURE 4.10 Stratifiaction chart.

TABLE 4.2
Management and Planning Tools

Serial Number	Name of Quality Tool	Usage
Tool 1	Activity network diagram (arrow diagram)/critical path method	Used when scheduling or monitoring task is complex or lengthy and has schedule constraints.
Tool 2	Affinity diagram	Used to organize a large group of items in smaller categories that are easier to understand and deal with.
Tool 3	Interrelationship digraph (relations diagram)	Used to show logical relationship between ideas, process, cause, and effect.
Tool 4	Matrix diagram	Used to analyze the correlations between two or more groups of information.
Tool 5	Prioritization matrix	Used to choose one or two options that have important criteria from several options.
Tool 6	Process decision program chart	Used to help contingency plan.
Tool 7	Tree diagram	Used to break down or stratify ideas in progressively more detailed step.

Source: Abdul Razzak Rumane. (2013). *Quality Tools for Managing Construction Projects.* Reprinted with permission from Taylor & Francis Group.

4.2.2.1 Activity Network Diagram

Activity network diagram (AND) is a graphical representation chart showing interrelationship among activities (task) associated with a project. An AND was developed by the U.S. Department of Defense. It was first used as a management tool for military projects. It was adapted as an education tool for business managers. In AND, each activity is represented by only one arrow in the network and is associated with an estimated time to perform the activity. AND analyzes the sequences of tasks necessary to complete the project. The direction of arrow specifies the order in which the events must occur. The event represents a point in time that indicates the completion of one or more activities and beginning of new ones. Figure 4.11 illustrates AND.

There are two kinds of network diagrams, the "activity-on-arrow" (A-O-A) network diagram and the "activity-on-node" (A-O-N) network diagram.

Arrow diagram or A-O-A is a diagramming method to represent the activities on arrows and connect them at nodes (circles) to show the dependencies. With A-O-A method, the detailed information about each activity is placed on arrow or as footnotes at the bottom.

Figure 4.12 illustrates arrow diagramming method for concrete foundation work.

Activities originating from a certain event cannot start until the activities terminating the same event have been completed. This is known as precedence relations. These relationships are drawn using precedence diagramming method (PDM). The PDM technique is also referred to as "A-O-N" because it shows the activities in a node (box) with arrows showing dependencies. "A-O-N" network diagram has the activity information written in a small box that are the nodes of the diagram. Arrows connect the boxes to show the logical relationships between pairs of the activities.

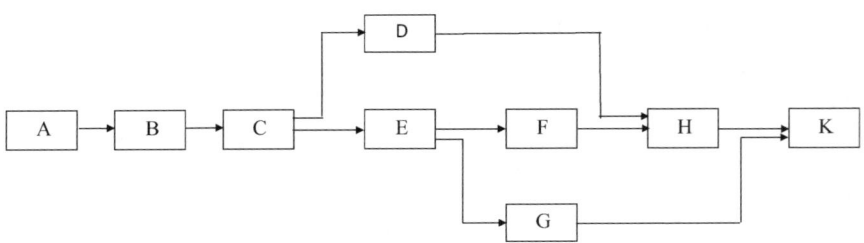

FIGURE 4.11 Activity network diagram. (Abdul Razzak Rumane. (2013). *Quality Tools for Managing Construction Projects*. Reprinted with permission from Taylor & Francis Group.)

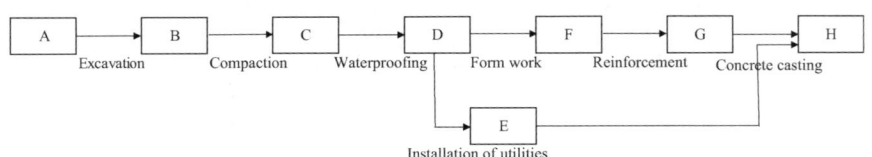

FIGURE 4.12 Arrow diagraming method for concrete foundation. (Abdul Razzak Rumane. (2013). *Quality Tools for Managing Construction Projects*. Reprinted with permission from Taylor & Francis Group.)

In networking diagram, all activities are related in some direct way and may be further constrained by indirect relationship. The following are direct logical relationships or dependencies among project-related activities:

1. Finish-to-start—Activity A must finish before activity B can begin.
2. Start-to-start—Activity A must begin before activity B can begin.
3. Start-to-finish—Activity A must begin before activity B can finish.
4. Finish-to-finish—Activity A must finish before activity B can finish.

Apart from these, there are other dependencies such as follows:

1. Mandatory
2. Discretionary
3. External

Figure 4.13 illustrates dependency relationship diagrams, and Figure 4.14 illustrates PDM.

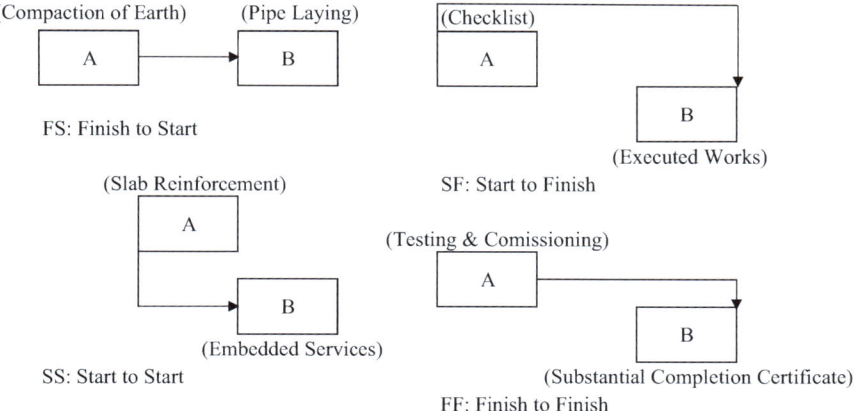

FIGURE 4.13 Dependency relationship diagram. (Abdul Razzak Rumane. (2013). *Quality Tools for Managing Construction Projects*. Reprinted with permission from Taylor & Francis Group.)

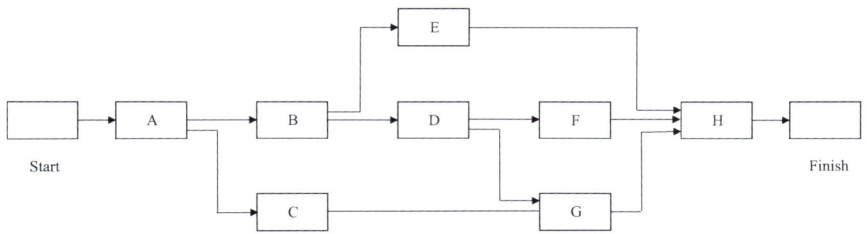

FIGURE 4.14 PDM diagramming method. (Abdul Razzak Rumane. (2013). *Quality Tools for Managing Construction Projects*. Reprinted with permission from Taylor & Francis Group.)

AND or arrow diagram is a tool used for detailed planning, a tool for analyzing schedule during execution, and a tool for controlling of complex or large-scale project. A network diagram uses nodes and arrows. Date information is added to each activity node.

Critical path method (CPM) chart is an expanded AND showing an estimate time to complete each activity and connecting these activities based on the task to be performed. Critical path is sequence of interrelated predecessor/successor activities that determines the minimum completion time for a project. The duration of critical path is the sum of the activities' duration along the path. The activities in the critical path have the least scheduling flexibility. Any delays along the critical path would imply that additional time would be required to complete the project. There may be more than one critical path among all the project activities, so completion of the entire project could be affected due to delaying activity along any of the critical paths. Table 4.3 illustrates activity relationship for pumping station project, and Figure 4.15 illustrates CPM diagram for construction of pumping station.

AND is also known as program evaluation and review technique (PERT). PERT is used to schedule, organize, and coordinate tasks within a project.

PERT planning involves the following steps:

1. Identify specific activity
2. Identify milestones of each activity
3. Determine proper sequence of each activity
4. Construct a network diagram
5. Estimate time required to complete each activity. Three-point estimation method using the following formula can be used to determine approximated estimate time for each activity:
6. Compute the early start (ES), early finish (EF), late start (LS), and late finish (LF) times for each activity in the network.
7. Determine critical path.
8. Identify the critical path for possible schedule compression.
9. Evaluate the diagram for milestones and target dates in the overall project.

Figure 4.16 illustrates Gantt chart for pumping station.

4.2.2.2 Affinity Diagram

Affinity diagram is a tool that gathers a large group of ideas/items and organizes into a smaller grouping based on their natural relationships. Affinity diagram is refinement of brainstorming ideas into something smaller groups, which can be dealt more easily and satisfy the team members. The affinity process is often used to group ideas generated by brainstorming. Affinity diagram is created as per the following steps:

1. Generate ideas and list the ideas without criticism.
2. Display the ideas in a random manner.
3. Sort the ideas and place them into multiple groups.
4. Continue until smaller groups satisfy all the members.
5. Draw affinity diagram.

Figure 4.17 illustrates affinity diagram for concrete slab.

TABLE 4.3
Activities to Construct Pumping Station

Activity Number	Description of Activity	Duration in Days	Preceding Activity
1	Start	0	
2	Mobilization	21	1
3	Preparation of site	15	2
4	Staff approval	15	2
5	Material approval	15	2,4
6	Shop drawing approval	15	2,5
7	Procurement (structural work)	15	5
8	Procurement (conduits, ducts, sleeves)	7	5
9	Procurement (pumps, valves, piping material, control panels, security system)	60	5
10	Procurement (civil, MEP, furnishing)	30	5
11	Excavation	4	3,6
12	Blinding concrete	4	5,7,11
13	Raft foundation	7	5,6,7,12
14	Utility services (embedded)	4	8,12
15	Concrete (floor)	1	6,13,14
16	Trenches	7	13
17	Embedded services/ducts in trench	2	8,13
18	Concrete, foundation (pump area)	1	16,17
19	Walls and columns	7	15,18
20	Formwork for slab	3	6,7,19
21	Reinforcement	2	7,20
22	Embedded services	1	8,20
23	Concrete (roof slab)	1	21,22
24	Masonry work	14	6,10,23
25	Installation of pumps, valves, equipment	7	9,18,24
26	Installation of panels, electromechanical items	14	6,10,23,24
27	Installation of ventilation system	4	10,23,24
28	Finishes	7	10,24
29	Installation of isolators, starters, electrical fixes	3	10,28
30	Furnishing	2	10,28
31	Testing of pumps, valves, equipment	2	25
32	Testing of HVAC, firefighting, electrical system	4	25,26,27
33	Handing over	1	29,30,31,32
34	End	0	

HVAC, heating, ventilation, and air conditioning.
Note: The duration is indicative only to understand sequencing and relationship of activities.

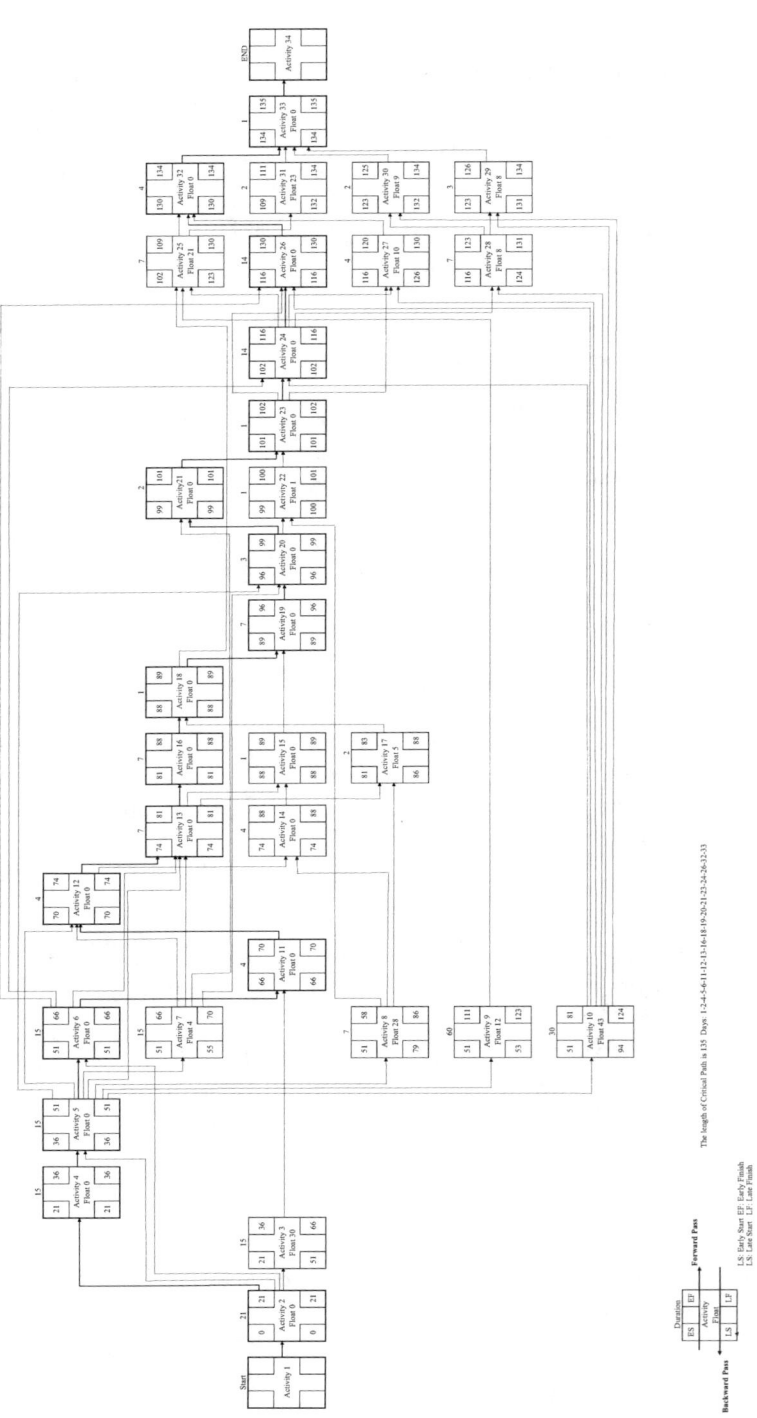

FIGURE 4.15 Critical path method.

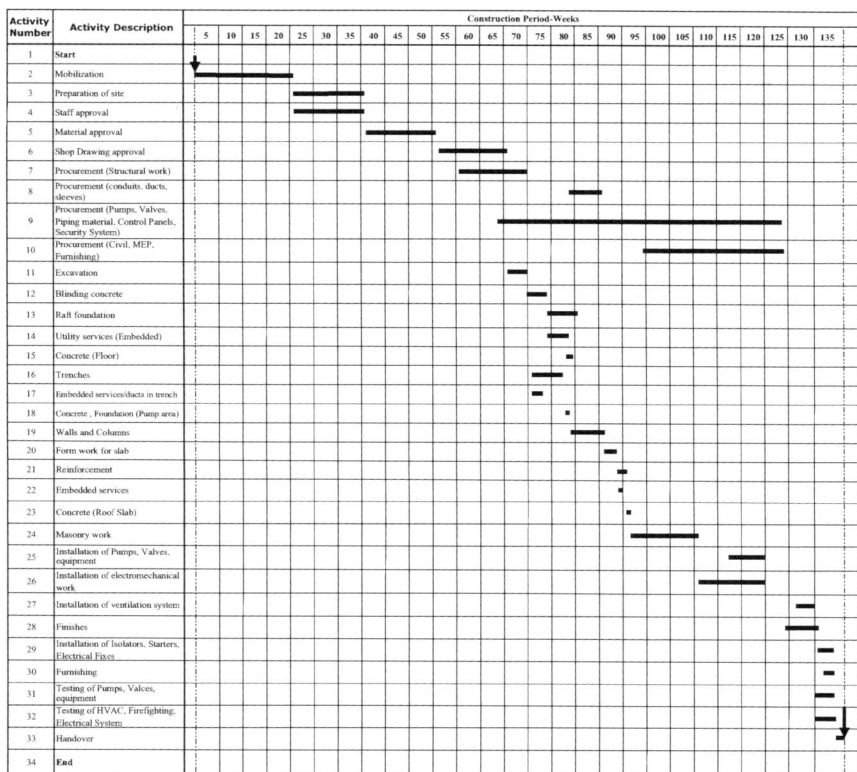

FIGURE 4.16 Gantt chart for pumping station.

4.2.2.3 Interrelationship Digraph

Interrelationship digraph (di is for directional) is an analysis tool that allows the team members to identify logical cause-and-effect relationship between the ideas. It is drawn to show all the different relationships between factors, areas, or process. They make it easy to pick out the factors in a situation, which are the ones that are driving many of the other symptoms or factors. While affinity diagrams organize and arrange the ideas into groups, interrelationship digraph identifies problems to define the ways in which ideas influence one and another. Interrelation digraph is used to identify cause-and-effect relationship with the help of directional arrows among critical issues. The number of arrows coming into the node determines outcome (Key Indicator) while the outgoing arrows determine the cause (driver) of the issue.

Figure 4.18 illustrates interrelationship digraph for causes of bridge collapse.

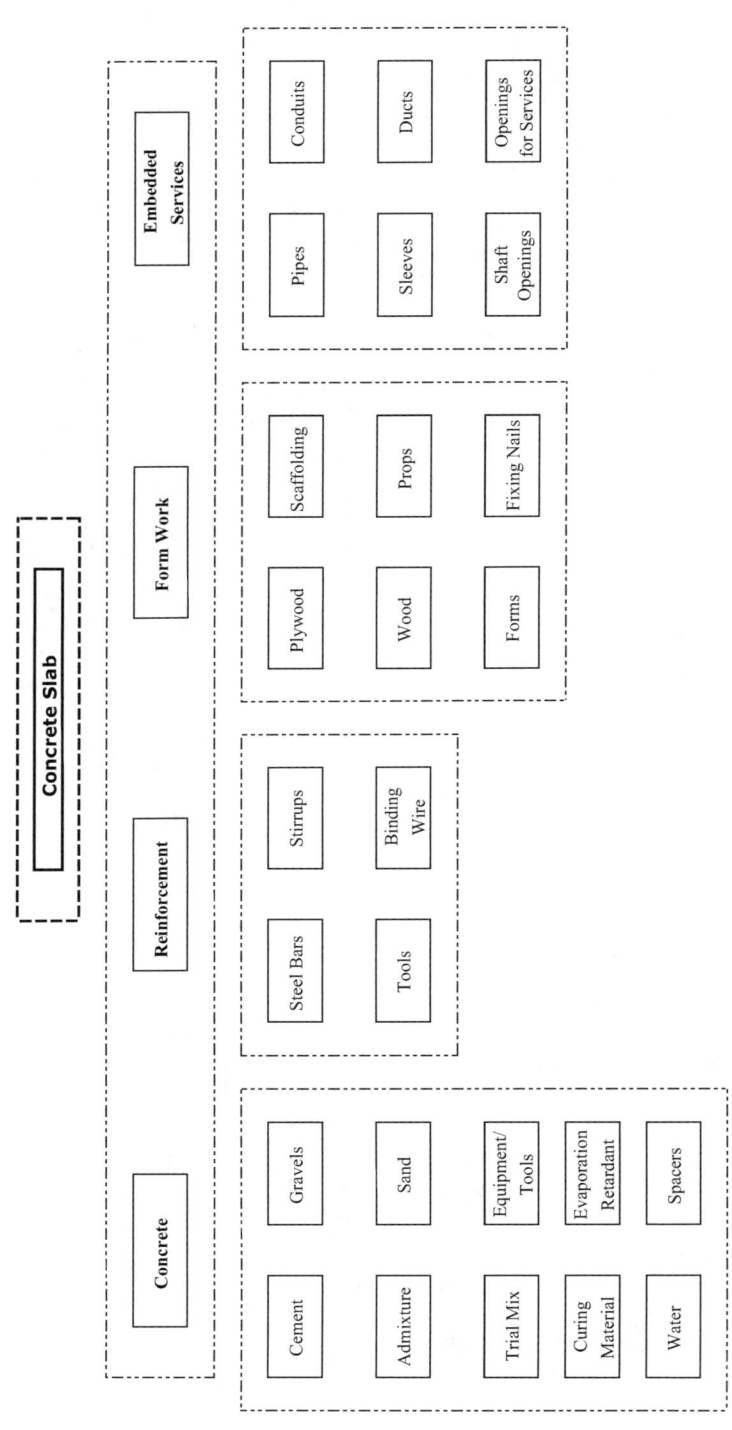

FIGURE 4.17 Affinity diagram for concrete slab. (Abdul Razzak Rumane. (2013). *Quality Tools for Managing Construction Projects*. Reprinted with permission from Taylor & Francis Group.)

Quality Tools for Oil and Gas Industry

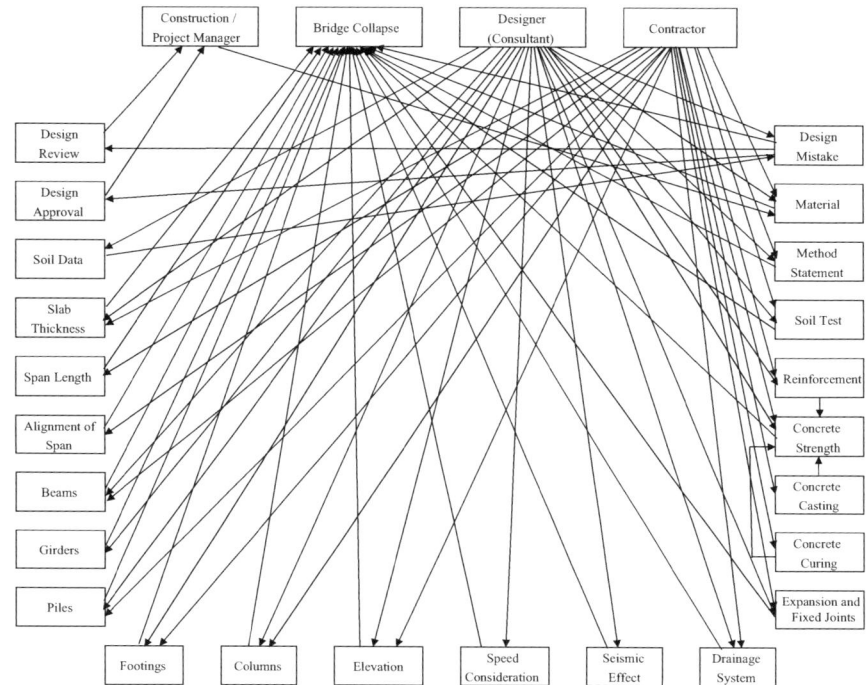

FIGURE 4.18 Interrelationship digraph. (Abdul Razzak Rumane. (2013). *Quality Tools for Managing Construction Projects*. Reprinted with permission from Taylor & Francis Group.)

4.2.2.4 Matrix Diagram

Matrix diagram is constructed to analyze systematically the correlations between two or more group of items or ideas. The matrix diagram can be shaped in the following ways:

1. L-shaped
2. T-shaped
3. X-shaped
4. C-shaped
5. Inverted Y-shaped
6. Roof-shaped

Each shape has its own purpose.

1. L-shaped is used to show interrelationships between two groups or processes.
2. T-shaped is used to show relation between three groups. For example, consider there are three groups A, B, and C. In T-shaped matrix, groups A and B are each related to group C, whereas groups A and B are not related to each other.

3. X-shaped matrix is used to show relationship among four groups. Each group is related to two other groups in a circular fashion.
4. C-shaped matrix interrelates three groups of processes or ideas in three-dimensional ways.
5. Inverted Y-shaped matrix is used to show relation between three groups. Each group is related with other groups in a circular fashion.
6. Roof-shaped matrix relates one group of items to itself. Roof-shaped matrix is used with L- or T-shaped matrix.

Table 4.4 illustrates L-shaped matrix for customer requirements of valves.

Figure 4.19 illustrates T-shaped matrix for pressure vessels.

Figure 4.20 illustrates roof-shaped matrix for different divisions of construction activities.

TABLE 4.4
L-Shaped Matrix

Customer Requirements of Valves

Serial Number	Valve Details	Customer A	Customer B	Customer C	Customer D
1	Type	Gate	Globe	Ball	Check
2	Body type	Cast steel	Forged steel	Cast steel	Forged steel
3	Size	250 mm	300 mm	50 mm	150 mm
4	Actuation	Manual	Manual	Power	Power
5	End connection	Flanged	Flanged	Flanged	Flanged
6	Opening and closing	Linear motion	Linear motion	Rotary motion	Linear motion
7	Standard	API/ANSI/ASME	API/ASME	API/ANSI	API/ANSI
8	Pressure	50 bar max	50 bar max	50 bar max	50 bar max
9	Flow of material	Oil	Steam	Oil	Oil
10	Operating temperature	90°C	90°C	90°C	90°C

Manufacturing Plant	Products (Pressure Vessels)				
Customer					
International				#	#
European Manufacturer	≠	#	#		
Local Plant	#	≠	≠		
# Large Capacity (20,000–30,000) ≠ Small Capacity (5,000–15,000)	Pressure Vessel (5,000 Gallons)	Pressure Vessel (10,000 Gallons)	Pressure Vessel (20,000 Gallons)	Pressure Vessel (25,000 Gallons)	Pressure Vessel (3,000 Gallons)
ABC Company	#	≠	≠		
XYZ Company	≠	#	#		
Others				#	#

FIGURE 4.19 T-shaped matrix.

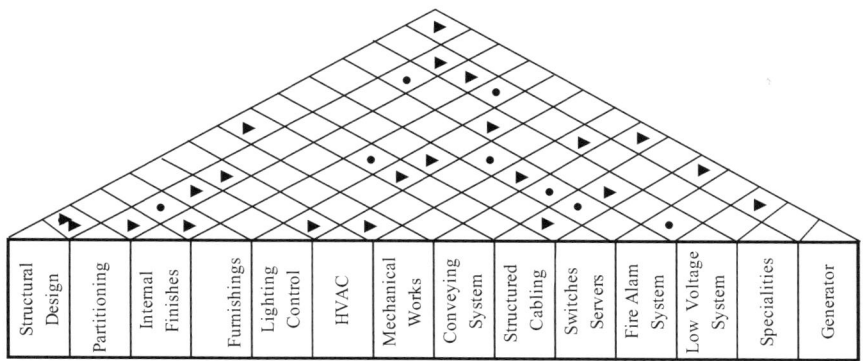

FIGURE 4.20 Roof-shaped matrix. (Abdul Razzak Rumane. (2013). *Quality Tools for Managing Construction Projects.* Reprinted with permission from Taylor & Francis Group.)

4.2.2.5 Prioritization Matrix

The prioritization matrix assists in choosing between several options in order of importance and priority. It helps decision-makers determine the order of importance considering the relative merit of each of the activities or goal being considered. Prioritization matrix focuses the attention of team members to those key issues and options, which are more important for the organization or project.

Figure 4.21 illustrates prioritization matrix.

4.2.2.6 Process Decision Program

The process decision program is a technique used to help prepare contingency plans. The process decision program systematically identifies what might go wrong in a project plan or project schedule and describes specific actions to be taken to prevent the problems from occurring in first place and to mitigate or avoid the impact of the problems if they occur.

Figure 4.22 illustrates process decision program for submission of contract documents.

4.2.2.7 Tree Diagram

Tree diagrams are used to break down or stratify ideas progressively into more detailed steps. Tree diagram breaks broader ideas into specific details and help make decision easier to select the alternative. It is used to figure out all the various tasks that must be undertaken to achieve a given objective.

Figure 4.23 illustrates tree diagram for oil in the storage tank.

4.2.3 Process Analysis Tools

Table 4.5 illustrates process analysis tools.

4.2.3.1 Benchmarking

Benchmarking is the process of measuring the actual performance of the organization's products, processes, and services and comparing them with the best known

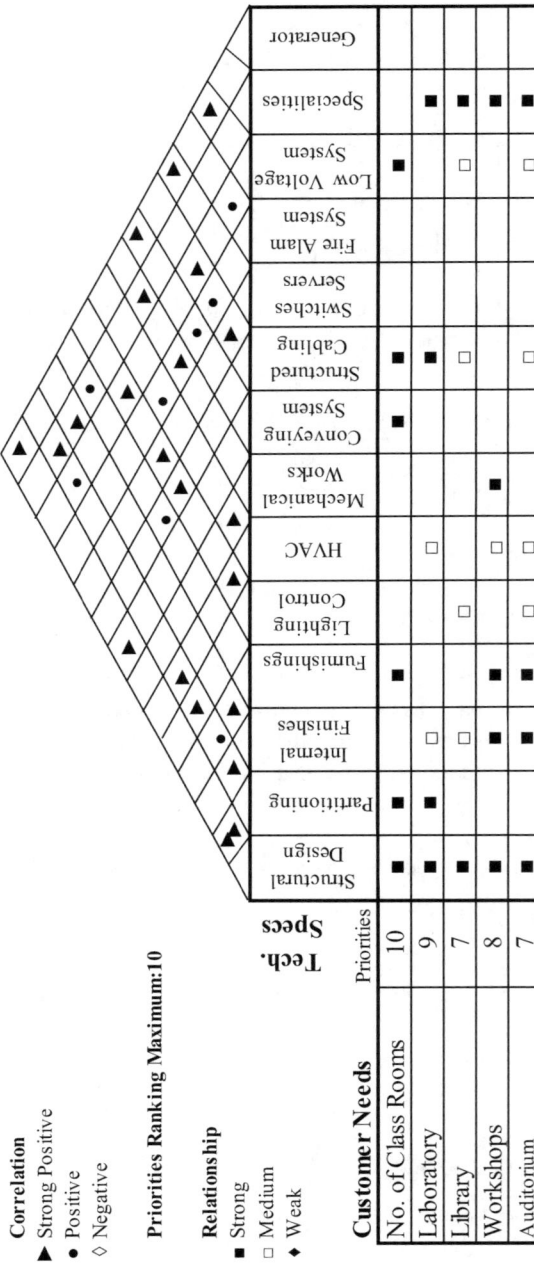

FIGURE 4.21 Prioritization matrix. (Abdul Razzak Rumane. (2013). *Quality Tools for Managing Construction Projects*. Reprinted with permission from Taylor & Francis Group.)

Quality Tools for Oil and Gas Industry

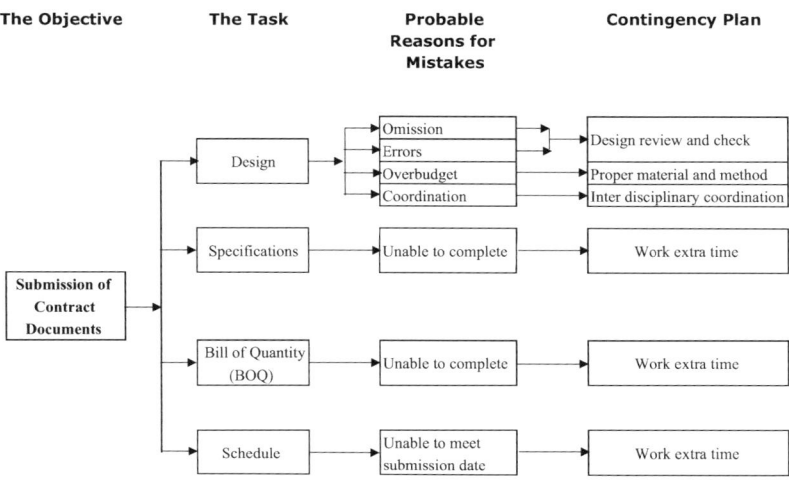

FIGURE 4.22 Process decision program chart. (Abdul Razzak Rumane. (2013). *Quality Tools for Managing Construction Projects*. Reprinted with permission from Taylor & Francis Group.)

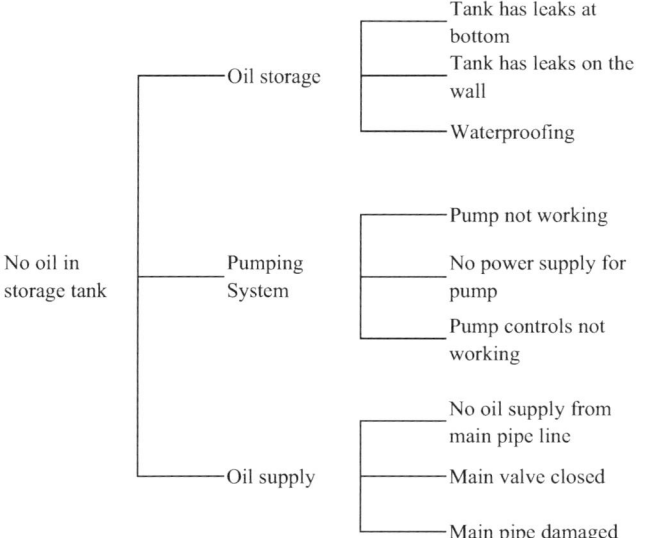

FIGURE 4.23 Tree diagram for no oil in storage tank.

industry standards to assist the organization in improving the performance of their products, processes, and services. Benchmarking involves analyzing an existing situation, identifying and measuring factors critical to the success of the product or services, comparing them with other businesses, analyzing the results, and implementing

TABLE 4.5
Process Analysis Tools

Serial Number	Name of Quality Tool	Usage
Tool 1	Bench marking	To identify the best practices in the industry and improve the process or project.
Tool 2	Cause and effect	To identify possible cause and its effect in the process.
Tool 3	Cost of quality	To identify hidden or indirect cost affecting the overall cost of product/project.
Tool 4	Critical to quality	To identify quality features or characteristics most important to the client.
Tool 5	Failure mode and effects analysis (FMEA)	To identify and classify failures according to their effect.
Tool 6	5 Why analysis	Used to analyze and solve any problem where the root cause is unknown.
Tool 7	5W2H	The questions used to understand why the things happen the way they do.
Tool 8	Process mapping/flowcharting	It is a technique used for designing, analyzing, and communicating work processes.

Source: Abdul Razzak Rumane. (2013). *Quality Tools for Managing Construction Projects.* Reprinted with permission from Taylor & Francis Group.

an action plan to achieve better performance. The following are the steps involved in the process of benchmarking:

1. Collect internal and external data on work, process, method, product characteristics, and system selected for benchmarking.
2. Analyze data to identify performance gaps and determine cause and differences.
3. Prepare action plan to improve the process in order to meet or exceed the best practices in the industry.
4. Search for the best practices among market leaders, competitors, and non-competitors that lead to their superior performance.
5. Improve the performance by implementing these practices.

Figure 4.24 illustrates benchmarking process.

4.2.3.2 Cause-and-Effect Diagram

It is one of the quality classic tools. It is used to analyze the cause and effect of defects or nonconformance and effect on the process due to these causes.

Figure 4.25 illustrates cause-and-effect diagram for rejection of storage tank work.

4.2.3.3 Cost of Quality

Cost of quality is discussed in detail under Section 4.2.7 of this chapter.

Quality Tools for Oil and Gas Industry

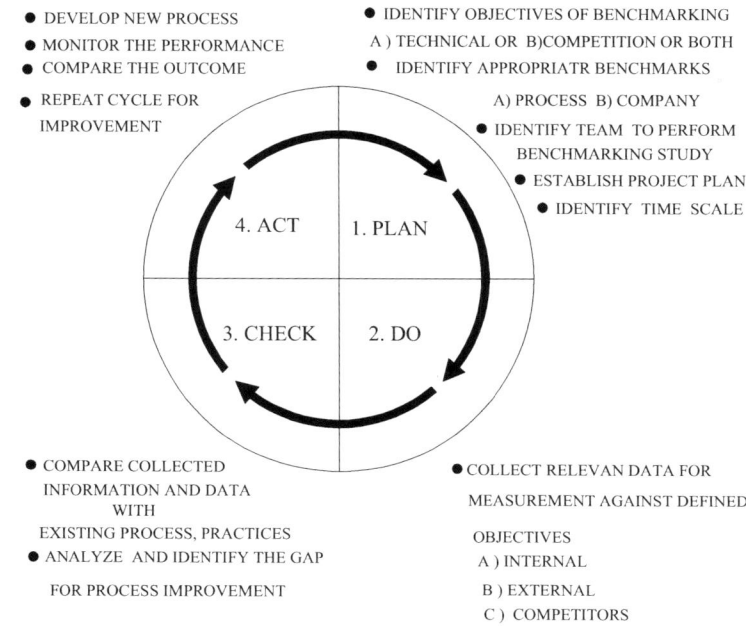

FIGURE 4.24 Benchmarking process. (Abdul Razzak Rumane. (2013). *Quality Tools for Managing Construction Projects.* Reprinted with permission from Taylor & Francis Group.)

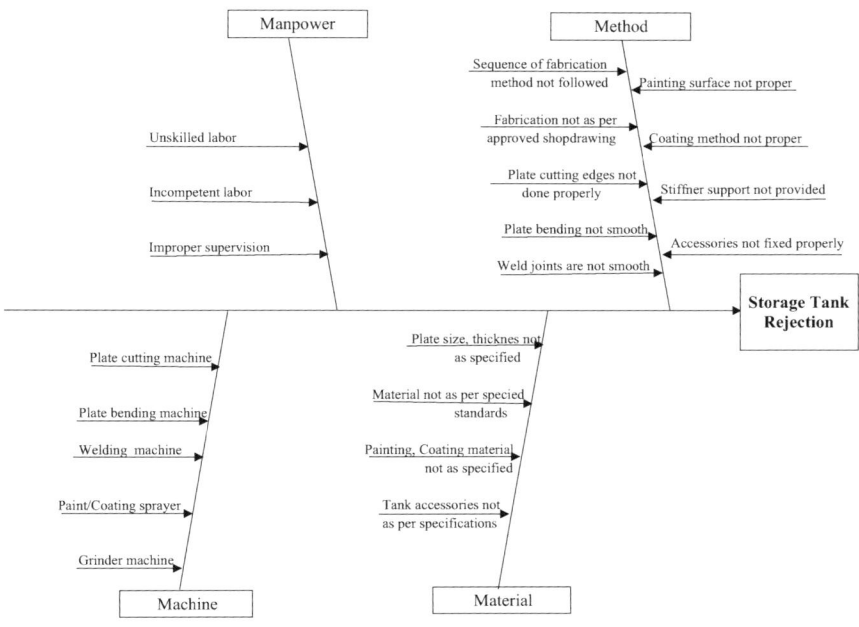

FIGURE 4.25 Cause-and-effect diagram for rejection of storage tank.

4.2.3.4 Critical to Quality

Critical to quality is a significant step in the design process of a product or service to identify the customer's/client's expectation to fulfill their needs and requirements.

4.2.3.5 Failure Mode and Effects Analysis

Failure mode and effects analysis (FMEA) is to identify all the possible failures in the design of a product, process, and service and their effects on the product, process, and service. Its aim is to reduce risk of failure and improve the process.

Figure 4.26 illustrates FMEA process, and Figure 4.27 illustrates an example form used to record FMEA readings.

4.2.3.6 5 Whys Analysis

It is used to analyze and solve any problem where the root cause is unknown.

Table 4.6 illustrates 5 Whys Analysis chart for burning of cable.

4.2.3.7 5W2H

5W2H is about asking the questions to understand about a process or problem.

The 5 Ws are:

1. Why
2. What
3. When
4. Where
5. Who

And 2 Hs are:

1. How
2. How much

Table 4.7 illustrates 5W2H for slab collapse.

4.2.3.8 Process Mapping/Flowcharting

Process mapping/flowcharting is a graphical representation of workflow giving clear understanding of a process or services of parallel processes. Process mapping/flowcharting is a technique that can be employed to not only procedure visual representation of the production processes but also processes related to other departments.

Figure 4.28 illustrates process mapping/flowcharting diagram for approval of variation order.

4.2.4 PROCESS IMPROVEMENT TOOLS

Table 4.8 illustrates process improvement tools.

Quality Tools for Oil and Gas Industry 153

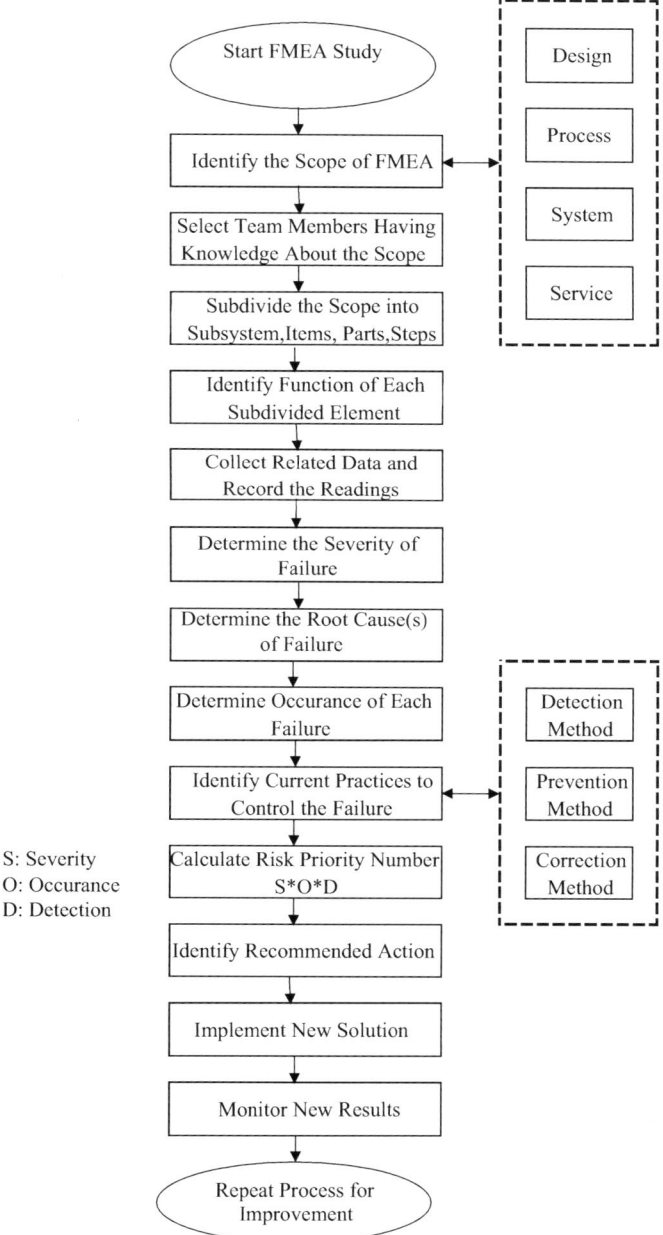

FIGURE 4.26 Failure mode and effects analysis (FMEA) process. (Abdul Razzak Rumane. (2013). *Quality Tools for Managing Construction Projects*. Reprinted with permission from Taylor & Francis Group.)

Failure Mode and Effect Analysis

Product Name:

Drawing Reference:
Revision:

EXAMPLE ANALYSIS

Team Members
1
2
3
4

Operation Number	Scope Description	Subdivided Elements	Failure Mode	Effects of Failure	Severity Rating	Cause of Failure	Occurance Rating	Current Practice of Controls			Detection Rating	Current Status of Product				Recommended Corrective Action	Action By	Action Taken Date	Revise Status			
								Detection	Prevention	Correction		S	O	D	RPN				S	O	D	RPN
1	Emergency generator system	Operation	Generator failed to start	1. No lights 2.Life support equipment stopped functioning 3. No power supply fot IT System 4. Water supply pumps stopped 5. No power supply for lift 6. Fire mode operation equipment will not operate 5. HVAC system stopped	7 10 9 6 4	1. No signal from ATS 2. Automatic stsrting system failed 3. Low Battery voltage for Starter motor 4. Circuit breaker in off position 5. No diesel in day tank		1.Through BMS 2. No regular check 3. Manual 4. Manual	1. Not in practice	1. Manual						1. Regular check of starting system 2. Check starter regularly 3.Check diesel level regularly. Interface level indicator with BMS 4. Check breaker position regularly. Interface with BMS	Maintenance Engineer					

Legend:-
RPN: Risk Priority Number
S: Severity
O: Occurance
D: Dtetection

ATS: Automatic Transfer Switch
BMS: Building Management System

FIGURE 4.27 FMEA recording form. FMEA, failure mode and effects analysis. (Abdul Razzak Rumane. (2013). *Quality Tools for Managing Construction Projects*. Reprinted with permission from Taylor & Francis Group.)

TABLE 4.6
5 Why Analysis for Cable Burning

Serial Number	Why	Related Analyzing Question
1	Why	Why the cable burned
2	Why	Why the earth leakage relay not tripped
3	Why	Why circuit breaker not tripped
4	Why	Why poor insulation of cable was not noticed
5	Why	Why under size rating of breaker with respect to current carrying capacity of cable was not noticed

Source: Abdul Razzak Rumane. (2013). *Quality Tools for Managing Construction Projects.* Reprinted with permission from Taylor & Francis Group.

TABLE 4.7
5W2H Analysis for Slab Collapse

Serial Number	Why	Related Analyzing Question
1	Why	The slab collapse
2	What	What is the reason for collapse
3	Who	Who is responsible
4	Where	Where is the mistake
5	When	When did the slab collapse
6	How many	How many persons affected (injured or died)
7	How much	How much loss in terms of time and cost

Source: Abdul Razzak Rumane. (2013). *Quality Tools for Managing Construction Projects.* Reprinted with permission from Taylor & Francis Group.

4.2.4.1 Root Cause Analysis

It is used to analyze root causes of problems. The analysis is generally performed by using Ishikawa diagram or cause-and-effect diagram.

Figure 4.29 illustrates root cause analysis for rejection of backfilling work.

4.2.4.2 PDCA Cycle

PDCA (plan–do–check–act) is mainly used for continuous improvement. It consists of a four-step model for carrying changes. PDCA cycle model can be developed as a process improvement tool to reduce cost of quality.

Figure 4.30 illustrates PDCA cycle for preparation of shop drawings.

4.2.4.3 SIPOC Analysis

It is used to identify supplier–input–process–output–customer (SIPOC) relationship. The purpose of SIPOC analysis is to show the process flow by defining and documenting the suppliers, inputs, process steps, outputs, and customers.

Table 4.9 illustrates SIPOC Analysis for fuel storage tank.

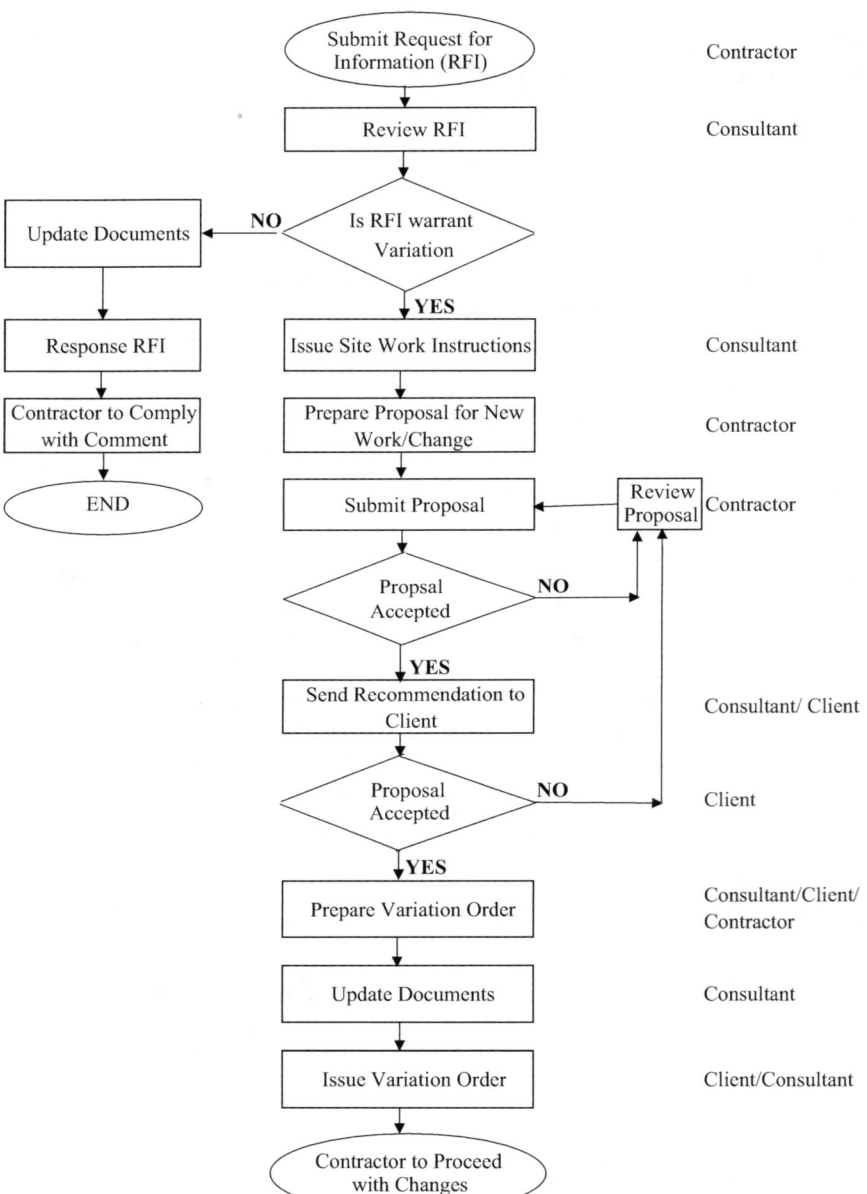

FIGURE 4.28 Process mapping/flowcharting for approval of variation order. (Abdul Razzak Rumane. (2013). *Quality Tools for Managing Construction Projects*. Reprinted with permission from Taylor & Francis Group.)

TABLE 4.8
Process Improvement Tools

Serial Number	Name of Quality Tool	Usage
Tool 1	Root cause analysis	To identify root causes that caused the problem to occur
Tool 2	PDCA cycle	Used to plan for improvement followed by putting into action
Tool 3	SIPOC analysis	Used to identify supplier–input–process–output–customer (SIPOC) relationship
Tool 4	Six Sigma DMAIC	Used as analytic tool for improvement
Tool 5	Failure mode and effects analysis	To identify and classify failures according to effect and prevent or reduce failure
Tool 6	Statistical process control	Used to study how the process changes over a time

DMAIC, define–measure–analyze–improve–control; PDCA, plan–do–check–act.

Source: Abdul Razzak Rumane. (2013). *Quality Tools for Managing Construction Projects.* Reprinted with permission from Taylor & Francis Group.

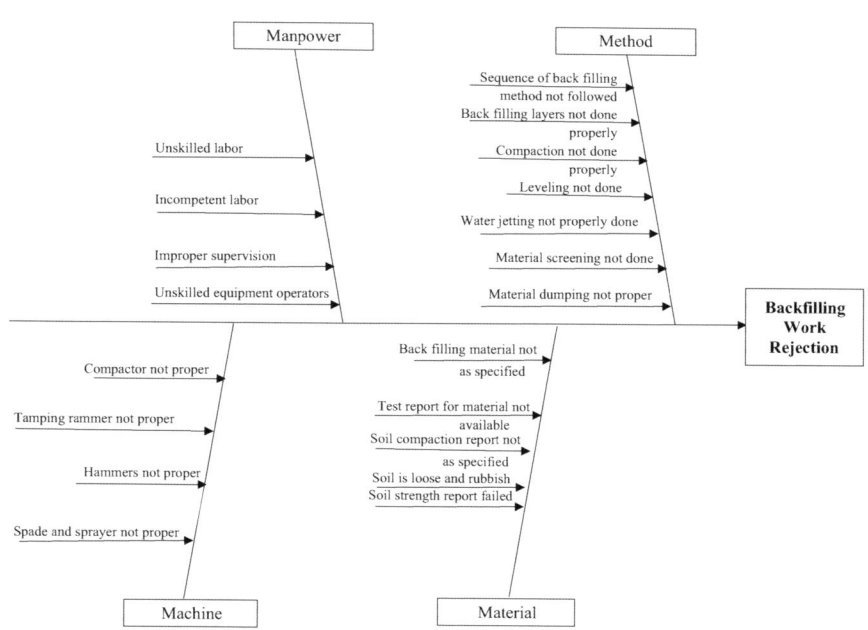

FIGURE 4.29 Root cause analysis for rejection of backfilling work.

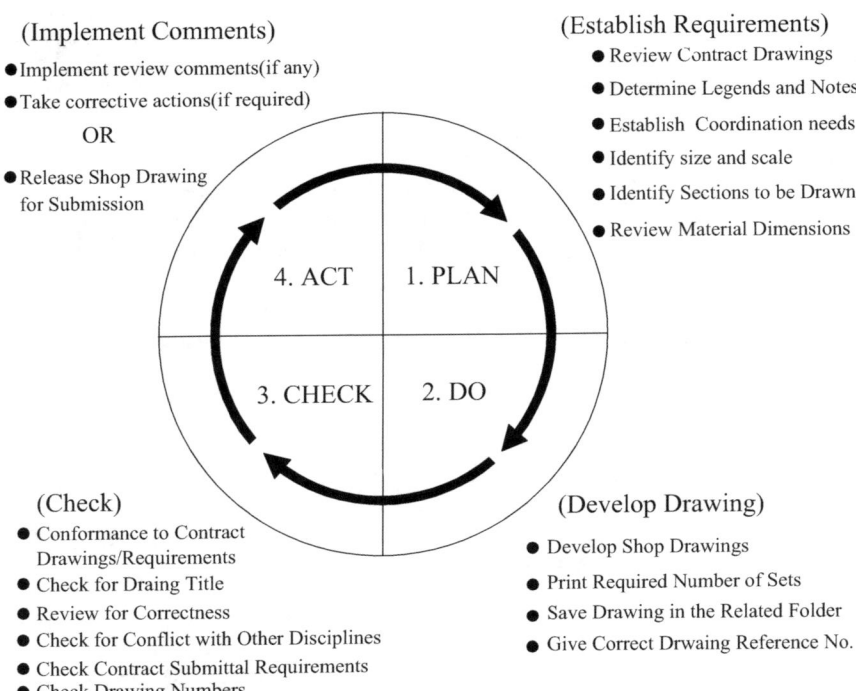

FIGURE 4.30 PDCA cycle for preparation of shop drawing. PDCA, plan–do–check–act. (Abdul Razzak Rumane. (2013). *Quality Tools for Managing Construction Projects*. Reprinted with permission from Taylor & Francis Group.)

4.2.4.4 Six Sigma DMAIC

Six Sigma is discussed in detail under Section 4.2.9 of this chapter.

4.2.4.5 Failure Mode and Effects Analysis

FMEA is also used as a process improvement tool. It identifies all the possible failures in the design of a product, process, and service and their effects on the product, process, and service. Its aim is to reduce risk of failure and improve the process. Refer Figure 4.26 under Section 4.2.3.5 that illustrates FMEA process.

4.2.4.6 Statistical Process Control

SPC is a quantitative approach based on the measurement of process control. Dr. Walter A. Shewhart developed the control charts as early as 1924. SPC charts are used for identification of common cause and special (or assignable) cause of variations and assisting diagnosis of quality problems. SPC charts reveal whether a process is "in control"-stable and exhibiting only random variation or "out of control" and needing attention. Control chart is one of the key tools of SPC. It is used to monitor processes that are not in control, using measured ranges. There are two types of process control charts. These are as follows:

TABLE 4.9
SIPOC Analysis for Storage Tank

(Who Are Suppliers)	(What the Suppliers are Providing)	(What Is the Process)	(What Is the Output of Process)	(Who Are the Customers)
Supplier	**Inputs**	**Process**	**Outputs**	**Customer**
Storage tank fabrication shop	Storage tank	Fuel/oil storage	Fuel/oil	Fuel/oil consumers
Subcontractor	Pumping system	Fuel supply		
	Control valve			
	Relief valve			
	Gauging system, level radar, float valve			
	Electrical/control panel			
	Depth/measuring gauge			
	Spiral stair			
	Fire suppression system			
	Security/surveillance system			

1. Variable charts
2. Attributes charts

Variable charts relate to variable measurements such as length, width, temperature, and weight.

Attributes charts relate to the characteristics possessed (or not possessed) by the process or the product.

Figure 4.31 illustrates SPC charts for generator frequency.

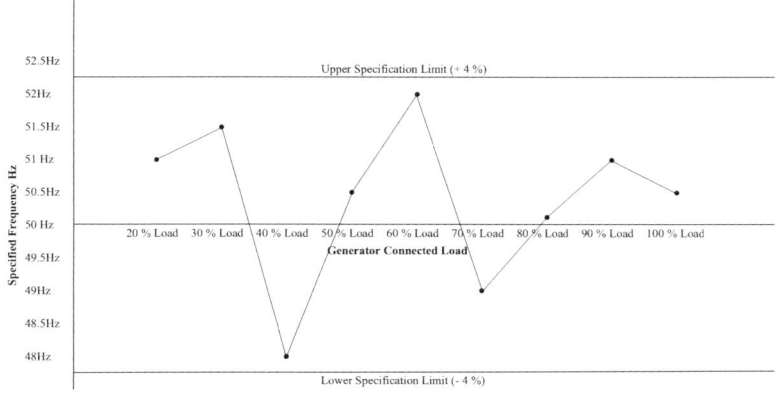

FIGURE 4.31 Statistical process control chart for generator frequency. (Abdul Razzak Rumane. (2013). *Quality Tools for Managing Construction Projects*. Reprinted with permission from Taylor & Francis Group.)

4.2.5 INNOVATION AND CREATIVE TOOLS

Table 4.10 illustrates innovation and creative tools.

4.2.5.1 Brainstorming

Brainstorming is listing of all the ideas put forth by a group in response to a given question or problem. It is a process of creating ideas by storming some objective. In 1939, a team led by advertising executive Alex Osborn coined the term "brainstorm". According to Osborn, brainstorm means using the brain to storm a creative problem. Classical brainstorming is the most well-known and often used technique for idea generation in a short period of time. It is based on the fundamental principles of deferment of judgment, and that quantity breeds quality. It involves questions such as:

- Does the item have any design features that are not necessary?
- Can two or more parts be combined together?
- How can we cut down the weight?
- Are these nonstandard parts that can be eliminated?

There are four rules for successful brainstorming:

1. Criticism is ruled out.
2. Freewheeling is welcomed.
3. Quantity is wanted.
4. Contribution and improvement are sought.

TABLE 4.10
Innovation and Creative Tools

Serial Number	Name of Quality Tool	Usage
Tool 1	Brainstorming	Used to generate multiple ideas.
Tool 2	Delphi technique	Used to get ideas from select group of experts.
Tool 3	5W2H	The questions used to understand why the things happen the way they do.
Tool 4	Mind mapping	Used to create a visual representation of many issues that can help get more understanding of the situation.
Tool 5	Nominal group technique	Used to enhance brainstorming by ranking the most useful ideas.
Tool 6	Six Sigma DMADV	Used primarily for the invention and innovation of modified or new product, services, or process.
Tool 7	Triz	Used to provide systematic methods and tools for analysis and innovative problem-solving.

DMADV, design–measure–analyze–design–verify.

Source: Abdul Razzak Rumane. (2013). *Quality Tools for Managing Construction Projects.* Reprinted with permission from Taylor & Francis Group.

A classical brainstorming session has the following basic steps:

- *Preparation.* The participants are selected, and a preliminary statement of the problem is circulated.
- *Brainstorming.* A warm-up session with simple unrelated problems is conducted, the relevant problem and the four rules of brainstorming are presented, and ideas are generated and recorded using checklists and other techniques if necessary.
- *Evaluation.* The ideas are evaluated relative to the problem.

Generally, a brainstorming group should consist of four to seven people, although some suggest larger group.

Figure 4.32 illustrates brainstorming process.

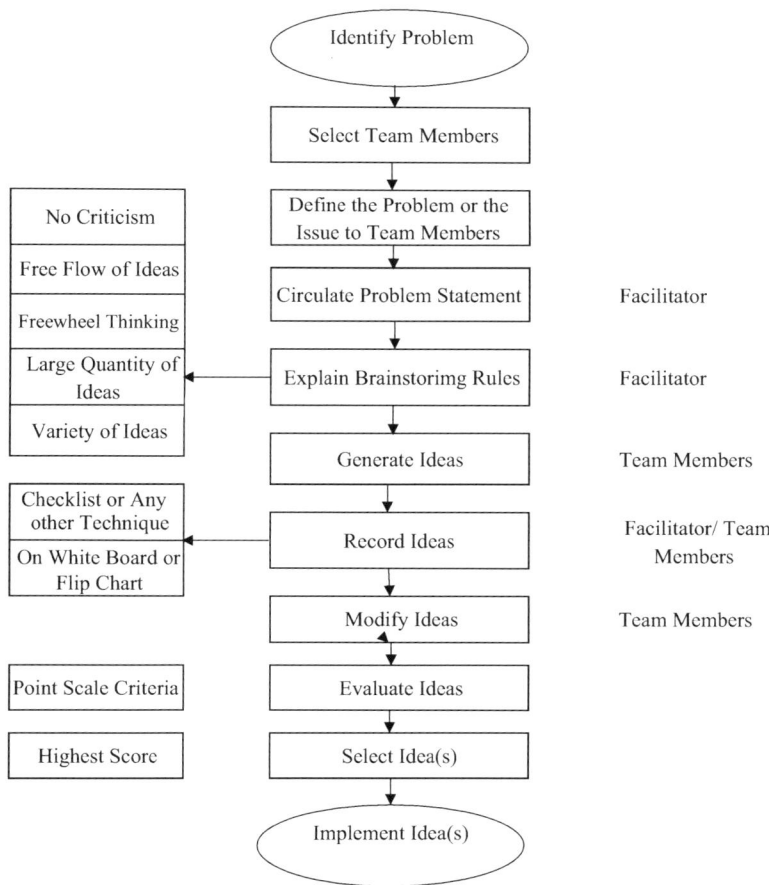

FIGURE 4.32 Brainstorming process. (Abdul Razzak Rumane. (2013). *Quality Tools for Managing Construction Projects.* Reprinted with permission from Taylor & Francis Group.)

4.2.5.2 Delphi Technique

Delphi technique is intended to determine a consensus among experts on a subject matter. The goal of Delphi technique is to pick brains of experts in the subject area, treating them as contributors to create ideas. It is a measure and method for consensus building by using questionnaire and obtaining responses from the panel of experts in the selected subjects. Delphi technique employs multiple iterations designed to develop consensus opinion about the specific subject. The selected expert group answers questions by facilitator. The responses are summarized and further circulated for group comments to reach the consensus. The iteration/feedback process allows the team members to reassess their initial judgment and change or modify the earlier suggestions.

Figure 4.33 illustrates Delphi technique process.

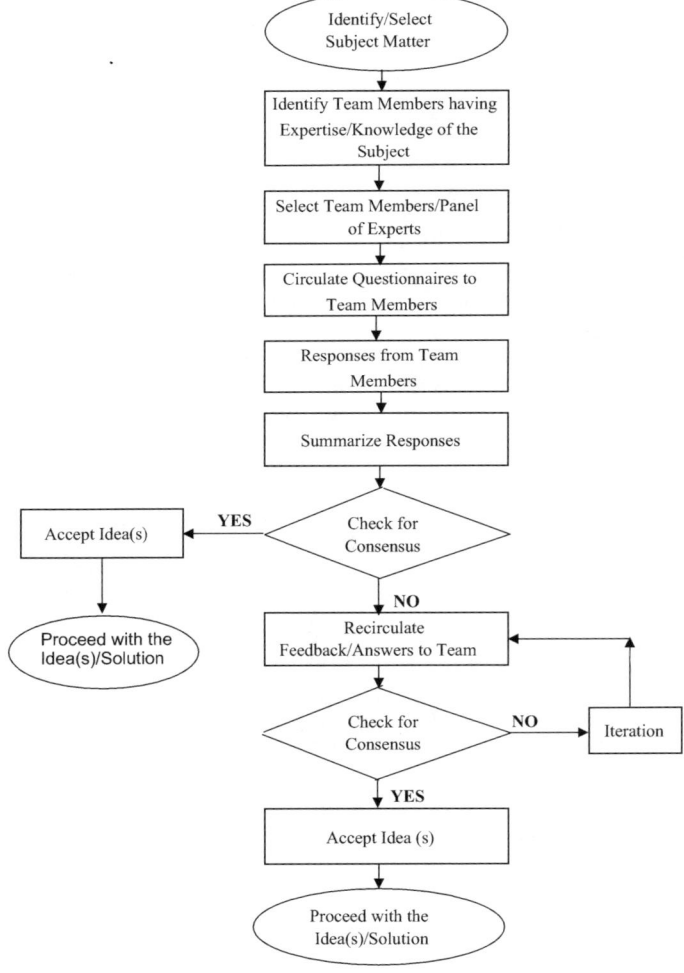

FIGURE 4.33 Delphi technique process. (Abdul Razzak Rumane. (2013). *Quality Tools for Managing Construction Projects*. Reprinted with permission from Taylor & Francis Group.)

4.2.5.3 5W2H

5W2H is also used as an innovation and creative tool. 5W2H is about asking the questions to understand about a process or problem.

The 5 Ws are:

1. Why
2. What
3. When
4. Where
5. Who

And 2 Hs are:

1. How
2. How much

Table 4.11 illustrates 5W2H for construction of new refinery.

4.2.5.4 Mind Mapping

Mind mapping is a graphical representation of ideas which can help get more understanding of the situation and create the solution or improve the task. Figure 4.34 illustrates mind mapping sketch to improve site safety.

4.2.5.5 Nominal Group Technique

The nominal group technique (NGT) involves a structural group meeting designed to incorporate individual ideas and judgments into a group consensus. By correctly applying the NGT, it is possible for groups of people (preferably 5–10) to generate alternatives or other ideas for improving the competitiveness of the firm. The technique can be used to obtain group thinking (consensus) on a wide range of topics. The technique, when properly applied, draws on the creativity of the individual participants, while reducing two undesirable effects of most group meetings:

a. The dominance of one or more participants
b. The suppression of conflict ideas

TABLE 4.11
5W2H Analysis for New Refinery

Serial Number	Why	Related Analyzing Question
1	Why	New refinery
2	What	What advantage it will have over other similar refineries
3	Who	Who will be the customers for the products from this refinery
4	Where	Where can we market the products
5	When	When the products will be ready for sale
6	How many	How many barrels/tons of the products will be produced/sold per year
7	How much	How much market share we will get for this refinery

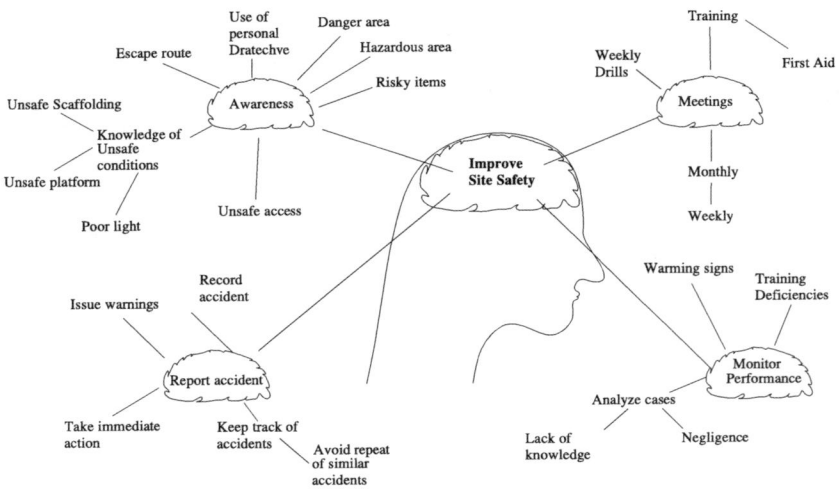

FIGURE 4.34 Mind mapping. (Abdul Razzak Rumane. (2013). *Quality Tools for Managing Construction Projects*. Reprinted with permission from Taylor & Francis Group.)

The basic format of an NGT session is as follows:

- Individual silent generation of ideas
- Individual round-robin feedback and recording the ideas
- Group's clarification of each idea
- Individual voting and ranking to prioritize ideas
- Discussion of group consensus results

The NGT session begins with an explanation of the procedure and a statement of question(s), preferably written by the facilitator.

4.2.5.6 Six Sigma DMADV
Six Sigma is discussed in detail under Section 4.2.9 of this chapter.

4.2.5.7 Triz
Triz is discussed in detail under Section 4.2.10 of this chapter.

4.2.6 Lean Tools

Table 4.12 illustrates lean tools.

4.2.6.1 Cellular Design
A self-contained unit dedicated to perform all the operational requirements to accomplish sequential processing. With cellular design, individual cells can be fabricated and assembled to give the same performance and saving time.

TABLE 4.12
Lean Tools

Serial Number	Name of Quality Tool	Usage
Tool 1	Cellular design	A self-contained unit dedicated to perform all the operational requirements to accomplish sequential processing
Tool 2	Concurrent engineering	It is used for product cycle reduction time. It is a systematic approach of creating a product design that simultaneously considers all elements of product life cycle.
Tool 3	5S	Used to eliminate waste that results from improper organization of work area.
Tool 4	Just-in-time (JIT)	Used to reduce inventory levels, improve cash flow, and reduce space requirements for storage of material.
Tool 5	Kanban	Used to signal that more material is required to be ordered. It is used to eliminate waste from inventory.
Tool 6	Kaizan	Used for continually eliminating waste from manufacturing processes by combining the collective talent of company.
Tool 7	Mistake proofing	Used to eliminate the opportunity for error by detecting the potential source of error.
Tool 8	Out sourcing	It is contracting out certain works, processes, and services to specialist in the discipline area.
Tool 9	Poka-Yoke	Used to detect the abnormality or error and fix or correct the error and take action to prevent the error.
Tool 10	Single minute exchange of die	Used to reduce setup time for change over to new process.
Tool 11	Value stream mapping	Used to establishing flow of material or information and eliminating waste and adding value.
Tool 12	Visual management	It addresses both visual display and control. It exposes waste elimination/prevention.
Tool 13	Waste reduction	It focusses on reducing waste.

Source: Abdul Razzak Rumane. (2013). *Quality Tools for Managing Construction Projects*. Reprinted with permission from Taylor & Francis Group

An electrical main switch board may consist of a number of cells assembled together to perform the desired operations. It helps easy maneuvering and assembling at workplace for proper functioning.

Figure 4.35 illustrates cellular design for electrical panel.

4.2.6.2 Concurrent Engineering

Product life cycle begins with need and extends through concept design, preliminary design, detail design, production or construction, product use, phase-out, and disposal. Concurrent engineering is defined as a systematic approach to create a product design that simultaneously considers all the elements of product life cycle,

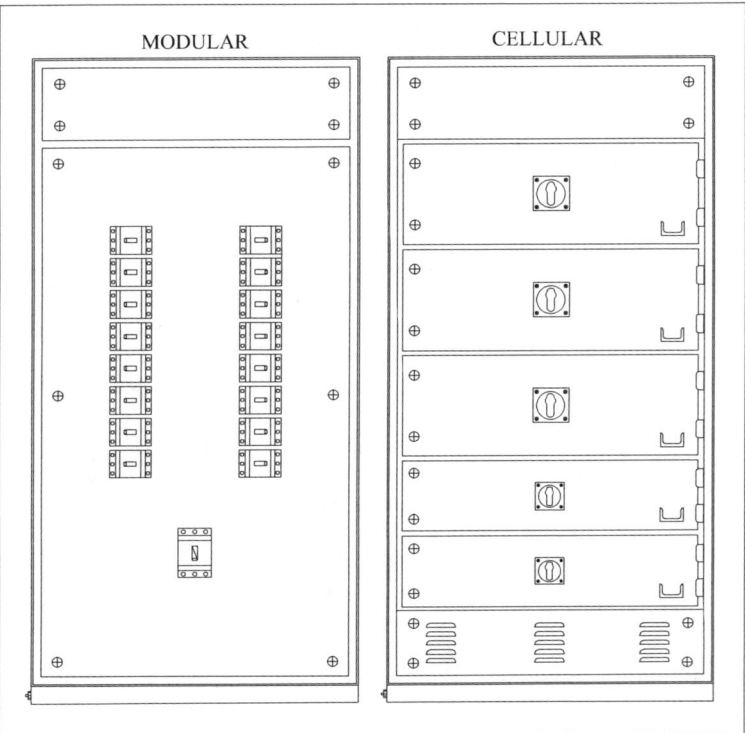

FIGURE 4.35 Cellular main switch board.

thus reducing the product life cycle time. It is used to expedite the development and launch of new product. In construction projects, construction can simultaneously start while the design is under development.

Figure 4.36 illustrates concurrent engineering for construction project life cycle.

4.2.6.3 5S

5S is a systematic approach for improvement of quality and safety by organizing a workplace. It is a methodology which advocates the following:

- What should be kept
- Where should be kept
- How should be kept

5S is a Japanese concept of housekeeping having reference to five Japanese words starting with letter "S". Table 4.13 illustrates 5S for construction projects.

4.2.6.4 Just in Time

It is used to reduce inventory levels, improve cash flow, and reduce storage space requirements for material. For example:

Quality Tools for Oil and Gas Industry

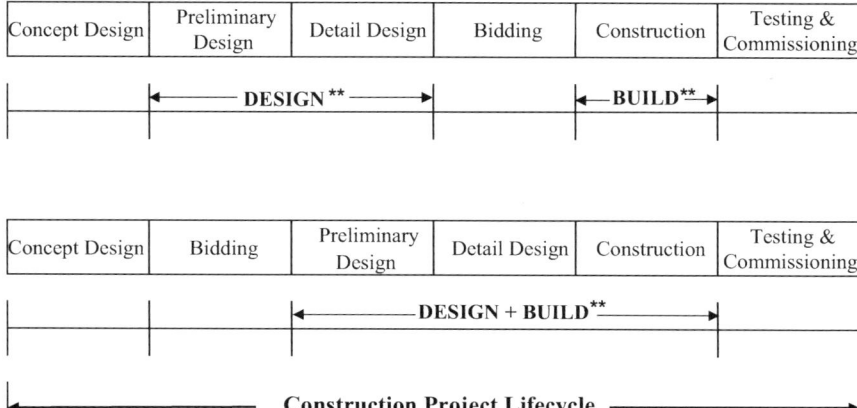

FIGURE 4.36 Concurrent engineering for construction project life cycle. (Abdul Razzak Rumane. (2013). *Quality Tools for Managing Construction Projects*. Reprinted with permission from Taylor & Francis Group.)

1. Concrete block can be received at site just before start of block work and can be stacked near the work area where masonry work is in progress.
2. Valves can be received at site and directly placed in the pipeline system without storing in the storage yard.

4.2.6.5 Kanban

It is used to signal when more material is required to be ordered. It is used to eliminate waste from inventory and inventory control thus to avoid extra storage required for large inventory. In construction projects, electrical wires for circuiting can be ordered to receive at site when the wire pulling work is under progress. Similarly concrete blocks, false ceiling tiles can be ordered and received as and when required.

4.2.6.6 Kaizan

It is used for continually improving through small changes to eliminate waste from manufacturing process by combining the collective talent of every employee of the company.

4.2.6.7 Mistake Proofing

Mistake proofing is used to eliminate the opportunity for error by detecting the potential source of error. Mistakes are generally categorized as follows:

1. Information
2. Mismanagement
3. Omission
4. Selection

Table 4.14 illustrates mistake proofing chart for eliminating design error.

TABLE 4.13
5S for Construction Projects

Serial Number	5S	Related Action
1	Sort	• Determine what is to be kept in open and what under shed. • Allocate area for each type of construction equipment and machinery. • Allocate area for electrical tools. • Allocate area for hand tools. • Allocate area for construction material/equipment to be used/installed in the project. • Allocate area for hazardous, inflammable material. • Allocate area for chemicals and paints. • Allocate area for spare part for maintenance.
2	Set in order	• Keep/arrange equipment in such a way that their maneuvering/movement shall be easy. • Vehicles to be parked in the yard in such a way that frequently used vehicles are parked near the gate. • Frequently used equipment/machinery to be located near the workplace. • Set boundaries for different types of equipment and machinery. • Identify and arrange tools for easy access. • Identify and store material/equipment as per relevant division/section of contract documents. • Identify and store material in accordance with their usage as per construction schedule. • Determine items which need special conditions. • Mark/tag the items/material. • Display route map and location. • Put the materials in sequence as per their use. • Frequently used consumables have to be kept near workplace. • Label on the drawer with list of contents. • Keep shuttering material at one place. • Determine inventory level of consumable items.

(Continued)

TABLE 4.13 (Continued)
5S for Construction Projects

Serial Number	5S	Related Action

3	Sweeping	• Clean site on daily basis by removing
		  • cut pieces of reinforced bars;
		  • cut pieces of plywood;
		  • left-out concrete;
		  • cut pieces of pipes;
		  • cut pieces of cables and wires; and
		  • used welding rods.
		• Clean equipment and vehicles.
		• Check electrical tools after return by the technician.
		• Attend to breakdown report.

4	Standardize	• Standardize the store by allocating separate areas for material used by different divisions/sections. • Standardize area for long lead items. • Determine regular schedule for cleaning the workplace. • Make available standard tool kit/box for a group of technicians. • Make every one informed of their responsibilities and related area where the things are to be placed and are available. • Standardize the store for consumable items. • Inform suppliers/vendors in advance the place for delivery of material.
5	Sustain	Follow the system till the end of project.

Source: Abdul Razzak Rumane. (2010). *Quality Management in Construction Projects*. Reprinted with permission from Taylor & Francis Group.

TABLE 4.14
Mistake Proofing for Eliminating Design Errors

Serial Number	Items	Points to be Considered to Avoid Mistakes
1	Information	1. Terms of reference (TOR)
		2. Client's preferred requirements matrix
		3. Data collection
		4. Regulatory requirements
		5. Environmental considerations
		6. Codes and standards
		7. Historical data
		8. Organizational requirements
2	Mismanagement	1. Compare production with actual requirements
		2. Interdisciplinary coordination
		3. Application of different codes and standards
		4. Drawing size of different trades/specialist consultants
3	Omission	1. Review and check design with TOR
		2. Review and check design with client requirements
		3. Review and check design with regulatory requirements
		4. Review and check design with codes and standards
		5. Check for all required documents
		6. Review and check design with FEED/contract documents (EPC contracting system)
4	Selection	1. Qualified team members
		2. Available material
		3. Installation methods

EPC, engineering, procurement, and construction; FEED, front-end engineering design.

Source: Abdul Razzak Rumane. (2013). *Quality Tools for Managing Construction Projects.* Reprinted with permission from Taylor & Francis Group.

4.2.6.8 Outsourcing

It is contracting out certain works, processes, and services to a specialist in a particular discipline area. For example, in construction projects, the following is the list of some of the works which are outsourced (subcontracted):

1. Excavation and backfilling works
2. Shoring works
3. Structural concrete
4. Piping work
5. Tank fabrication works
6. Fire suppression work
7. Electrical work
8. Instrumentation work
9. Automation/low-voltage work

10. Landscape work
11. External work

4.2.6.9 Poka-Yoke

Poka-Yoke is a quality management concept developed by Shigeo Shino to prevent human errors occurring in the production line. The main objective of Poka-Yoke is to achieve zero defects.

4.2.6.10 Single Minute Exchange of Die

It is used to reduce the setup time for changeover to new process.

For example, a spare circuit breaker of similar rating can be used as immediate replacement to damaged circuit breaker in the electrical distribution board to avoid breakdown of electrical supply for long duration. Subsequently, a new circuit breaker can be fixed in place of spare breaker.

4.2.6.11 Value Stream Mapping

It is used to establish flow of material or information, eliminate waste, and add value. Value stream mapping is used to identify areas for improvement.

Figure 4.37 illustrates value stream mapping diagram for emergency power system.

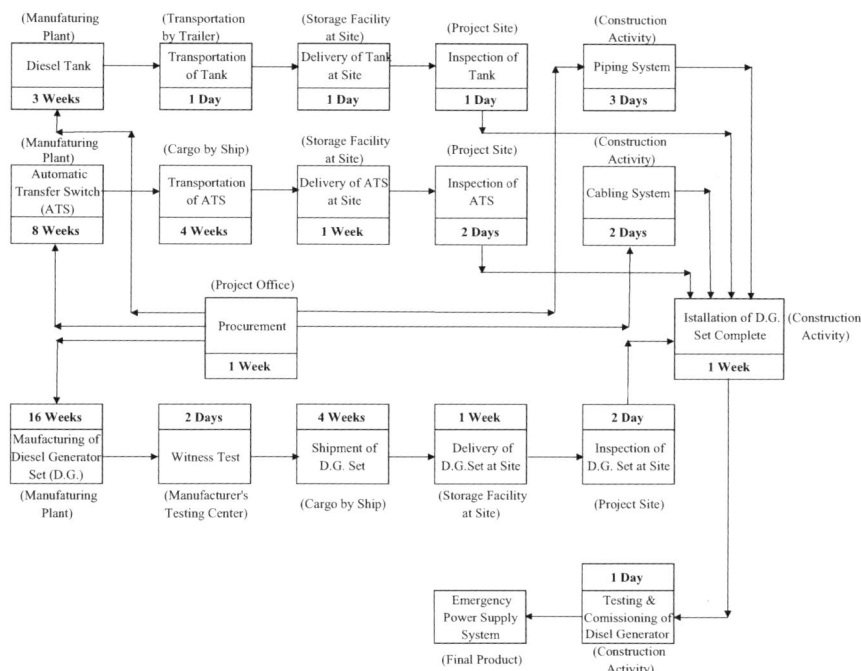

FIGURE 4.37 Value stream mapping for emergency power system. (Abdul Razzak Rumane. (2013). *Quality Tools for Managing Construction Projects*. Reprinted with permission from Taylor & Francis Group.)

4.2.6.12 Visual Management

It addresses both visual display and control. It exposes waste elimination/prevention. Visual displays present information, while visual control focuses on a need to act.

4.2.6.13 Waste Reduction

It focuses on reducing waste. The following are the general type of waste:

1. Defective parts
2. Delays, waiting
3. Excess inventory
4. Misused resources
5. Overproduction
6. Processing
7. Transportation
8. Untapped resources
9. Wasted motion

4.2.7 COST OF QUALITY

4.2.7.1 Introduction

Quality has an impact on the costs of products and services. The cost of poor quality is the annual monetary loss of products and processes that are not achieving their quality objective. The main components of the cost of low quality are as follows:

1. Cost of conformance
2. Cost of nonconformance

Table 4.15 illustrates elements of cost of quality.

4.2.7.2 Categories of Costs

Costs of poor quality are the costs associated with providing poor-quality products or services. These are incurred because of poor quality costs that would not be incurred if things were done right from the time and at every time to follow thereafter in order to achieve the quality objective. There are four categories of costs:

1. Internal failure costs (The costs associated with defects found before the customer receives the product or service. It also consists of cost of failure to meet customer satisfaction and needs and cost of inefficient processes.)
2. External failure costs (The cost associated with defects found after the customer receives the product or service. It also includes lost opportunity for sales revenue.)
3. Appraisal costs (The costs incurred to determine the degree of conformance to quality requirements)
4. Prevention costs (The costs incurred to keep failure and appraisal costs to minimum)

TABLE 4.15
Cost of Quality

Cost of Compliance	Cost of Noncompliance
• Quality planning	• Scrap
• Process control planning	• Rework
• Quality training	• Corrective action
• Quality audit	• Additional material/inventory cost
• Design review	• Expedition
• Product design validation	• Customer complains
• Work procedure	• Product recalls
• Method statement	• Warranty
• Process validation	• Maintenance service
• Field testing	• Field repairs
• Third party inspection	• Rectification of returned material
• Receiving inspection	• Reinspection or retest
• Prevention action	• Downgrading
• In-process inspection	• Loss of business
• Outside endorsement	
• Calibration of equipment	
• Laboratory acceptance testing	
• Regulatory requirements	

Source: Abdul Razzak Rumane. (2013). *Quality Tools for Managing Construction Projects.* Reprinted with permission from Taylor & Francis Group.

These cost categories allow the use of quality cost data for a variety of purpose. Quality costs can be used for measurement of progress, for analyzing the problem, or for budgeting. By analyzing the relative size of the cost categories, the company can determine if its resources are properly allocated.

4.2.7.3 Quality Cost in Construction

Quality of construction is defined as follows:

1. Scope of work
2. Time
3. Budget

Cost of quality refers to the total cost incurred during the entire life cycle of construction project in preventing nonconformance to owner requirements (defined scope). There are certain hidden costs which may not affect directly the overall cost of the project; however, it may cost consultant/designer to complete design within stipulated schedule to meet owner requirements and conformance to all the regulatory codes/standards, and contractor to construct the project within stipulated schedule meeting all the contract requirements. Rejection/nonapproval of executed/installed works by the supervisor due to noncompliance with specification will cause the contractor loss in terms of the following:

- Material
- Manpower
- Time

The contractor shall have to rework or rectify the work, which will need additional resources and will need extra time to do the work as specified. This may disturb contractor's work schedule and affect execution of other activities. The contractor has to emphasis upon "zero defect" policy, particularly for concrete works. To avoid rejection of works, contractor has to take the following measures:

1. Execute works as per approved shop drawings using approved material.
2. Follow approved method of statement or manufacturer's recommended method of installation.
3. Conduct continuous inspection during construction/installation process.
4. Employ properly trained workforce.
5. Maintain good workmanship.
6. Identify and correct deficiencies before submitting the check list for inspection and approval of work.
7. Coordinate requirements of other trades, for example, if any opening is required in the concrete beam for crossing of services pipe.

Timely completion of project is one of the objectives to be achieved. To avoid delay in completion schedule, proper planning and scheduling of construction activities is necessary. Since construction projects have involvement of many participants, it is essential that requirements of all the participants are fully coordinated. This will ensure execution of activities as planned, resulting in timely completion of project.

Normally the construction budget is fixed at the inception of project; therefore, it is required to avoid variations during construction process as it may take time to get approval of additional budget resulting time extension to the project. Quality costs related to construction projects can be summarized as follows.

4.2.7.3.1 Internal Failure Costs
- Rework
- Rectification
- Rejection of checklist
- Corrective action

4.2.7.3.2 External Failure Costs
- Breakdown of the installed system
- Repairs
- Maintenance
- Warranty

4.2.7.3.3 Appraisal Costs

- Design review/preparation of shop drawings
- Preparation of composite/coordination drawings
- On-site material inspection/test
- Off-site material inspection/test
- Prechecklist inspection

4.2.7.3.4 Prevention Costs

- Preventive action
- Training
- Work procedures
- Method statement
- Calibration of instruments/equipment

4.2.8 Quality Function Deployment

QFD is a technique to translate customer requirements into technical requirements. It was developed in Japan by Dr. Yoji Akao in 1960s to transfer the concepts of quality control from the manufacturing process into the new product development process. QFD is referred as "voice of customer" which helps in identifying and developing customer requirements through each stage of product or service development. It is a development process which utilizes a comprehensive matrix involving project team members.

QFD involves constructing one or more matrices containing information related to others. The assembly of several matrices showing the correlation with one another is called "the house of quality". The "house of quality" matrix is the most recognized form of QFD. QFD is being applied virtually in every industry and business from aerospace, communication, software, transportation, manufacturing, services industry, and construction industry. The house of quality is made up of the following major components:

1. Whats
2. Hows
3. Correlation matrix (roof)—technical requirements
4. Interrelationship matrix
5. Target value
6. Competitive evaluation

Whats is the first step in developing the house of quality. It is a structured set of needs/requirements ranked in terms of priority and the levels of importance being specified quantitatively. It is generated by using questions such as follows:

- What are the types of finishes needed for the building?
- What type of air conditioning system required for the building?
- What type of communication system is required for the building?

- What type of flooring material is required?
- Whether the building needs any security system?

Hows is the second step in which project team members translate the requirements (Whats) into technical design characteristics (specifications), which are listed across the columns.

Correlation matrix identifies technical interaction or physical relationship among the technical specifications.

Interrelationship matrix illustrates team member's perception of interrelationship between owner's requirements and technical specifications.

The bottom part allows for technical comparison between possible alternatives, target values for each technical design characteristics, and performance measurement.

The right side of the house of quality is used for planning purpose. It illustrates customer perceptions observed in market survey.

QFD technique can be used to translate owner's needs/requirements into development of a set of technical requirements during conceptual design.

Figure 4.38 illustrates the house of quality for offices building.

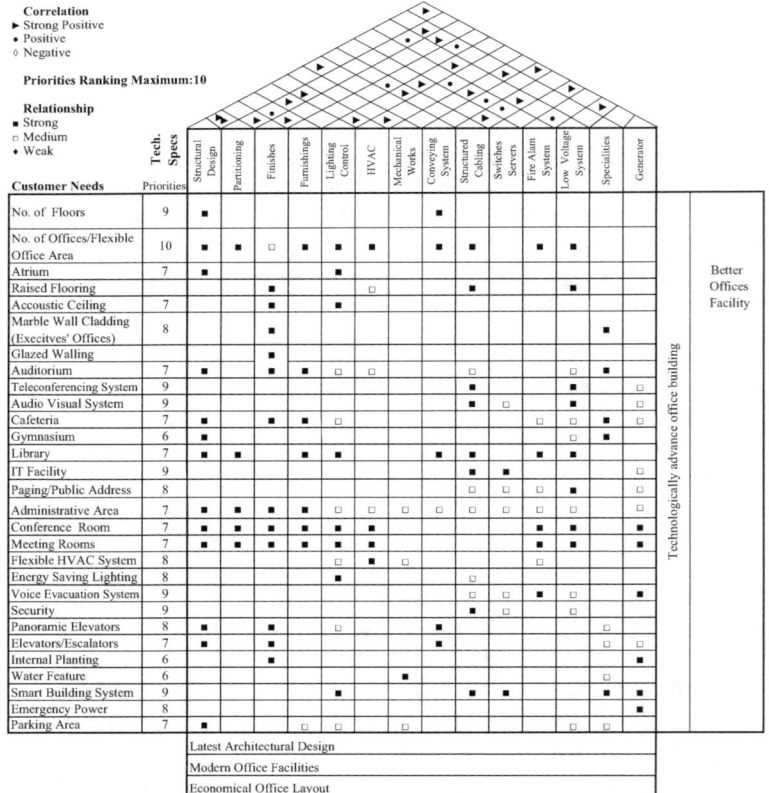

FIGURE 4.38 House of quality for office building.

4.2.9 SIX SIGMA

4.2.9.1 Introduction

Six sigma is, basically, a process quality goal. It is a process quality technique that focuses on reducing variation in process and preventing deficiencies in product. In a process that has achieved Six Sigma capability, the variation is small compared with the specification limits.

Sigma is a Greek letter σ that stands for standard deviation. Standard deviation is a statistical way to describe how much variation exists in a set of data, a group of items, or a process. Standard deviation is the most useful measure of dispersion. Six Sigma means that a process to be capable at Six Sigma level, and the specification limits should be at least 6 σ from the average point. So the total spread between upper specification (control) limit and lower specification (control) limit should be 12 σ. With Motorola's Six Sigma program, no more than 3.4 defects per million fall outside specification limits with process shift of not more than 1.5 σ from the average or mean. Six Sigma started as a defect reduction effort in manufacturing and was then applied to other business processes for the same purpose.

Six Sigma is a measurement of "goodness" using a universal measurement scale. Sigma provides a relative way to measure improvement. Universal means sigma can measure anything from coffee mug defects to missed chances to close a sales deal. It simply measures how many times a customer's requirements were not met (a defect), given a million opportunities. Sigma is measured in defects per million opportunities. For example, a level of sigma can indicate how many defective coffee mugs were produced when 1 million mugs were manufactured. Levels of sigma are associated with improved levels of goodness. To reach a level of Three Sigma, you can only have 66,811 defects, given a million opportunities. A level of Five Sigma only allows 233 defects. Minimizing variation is a key focus of Six Sigma. Variation leads to defects, and defects lead to unhappy customers. To keep customers satisfied, loyal, and coming back, you have to eliminate the sources of variation. Whenever a product is created or a service performed, it needs to be done the same way every time, no matter who is involved. Only then will you truly satisfy the customer. Figure 4.39 illustrates Six Sigma roadmap.

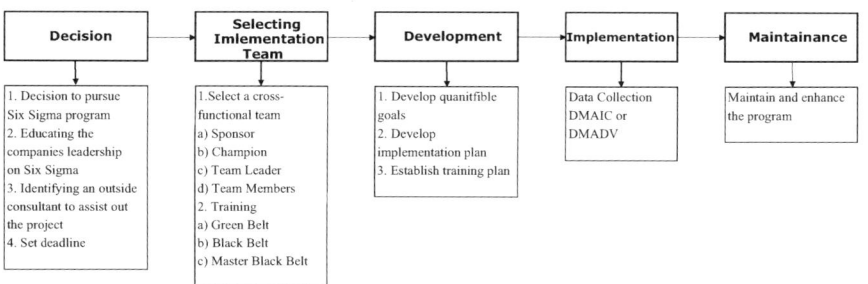

FIGURE 4.39 Six Sigma roadmap. (Abdul Razzak Rumane (2010). *Quality Management in Construction Projects*. Reprinted with permission from Taylor & Francis Group.)

4.2.9.2 Six Sigma Methodology

Six Sigma is overall business improvement methodology that focuses an organization on the following:

- Understanding and managing customer requirements
- Aligning key business process to achieve these requirements
- Utilizing rigorous data analysis to minimize variation in these processes
- Driving rapid and sustainable improvement in business process by reducing defects, cycle time, impact to the environment, and other undesirable variations
- Timely execution

As a management system, Six Sigma is a high-performance system for executing business strategy. It uses concept of fact and data to drive better solutions. Six Sigma is a top-down solution to help organizations:

- Align their business strategy to critical improvement efforts.
- Mobilize teams to attack high-impact projects.
- Accelerate improved business results.
- Govern efforts to ensure improvements are sustained.

Six Sigma methodology also focuses on the following:

- Leadership Principles
- Integrated approach to improvement
- Engaged teams
- Analytic tool
- Hard coded improvements

4.2.9.2.1 Leadership Principles

Six Sigma methodology has four leadership principles. These are as follows:

1. Align
2. Mobilize
3. Accelerate
4. Govern

Brief description of these leadership principles is as follows:

1. *Align*—Leadership should ensure that all improvement projects are in line with the organization's strategic goals.
   Alignment begins with the leadership team developing a scorecard. This vital tool, the cornerstone of the Six Sigma business improvement campaign, translates strategy into tactical operating terms. The scorecard also defines metrics that an organization can use to determine success. Just as a

scoreboard at a sporting event tells you who is winning, the scorecard tells leadership how well the company is meeting its goals.
2. *Mobilize*—Leadership should enable teams to take action by providing clear direction, feasible scope, a definition of success, and rigorous reviews.

    Mobilizing sets clear boundaries, lets people go to work, and trains them as required.

    The key to mobilizing is focus–lack of focused action was one of the downfalls of previous business improvement efforts. True focus means that the project is correctly aligned with the organization's scorecard. Mobilized teams have a valid reason for engaging in improvement efforts—they can see benefit for the customer. The project has strategic importance, and they know it. They know exactly what must be done and the criteria they can use to determine success.

3. *Accelerate*—Leadership should drive a project to rapid results through tight clock management, training as needed, and shorter deadlines

    More than 70% of all improvement initiatives fail to achieve the desired results in time to make a difference. For projects to make an impact, they must achieve results quickly, and that is what acceleration is all about.

    Accelerating leadership principle involves three main components:
    a. Action learning
    b. Clock management
    c. Effective planning

    Accelerate employs "action learning" methodology to quickly bridge from "learning" to "doing". Action learning mixes traditional training with direct application. Training is received while working on a real-world project, allowing plenty of opportunities to apply new knowledge. The instructor is not simply a trainer, but a coach as well, helping work with real-world project. Action learning accelerates improvement over traditional learning methods. It helps receiving training and also completing a worthwhile project at the same time. In addition to the 4-to-6-month time frame, *accelerate* requires teams to set deadlines that are reinforced through rigorous reviews.

4. *Govern*—Leadership must visibly sponsor projects and conduct regular and rigorous reviews to make critical midcourse corrections.

    The fourth leadership principle is to *govern*. Once leadership selects an improvement opportunity, their work is not done. They must remain ultimately responsible for the success of that project. *Govern* requires leaders to drive for results.

While governing a Six Sigma project, you need the following:

- A regular communications plan and a clear review process
- Actively sponsor teams and their projects
- To encourage proactive dialogue and knowledge sharing on the team and throughout the organization

4.2.9.2.2 Six Sigma Team

Team work is absolutely vital for complex Six Sigma projects. For teams to be effective, they must be engaged—involved, focused, and committed to meet their goals. Engaged teams must have leadership support. There are four types of teams. These are as follows:

1. Black Belts
2. Green Belts
3. Breakthrough
4. Blitz

Brief description of these teams is as follows:

1. Black Belt

    Black Belt teams are led by a Black Belt and may have Green Belts and functional experts assigned to complex, high-impact process improvement projects or designing new products, services, or complex processes. Black Belts are internal Six Sigma practitioners, skilled in the application of rigorous statistical methodologies, and they are crucial to the success of Six Sigma. Their additional training and experience provide them with the skills they need to tackle difficult problems. Black Belts have many responsibilities. They
    - function as a team leader on Black Belt projects;
    - integrate their functional discipline with statistical, project, and interpersonal skills;
    - serve as internal consultants;
    - tackle complex, high-impact improvement opportunities; and
    - mentor and train Green Belts.

2. Green Belt

    Led by a Green Belt and comprised of nonexperts, Green Belt teams tackle less complex, high-impact process improvement projects. Green Belt teams are often coached by Black Belts or Master Black Belts.

    Green Belts are also essential to the success of Six Sigma. They perform many of the same functions as Black Belts, but their work requires less complex analysis. Green Belts are trained in basic problem-solving skills and the statistical tools needed to work effectively as members of process improvement teams.

    Green Belt responsibilities include
    - acting as a team leader on business improvements requiring less complex analysis;
    - adding their unique skills and experiences to the team;
    - working with the team to come up with inventive solutions;
    - performing basic statistical analysis; and
    - conferring with a Black Belt as questions arise.

3. Breakthrough

    While creating simple processes, sophisticated statistical tools may not be needed. Breakthrough teams are typically used to define low-complexity, new processes.

4. Blitz

   Blitz teams are put in place to quickly execute improvements produced by other projects. These teams can also implement digitization for efficiency using a new analytic tool set.

For typical Six Sigma project, four critical roles exist.

1. Sponsor
2. Champion
3. Team leader
4. Team member

A sponsor typically

- remains ultimately accountable for a project's impact;
- provides project resources;
- reviews monthly and quarterly achievements, obstacles, and key actions; and
- supports the project champion by removing barriers as necessary.

A champion typically

- reviews weekly achievements, obstacles, and key actions;
- meets with the team weekly to discuss progress;
- reacts to changes in critical performance measures as needed;
- supports the team leader, removing barriers as necessary; and
- helps ensure project alignment.

A team leader typically

- leads improvement projects through an assigned, disciplined methodology;
- works with the champion to develop the team charter, review project progress, obtain necessary resources, and remove obstacles;
- identifies and develops key milestones, timelines, and metrics for improvement projects;
- establishes weekly, monthly, and quarterly review plans to monitor team progress; and
- supports the work of team members as necessary.

Team members typically

- Assist the team leader;
- follow a disciplined methodology;
- ensure the team charter and timeline are being met;
- accept and execute assignments; and
- add their views, opinions, and ideas

4.2.9.3 Analytic Tool Sets

The following are the analytic tools used in Six Sigma projects.

4.2.9.3.1 FORD Global 8D Tool

What Problem Needs Solving? →
Who should help solve problem? →
How do we quantify symptoms? →
How do we contain it? →
What is the root cause? →
What is the permanent corrective action? →
How do we implement? →
How can we prevent this in future? →
Who should we reward? →

Ford Global 8D Tool is primarily used to bring performance back to a previous level.

4.2.9.3.2 DMADV Tool Set Phases

Define → What is important?
Measure → What is needed?
Analyze → How will we fulfill?
Design → How do we build it?
Verify → How do we know it will work?

DMADV tool is used primarily for the invention and innovation of modified or new products, services, or process. Using this tool set, Black Belts optimize performance before production begins. DMADV is proactive, solving problems before they start. This tool is also called as DFSS (Design for Six Sigma).

Table 4.16 lists fundamental objectives of DMADV.

TABLE 4.16
Fundamental Objectives of Six Sigma DMADV Tool

DMADV	Phase	Fundamental Objective
1	**Define**—What is important?	Define the project goals and customer deliverables (internal and external)
2	**Measure**—What is needed?	Measure and determine customer needs and specifications
3	**Analyze**—How we fulfill?	Analyze process options and prioritize based on capabilities to satisfy customer requirements
4	**Design**—How we build it?	Design detailed process(es) capable of satisfying customer requirements
5	**Verify**—How do we know it will work	Verify design performance capability

Source: Abdul Razzak Rumane. (2013). *Quality Tools for Managing Construction Projects.* Reprinted with permission from Taylor & Francis Group.

4.2.9.3.2.1 DMADV PROCESS
Define phase: What is important?
(Define the project goals and customer deliverables.)
Key deliverables of this phase are as follows:

- Establish the goal.
- Identify the benefits.
- Select project team.
- Develop project plan.
- Project charter.

Measure phase: What is needed?
(Measure and determine customer needs and specifications)
Key deliverable in this phase is as follows:

- Identify specification requirements.

Analyze phase: How we fulfill?
(Analyze process options and prioritize based on capability to satisfy customer requirements.)
Key deliverables in this phase are as follows:

- Design generation (data collection).
- Design analysis.
- Risk analysis.
- Design model (prioritization of data under major variables).

Design phase: How we build it?
(Design detailed process(es) capable of satisfying customer requirements.)
Key deliverables in this phase are as follows:

- Construct a detail design.
- Convert CTQs (critical to quality) into CTPs (critical to process elements).
- Estimate the capabilities of the CTPs in the design.
- Prepare a verification plan.

Verify phase: How do we know it work?
(Verify design performance capability.)
Key deliverable in this phase is as follows:

- Design a control and transition plan.

4.2.9.3.3 DMAIC Tool

Define → What is important?
Measure → How are we doing?
Analyze → What is wrong?

TABLE 4.17
Fundamental Objectives of Six Sigma DMAIC Tool

DMAIC	Phase	Fundamental Objective
1	**Define**—What is important?	Define the project goals and customer deliverables (internal and external)
2	**Measure**—How are we doing?	Measure the process to determine current performance
3	**Analyze**—What is wrong?	Analyze and determine the root cause(es) of the defects
4	**Improve**—What needs to be done?	Improve the process by permanently removing the defects
5	**Control**—How do we guarantee performance?	Control the improved process's performance to ensure sustainable results

Source: Abdul Razzak Rumane. (2013). *Quality Tools for Managing Construction Projects.* Reprinted with permission from Taylor & Francis Group

Improve → What needs to be done?
Control → How do we guarantee performance?

DMAIC tool refers to a data-driven quality strategy and is used primarily for improvement of an existing product, service, or process.

Table 4.17 lists fundamental objectives of DMAIC.

4.2.9.3.3.1 The DMAIC Process The majority of the time, Black and Green Belts approach their projects with the DMAIC analytic tool set, driving process performance to never-before-seen levels.

DMAIC has the following fundamental objective:

1. Define: Define the project and customer deliverables.
2. Measure: Measure the process performance and determine current performance.
3. Analyze: Collect, analyze, and determine the root cause(s) of variation and process performance.
4. Improve: Improve the process by diminishing defects with alternative remedial.
5. Control: Control improved process performance.

The DMAIC process contains five distinct steps that provide a disciplined approach to improving existing processes and products through the effective integration of project management, problem-solving, and statistical tools. Each step has fundamental objectives and a set of key deliverables, so the team member will always know what is expected of him/her and his/her team.

DMAIC stands for the following:

- Define opportunities.
- Measure performance.

- Analyze opportunity.
- Improve performance.
- Control performance.

Define opportunities (What is important?)

The objective of this phase is as follows:

To identify and/or validate the improvement opportunities that will achieve the organization's goals and provide the largest payoff, develop the business process, define critical customer requirements, and prepare to function as an effective project team.

Key deliverables in this phase include the following:

- Team charter
- Action plan
- Process map
- Quick win opportunities
- Critical customer requirements
- Prepared team

Measure performance (How are we doing?)

The objectives of this phase are as follows:

- To identify critical measures that are necessary to evaluate the success or failure, meet critical customer requirements, and begin developing a methodology to effectively collect data to measure process performance.
- To understand the elements of the Six Sigma calculation and establish baseline sigma for the processes the team is analyzing.

Key deliverables in this phase include the following:

- Input, process, and output indicators
- Operational definitions
- Data collection format and plans
- Baseline performance
- Productive team atmosphere

Analyze opportunity (What is wrong?)

The objectives of this phase are as follows:

- To stratify and analyze the opportunity to identify a specific problem and define an easily understood problem statement.
- To identify and validate the root causes and thus the problem the team is focused on.
- To determine true sources of variation and potential failure modes that lead to customer dissatisfaction.

Key deliverables in this phase include the following:

- Data analysis
- Validated root causes
- Sources of variation
- FMEA
- Problem statement
- Potential solutions

Improve performance (What needs to be done?)
The objectives of this phase are as follows:

- To identify, evaluate, and select the right improvement solutions.
- To develop a change management approach to assist the organization in adapting to the changes introduced through solution implementation.

Key deliverables in this phase include the following:

- Solutions
- Process maps and documentation
- Pilot results
- Implementation milestones
- Improvement impacts and benefits
- Storyboard
- Change plans

Control performance (How do we guarantee performance?)
The objectives of this phase are as follows:

- To understand the importance of planning and executing against the plan and determine the approach to be taken to ensure achievement of the targeted results.
- To understand how to disseminate lessons learned, identify replication and standardization opportunities/processes, and develop related plans.

Key deliverables in this phase include the following:

- Process control systems
- Standards and procedures
- Training
- Team evaluation
- Change implementation plans
- Potential problem analysis
- Solution results
- Success stories
- Trained associates

- Replication opportunities
- Standardization opportunities

Six Sigma methodology is not so commonly used in construction projects; however, DMAIC tool can be applied at various stages in construction projects. These are as follows:

1. Detailed design stage—To enhance coordination method in order to reduce repetitive work
2. Construction stage—Preparation of builders workshop drawings and composite drawings, as it needs lot of coordination among different trades
3. Construction stage—Preparation of contractor's construction schedule
4. Execution of works

4.2.9.3.4 DMADDD Tool

Define → Where must we be leaner?
Measure → What is our baseline?
Analyze → Where can we free capacity and Improve yields?
Design → How should we implement?
Digitize → How do we execute?
Drawdown → How do we eliminate parallel paths?

DMADDD tool is primarily used to drive the cost out of a process by incorporating digitization improvements. These improvements can drive efficiency by identifying non—value-added tasks and use simple web-enabled tools to automate certain tasks and improve efficiency. In doing so, employees can be freed up to work on more value-added tasks.

Table 4.18 lists fundamental objectives of DMADDD.

Impact of Six Sigma Strategy

The Six Sigma strategy affects five fundamental areas of business. These are as follows:

1. Process improvement
2. Product and service improvement
3. Customer satisfaction
4. Design methodology
5. Supplier improvement

(*Source*: Abdul Razzak Rumane (2010). *Quality Management in Construction Projects*. Reprinted with permission from Taylor & Francis Group).

4.2.10 TRIZ

TRIZ is a short form for *teirija rezhenijia izobretalenksh zadach* (theory of inventive problem solving), developed by the Russian Scientist Genrish Altshuller. TRIZ

TABLE 4.18
Fundamental Objectives of Six Sigma DMADDD Tool

DMADDD	Phase	Fundamental Objective
1	**Define**—Where must we be learner?	Identify potential improvements
2	**Measure**—What's our baseline?	Analog touch points
3	**Analyze**—Where can we free capacity and improve yields?	Task elimination and consolidated ops. Value-added/non–value-added tasks Free capacity and yield
4	**Design**—How should we implement?	Future state vision Define specific projects Define drawdown timing Define commercialization plans
5	**Digitize**—How do we execute?	Execute project
6	**Draw Down**—How do we eliminate parallel paths?	Commercialize new process Eliminate parallel path

Source: Abdul Razzak Rumane. (2013). *Quality Tools for Managing Construction Projects.* Reprinted with permission from Taylor & Francis Group.

provides systematic methods and tools for analysis and innovative problem-solving to support decision-making process.

Continuous and effective quality improvement is critical for organization's growth, sustainability, and competitiveness. The cost of quality is associated with both chronic and sporadic problems. Engineers are required to identify, analyze the causes, and solve these problems by applying various quality improvement tools. Any of these quality tools taken individually does not allow a quality practitioner to carry out whole problem-solving cycle. These tools are useful for solving a particular phase of problem and need combination of various tools and methods to find problem solution. TRIZ is an approach that starts at a point where fresh thinking is needed to develop a new or redesign a process. It focuses on method for developing ideas to improve a process, get something done, design a new approach, or redesign an existing approach. TRIZ offers a more systematic although still universal approach to problem-solving. TRIZ has advantages over other problem-solving approaches in terms of time efficiency and has low cost quality improvement solution. The pillar of TRIZ is the realization that contradictions can be methodically resolved through the application of innovative solutions. Altshuller defined an inventive problem as one containing a contradiction. He defined contradiction as a situation where an attempt to improve one feature of system detracts from another feature.

4.2.10.1 TRIZ Methodology

Traditional processes for increasing creativity have a major flaw in that their usefulness decreases as the complexity of the problem increases. At times, trial-and-error method is used in every process, and the number of trials increases with the complexity of the inventive problem. In 1946, Altshuller determined to improve the

TABLE 4.19
Level of Inventives

Level	Degree of Inventiveness	% of Solutions	Source of Solution
1	Obvious solution	32	Personal skill
2	Minor improvement	45	Knowledge within existing systems
3	Major improvement	18	Knowledge within the industry
4	New concept	4	Knowledge outside industry are found in science, not in technology
5	Discovery	1	Outside the confines of scientific knowledge

Source: Abdul Razzak Rumane. (2010). *Quality Management in Construction Projects.* Reprinted with permission from Taylor & Francis Group.

inventive process by developing the "Science" of creativity, which led to creation of TRIZ. TRIZ was developed by Altshuller as a result of analysis of many thousands of patents. He reviewed over 200,000 patents looking for problems and how they are solved. He selected 40,000 as a representative of inventive solutions, the rest were direct improvements easily recognized within the system. Altshuller recognized a pattern where some fundamental problems were solved with solutions that were repeatedly used from one patent to another, although the patent subject, applications, and timings varied significantly. Altshuller categorized these patterns into five level of inventiveness. Table 4.19 summarizes Altshuller's findings.

He noted that with each succeeding level, the source of the solution required broader knowledge and more solutions to consider before an ideal solution could be found.

TRIZ is a creative thinking process, which provides a highly structured approach for generating innovative ideas and solutions for problem-solving. It provides tools and methods for use in problem formulation, system analysis, failure analysis, and pattern of system evolution. TRIZ, in contrast to techniques such as brainstorming, aims to create algorithmic approach to invention of new systems and refinement of old systems. Using TRIZ requires some training and good deal of practice.

TRIZ Body of Knowledge contains 40 creative principles drawn from analysis of how complex problems have been solved.

- The laws of systems solution
- The algorithm of inventive problem-solving
- Substance-field analysis
- 76 standard solutions

4.2.10.2 Application of TRIZ

Engineers can apply TRIZ for solving the following problems in construction projects:

- Nonavailability of specified material
- Regulatory changes to use certain type of material

- Failure of dewatering system
- Casting of lower grade of concrete to that of specified higher grade
- Collapse of trench during excavation
- Collapse of formwork
- Collapse of roof slab while casting in progress
- Chiller failure during peak hours in the summer
- Modifying method statement
- Quality auditor can use TRIZ to develop corrective actions to the audit findings during auditing.

4.2.10.3 TRIZ Process

Altshuller has recommended four steps to invent new solution to a problem. These are as follows:

Step 1—Identify the problem.
Step 2—Formulate the problem.
Step 3—Search for precisely well-solved problem.
Step 4—Generate multiple ideas and adapt a solution.

The aforementioned methods are primarily used for low-level problems. To solve more difficult problems, more precise tools are used. These are as follows:

1. ARIZ (algorithm for inventive problem-solving)
2. Separation principles
3. Substance-field analysis
4. Anticipator failure determination
5. Direct product evaluation

QFD matrix is also used to identify new functions and performance levels to achieve truly exiting level of quality by eliminating technical bottlenecks at the conceptual stage. QFD may be used to feed data into TRIZ, especially using the "rooftop" to help develop contradictions.

The different schools for TRIZ and individual practitioners have continued to improve and add to the methodology.

4.3 TOOLS FOR CONSTRUCTION PROJECT DEVELOPMENT

Project development is a process spanning from the commencement of project initiation and end with closeout and finalizing project records after project construction and handing over of the project. The project development process is initiated in response to an identified need. It covers a range of time-framed activities extending from identification of a project need to a finished set of contract documents and to construction.

Oil and gas construction project development has four major elements/stages. These are as follows:

1. Study
2. Design
   2.1 Concept design
   2.2 Front-end engineering design (FEED)
3. Bidding and tendering
4. EPC construction
   4.1 Detailed engineering
   4.2 Procurement
   4.3 Construction
5. Testing and commissioning

As the project develops, more information and specifications are developed.
Table 4.20 illustrates oil and gas construction project development stages.

TABLE 4.20
Oil and Gas Construction Project Development Stages

Serial Number	Stages	Elements	Description
1	Study	Problem statement/need identification	Project needs, goals, and objectives.
		Need assessment	Identification of needs. Prioritization of needs. Leveling of needs. Deciding what needs to be addressed.
		Need analysis	Perform project need analysis/study to outline the scope of issues to be considered in the planning phase.
		Need statement	Develop project need statement
		Establish project goals and objectives	Scope, time, cost, quality.
		Feasibility study	Technical studies, economics assessment, financial assessment, scheduling, market demand, risk, and environmental and social assessment.
		Identify and evaluate alternatives/options	Identify alternatives based on a predetermined set of performance measures.
		Identify regulatory approval requirements	Identify regulatory approval authorities.
		Identify project stakeholders	Identify project stakeholders.
		Estimate project schedule	Estimate the duration for completion of project/facility.

(*Continued*)

TABLE 4.20 (*Continued*)
Oil and Gas Construction Project Development Stages

Serial Number	Stages	Elements	Description
		Estimate project cost	Preliminary budget estimates of total project cost (life cycle cost) on the basis of any known research and development requirements. This will help arrange the finances. (Funding agency).
		Quality management/ HSE requirements	Identify quality codes and standards, and HSE requirements.
		Estimate project resources	Estimate resources.
		Identify project risk	Risk, constraints.
		Select preferred alternative/option and develop design basis	Assess technological and economical advantages to preferred alternative.
		Identify project delivery and contracting system	Establish project delivery and contracting/pricing system.
		Project launch	Project charter.
2	Design	Identify concept design requirements	Establish design process/requirements to be followed in the project development.
		Identify process technology	Identify process technology to be followed in the project development.
		Develop concept design	Report, drawings, models, presentation.
		Identify project stakeholders	Identify project stakeholders.
		Regulatory approvals	Obtain regulatory authority's approvals.
		Project scope	Establish project scope.
		Project schedule	Prepare project plan.
		Project cost	Estimate preliminary cost of the project.
		Finalize concept design	Finalize concept design
		Front-end engineering design (FEED)	FEED
		Establish process technology	Establish process technology to be followed for the project.
		Safety management system	Establish safety management for shutdown in case of malfunctioning.
		Project schedule/project plan	Determine project schedule.
		Project cost	Determine cost of the project.
		Project specifications/ construction documents	Project specifications and construction contract documents.

(*Continued*)

TABLE 4.20 (*Continued*)
Oil and Gas Construction Project Development Stages

Serial Number	Stages	Elements	Description
3	Bidding and tendering	Bid documents	Tendering documents.
		Selection of contractor	Advertisement for bidders.
			Prequalification of contractors
			Issuing tender documents/request for proposal.
			Receipt of bid documents.
			Tabulation of proposals.
			Analysis of proposal.
			Selection of contractor.
			Most competitive bidder (low bid, qualification based).
		Award contract	Signing of contract.
			Bonds and insurance.
			Notice to proceed.
4	Construction	**Detailed engineering**	
		Project schedule	Develop contractor's construction schedule.
		Project cost	Prepare planned S-curve.
		Detailed design	Detailed design.
		Procurement	
		Purchase of material, equipment, machinery	Identify approved supplier, procure project material, system, and equipment.
		Construction	Contractor to execute contracted works.
		Monitoring and control	Monitor and control scope, schedule, cost, quality, risk, and procurement of the project.
5	Testing and commissioning	Commissioning and handover	Develop testing and commission plan.
			Testing, commissioning, and handover.
		Project closeout	Close the project.

HSE, health, safety, and environmental.

4.3.1 Tools for Study Stage

Table 4.21 illustrates major elements of study stage and related tasks performed during this stage.

Figure 4.40 illustrates logic flowchart for activities in study stage.

Table 4.22 illustrates points to be considered for need assessment.

Table 4.23 illustrates major points to be considered for need analysis.

TABLE 4.21
Major Elements of Study Stage

Serial Number	Elements	Description	Related Task
1	Problem statement/need identification	Project needs, goals, and objectives	Strategic objectives, policies, and priorities (Please refer Figure 4.40.)
2	Need assessment	Identification of needs Prioritization of needs Leveling of needs Deciding what needs to be addressed	Ensure that owner's business case has been properly considered (Please refer Table 4.22.)
3	Need analysis	Perform project need analysis/study to outline the scope of issues to be considered in the planning phase	Perform need analysis (Please refer Table 4.23.)
4	Need statement	Develop project need statement	Develop need statement (Please refer Table 4.24.)
5	Establish project goals and objectives	Scope, time, cost, and quality	Project initiation documents developed on SMART concept
6	Feasibility study	Technical studies, economics assessment, financial assessment, scheduling, market demand, risk, and environmental and social assessment	Perform feasibility study considering points listed under Table 4.25
7	Identify and evaluate alternatives/options	Identify alternatives based on a predetermined set of performance measures	Select conceptual alternatives/preferred options
8	Authorities requirements/clearance	Identify issues, sustainability, impacts, and potential approvals (environmental, authorities, permits) required for subsequent design and authority approval processes	Establish requirements for statutory approvals and other regulatory authorities
9	Project stakeholders	Identify project stakeholders	Define responsibility matrix of stakeholders
10	Preliminary schedule	Estimate the duration for completion of project/facility	Establish project schedule

(Continued)

TABLE 4.21 (Continued)
Major Elements of Study Stage

Serial Number	Elements	Description	Related Task
11	Preliminary project cost	Preliminary budget estimates of total project cost (life cycle cost) on the basis of any known research and development requirements. This will help arrange the finances (funding agency)	Determine project budget
12	Quality management/HSE requirements	Identify quality codes and standards, and health, safety, and environment requirement	Define codes and standards to be followed
13	Preliminary resources	Estimate resources	Confirm availability of resources, manpower, material, and equipment
14	Project risk	Identify project risk and constraints	Establish risk response and mitigation plan
15	Select preferred alternative/option	Assess technological and economical feasibility and compare with the preferred option/alternative to prepare business case	Discuss relative merits of various alternative schemes and evaluate the performance measures to meet owner's needs/requirements. Consider social, economical, and environmental impacts, safety, reliability, and functional capability
16	Identify project delivery system	Establish project delivery system	Normally, oil and gas projects have design–build, turnkey type of project delivery system
17	Identify type of contracting system	Select contract pricing system such as firm fixed price or lump sum, unit price, cost reimbursement (cost plus), reimbursement, target price, time, and material	Select contracting/pricing most appropriate for the benefit of owner
18	Identify project team	Select designer for concept design and front-end engineering design.	Select project team for design–build/turnkey type of project delivery system
19	Project launch	Project charter	Prepare terms of reference

HSE, health, safety, and environmental.

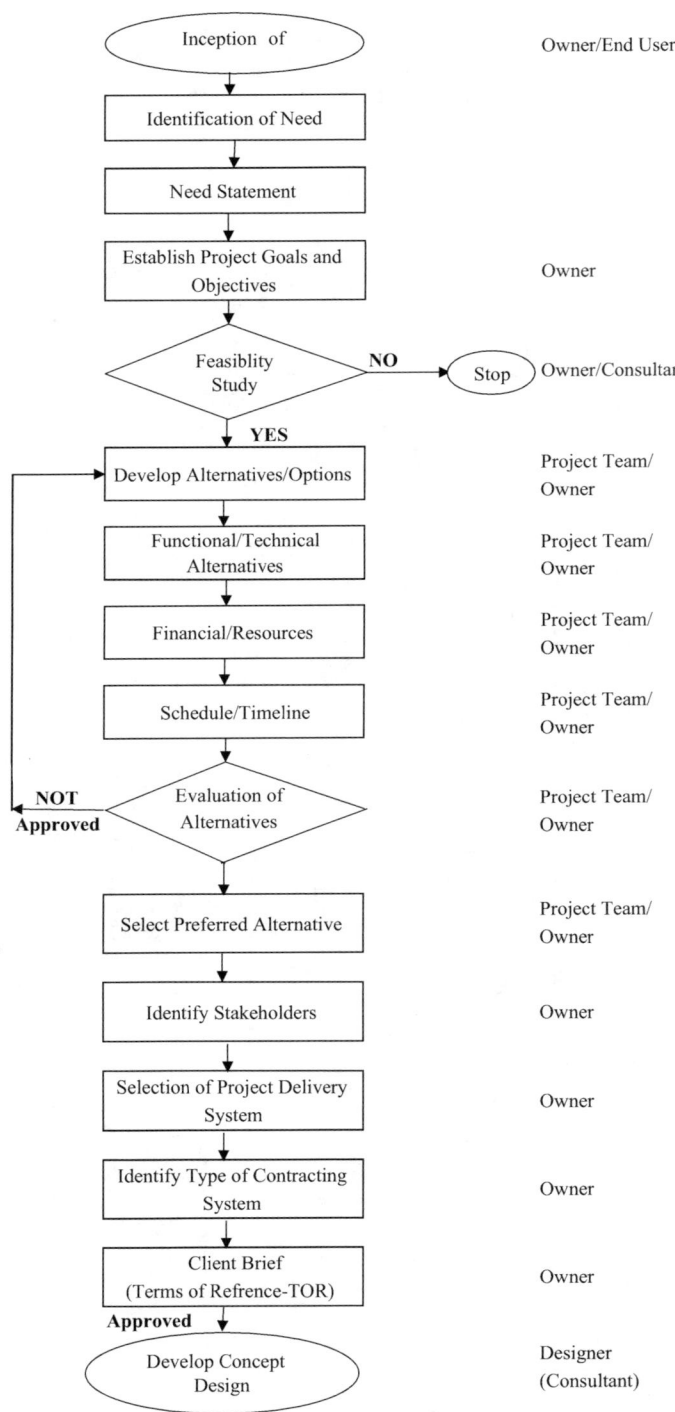

FIGURE 4.40 Logic flowchart of activities in the study stage.

Quality Tools for Oil and Gas Industry

TABLE 4.22
Need Assessment for a Construction Project

Serial Number	Points To Be Considered
1	Ensure the need is properly defined.
2	Confirm that the need outcome will benefit the owner/end user.
3	Gather and analyze owner/end-user requirements.
4	Set priorities and establish criteria for solutions to meet the targeted demand or requirements.
5	Identify and measure areas for improvement to achieve the objectives of the project.

TABLE 4.23
Major Considerations for Need Analysis of a Construction Project

Serial Number	Points To Be Considered
1	Is the project in line with organization's strategy/strategic plan and mandated by management in support of a specific objective?
2	Is the project a part of mission statement of the organization?
3	Is the project a part of vision statement of the organization?
4	Is the need mandated by regulatory body?
5	Is the need for meeting government regulations?
6	Is the need to fulfill the deficiency/gap of such type of project(s) in the market?
7	Is the need created to meet market demand?
8	Is the need to meet the research and development requirements?
9	Is the need for technical advances?
10	Is the need generated to construct a facility/project which is innovative in nature?
11	Does the need aim to improve the existing facility?
12	Is the need to develop infrastructure?
13	Will the need serve the community and fulfill social responsibilities?
14	Is the need created to resolve specific problem?
15	Will the need have effect on environment?
16	Is the need a part of mandatory investment?
17	Does the need have financial constraints?
18	Does the need have any time frame to implement?
19	Does the need have major risk?
20	Does the need comply with environmental protection agency requirements?
21	Does the need comply with government's health and safety regulations?
22	Is the need within capability of the owner/client, either alone or in cooperation with other organizations?
23	Will the need be beneficial?
24	Can the need be managed and implemented?
25	Is the need realistic and genuine?
26	Is the need measurable?

Source: Abdul Razzak Rumane. (2013). *Quality Tools for Managing Construction Projects.* Reprinted with permission from Taylor & Francis Group

Table 4.24 illustrates need statement.

Table 4.25 illustrates points to be considered for feasibility study of a construction project.

Table 4.26 illustrates contents of project charter—terms of reference.

TABLE 4.24
Need Statement

Serial Number	Points To Be Considered
1	Project purpose and need a. Project description
2	What is the purpose of project? a. Project justification
3	Why the project is needed now?
4	How did the need of the project determined? a. Supporting data
5	Is it important to have the needed project?
6	Whether such facility/project is required
7	What are the factors contributing to the need?
8	What is the impact of the need?
9	Will the need improve the existing situation and be beneficial?
10	What are the hurdles?
11	What is the timeline for the project
12	What are funding sources for the project?
13	What are the benefits of the projects?
14	What are the environmental impacts?

Source: Abdul Razzak Rumane. (2013). *Quality Tools for Managing Construction Projects.* Reprinted with permission from Taylor & Francis Group

TABLE 4.25
Major Considerations for Feasibility Study of a Construction Project

Serial Number	Points To Be Considered
1	Technical suitability of facility for intended use by the owner/end user
2	Economical feasibility to ascertain value of benefit that results from the project exceeds the cost that results from the project
3	Financial payback period
4	Market demand
5	Environmental impact
6	Social and cultural assessment
7	Legal and regulatory impacts
8	Political aspects
9	Resources availability
10	Scheduling of the project
11	Operational
12	Risk analysis

TABLE 4.26
Typical Contents of Project Charter (Terms of Reference)

Serial Number	Topics
1	Project objectives
	1.1 Background
	1.2 Project information
	1.3 General requirements
	1.4 Special considerations
2	Project requirements
	2.1 Scope of work
	2.2 Functional requirements (design life)
	2.3 Work program
	2.3.1 Feasibility study phase
	2.3.2 Design phase (concept design, FEED)
	2.3.3 Tender stage
	2.3.4 Construction phase
	2.3.5 Testing and commissioning
	2.4 Reports and presentations
	2.5 Schedule of requirements
	2.6 Time program
	2.7 Cost estimate
	2.8 Energy conservation considerations
	2.9 Drawings
	2.10 Process equipment/machinery
	2.11 Storage tank
	2.12 Mechanical/piping
	2.13 Instrumentation and control, and automation
	2.14 Supervisory control system
	2.15 Fire and gas detection system
	2.16 Flaring and venting system
	2.17 Buildings
	2.18 Conveying system
	2.19 HVAC system
	2.20 Electrical
	2.21 Engineering systems, pipeline management, and supervisory control
	2.22 Information management system, data security
	2.23 Telecom and internal communication
	2.24 Security and safety management system
	2.25 Fire suppression system
	2.26 Loss prevention system
	2.27 Leak detection system
	2.28 Cathodic protection system, corrosion monitoring system
	2.29 External works
	2.30 Landscape
	2.31 Parking
	2.32 Parking control system

(*Continued*)

TABLE 4.26 (*Continued*)
Typical Contents of Project Charter (Terms of Reference)

Serial Number	Topics
3	Opportunities and constraints
	3.1 Site location
	3.2 Site conditions
	3.3 Land size and access
	3.2 Climate
	3.3 Time
	3.4 Budget
4	Performance target
	4.1 Financial performance
	4.1.1 Performance bond
	4.1.2 Insurance
	4.1.3 Delay penalty
	4.2 Energy performance target
	4.2.1 Energy conservation
	4.3 Work program schedule
	4.4 Processing technology and method
5	Environmental considerations
6	Design approach
	6.1 Design parameters
	6.1.1 Processing system/equipment
	6.1.2 Mechanical and piping
	6.1.3 Instrumentation
	6.1.4 Automation and control
	6.1.5 Storage of fuel
	6.1.6 Supervisory control system
	6.1.7 Fire and gas detection systems
	6.1.8 Architectural design
	6.1.9 Structural design
	6.1.10 Civil engineering
	6.1.11 Conveying system
	6.1.12 Mechanical design
	6.1.13 HVAC design
	6.1.14 Electrical design
	6.1.15 Information and communication technology
	6.1.16 Security and safety system
	6.1.17 Fire suppression system
	6.1.18 Loss prevention system
	6.1.19 Leak detection system
	6.1.20 Landscape
	6.1.21 External works
	6.1.22 Parking
	6.2 Sustainable design
	6.3 Engineering systems
	6.4 Value engineering study
	6.5 Design review by client

(*Continued*)

TABLE 4.26 (Continued)
Typical Contents of Project Charter (Terms of Reference)

Serial Number	Topics
	6.6 Selection of products/systems
	6.7 Codes and standards
7	Specifications and contract documents
8	Project control guidelines
9	Submittals
	9.1 Reports
	9.2 Drawings
	9.3 Specifications
	9.4 Material selection criteria
	9.5 Models
	9.6 Sample boards
	9.7 Mock up
10	Presentation
11	Project team members
	8.1 Number of project personnel
	8.2 Staff qualification
	8.3 Selection of specialists
12	Visits

FEED, front-end engineering design; HVAC, heating, ventilation, and air conditioning.

TABLE 4.27
5W2H Analysis for Project Need

Serial Number	Why	Related Analyzing Question
1	Why	New project
2	What	What advantage it will have over other similar projects
3	Who	Who will be the customer for this project
4	Where	Where can we find the market for the project
5	When	When the project will be ready
6	How many	How many such projects are in the market
7	How much	How much market share we will have by this project

Some of the tools discussed earlier can be applied during study stage. Table 4.27 is an example how 5W2H (under Innovation and Creative Tool) can be used to analyze the project need.

4.3.2 Tools for Design Stage

Table 4.28 illustrates major elements of design stage and related tasks performed during this stage.

TABLE 4.28
Major Elements of Design Stage

Serial Number	Elements	Description	Related Task
1	Develop concept design	Report, drawings, models, and presentation	Data collection (Please refer Table 4.29.) Design development points (Please refer Table 4.30.)
2	Regulatory approvals	Obtain regulatory approvals	Submission of drawings and related documents to authorities and obtain their approvals
3	Project planning	Prepare project plan	Develop project plan (Please refer Figure 4.41.)
4	Front-end engineering design	Front-end engineering design and value engineering	Develop front-end engineering design considering accepted process technology. Preliminary (outline) specifications, authority approvals, value engineering (Please refer Table 4.31.)
5	Construction documents and tender documents	Construction contract documents and tender documents	Preparation of particular specifications and contract drawings
6	EPC contract	Detailed engineering	Develop detailed engineering based on contract documents (Please refer Figure 4.42.)

EPC, engineering, procurement, construction.

Table 4.29 illustrates points to be considered for data collection to facilitate development of concept design.

Table 4.30 illustrates points to be considered for development of concept design.

Figure 4.41 illustrates sample preliminary project plan.

Table 4.31 illustrates points to be considered for development of FEED for oil and gas project.

Figure 4.42 illustrates major activities in the detailed engineering review steps.

Some of the tools discussed earlier can be applied during design stage. These are as follows:

- Innovative and creative tool
  - Six Sigma DMADV (Please refer Table 4.16)
- Process analysis tool
  - Cost of quality during design (Figure 4.43)

TABLE 4.29
Major Items for Data Collection during Concept Design Phase

Serial Number	Items To Be Considered
1	Certificate of title
	a. Site legalization
	b. Historical records
2	Topographical survey
	a. Location plan
	b. Site visits
	c. Site coordinates
	d. photographs
3	Geotechnical investigations
4	Field and laboratory test of soil and soil profile
5	Existing structures in/under the project site
6	Existing utilities/services passing through the project site
7	Existing roads and structure surrounding the project site
8	Shoring and underpinning requirements with respect to adjacent area/structure
9	Requirements to protect neighboring area/facility
10	Environmental studies
11	Energy saving requirements
12	Wind load, seismic load, dead load, and live load
13	Site access/traffic studies
14	Ease for delivery of product
15	Applicable codes, standards, and regulatory requirements
16	Usage and space program
17	Design protocol
18	Scope of work/client requirements

TABLE 4.30
Development of Concept Design

Serial Number	Points To Be Considered
1	Project goals
2	Usage
3	Incorporate requirements from collected data
4	Technical and functional capability
5	Process technology
6	Codes and standards
7	Aesthetics
8	Constructability
9	Sustainability (environmental, social, and economical)
10	Occupational health and safety
11	Reliability
12	Environmental compatibility
13	Sustainability
14	Fire and gas detection system
15	Leak detection system
16	Fire suppression/protection measures
17	Supportability during maintenance/maintainability
18	Cost-effective over the entire life cycle (economy)
19	LEED (equivalent) compliance
20	Reports, drawings, models

LEED, leadership in energy and environmental design.

- Process improvement tool
  - PDCA cycle (Figure 4.44)
- Lean tool
  - Mistake proofing (Table 4.14)

4.3.3 Tools for Bidding and Tendering Stage

Table 4.32 illustrates major activities during bidding and tendering stage and related tasks performed during this stage.

4.3.4 Tools for EPC Stage

Table 4.33 illustrates major elements during EPC stage and related tasks performed during this stage.

Table 4.34 illustrates major activities to be performed by EPC contractor during construction phase.

Quality Tools for Oil and Gas Industry   205

Activity ID	Activity Description	Duration	Start	Finish
Design & Construction Phases - Preliminary Project Schedule				
Design Phase I: Data Collection & Concept Design				
P1030	Kick Off Meeting	0	10-12-17	
P1040	Contract Signing	0	14-04-18	
P1050	Data Collection	94	24-04-18	30-07-18
P1060	Concept Design	250	01-05-18	05-01-19
P1070	Concept Design Submittal	0		05-01-19
P1080	Review & Approval of Concept Design	26	06-01-19	31-01-19
Design Phase II: Front-End Engineering design and Tender Documents				
P2010	Piping and Process Engineering	59	01-02-19	31-03-19
P2020	Electrical and Instrumentation	31	01-03-19	31-03-19
P2030	Infrastructure and External	17	15-03-19	31-03-19
P2040	Front-End Engineering Design submittal	0		31-03-19
P2050	Review & Approve Front-End Engineering Design and Construction Contract Documents	12	01-04-19	12-04-19
Design Phase III: Tendering, Awarding of Contract & Finalisation of Construction Contract Documents				
P3010	Finalization of Tender Documents & Tender Publishing	14	13-04-19	26-04-19
P3020	Receive Bids	21	27-04-19	17-05-19
P3030	Review of Bids and Awarding of Contract	7	18-05-19	24-05-19
P3040	Finalization of Construction Contract documents	7	25-05-19	31-05-19
Construction Phase				
P4100	Project Construction (EPC Contracting System)	730	01-06-19	31-05-21
P4111	1-Detailed Engineering	270	01-06-19	31-03-20
P421	2-Procurement	540	01-06-19	31-12-20
P431	3-Construction	730	01-06-19	31-05-21

FIGURE 4.41 Preliminary schedule for construction project.

TABLE 4.31
Development of Front-End Engineering Design for Oil and Gas Construction Project

Serial Number	Points To Be Considered
1	Concept design deliverables
2	Calculations to support the design
3	General plot plan
4	Unit plot plan
5	Process technology
6	System schematic for process flow
7	Piping general arrangement drawings
8	Pipeline management system
9	Instrumentation and control
10	Instrumentation, control, and automation
11	Electrical
12	Mechanical
13	Civil
14	Buildings
15	System schematic for MEP services
16	Tie-in drawings
17	Fire and gas detection system
18	Supervisory control system
19	Leak detection system
20	Information and communication technology, data security
21	Constructability
22	Selection of systems and products which support functional goals of the entire facility
23	Cause-and-effect diagrams
24	Requirements of all stakeholders
25	Authorities requirements
26	Project schedule
27	Project cost
28	Coordination with other members of the project team
29	Quality management requirements
30	Reliability
31	Energy conservation issues
32	Availability of resources
33	Occupational health and safety
34	Environmental issues
35	Safety management system (shutdown in case of malfunctioning)
36	Security system
37	Landscape
38	External works
39	Sustainability
40	Optimized life cycle cost (value engineering)
41	Material lists
42	Material datasheet
43	Reports and analysis

Quality Tools for Oil and Gas Industry

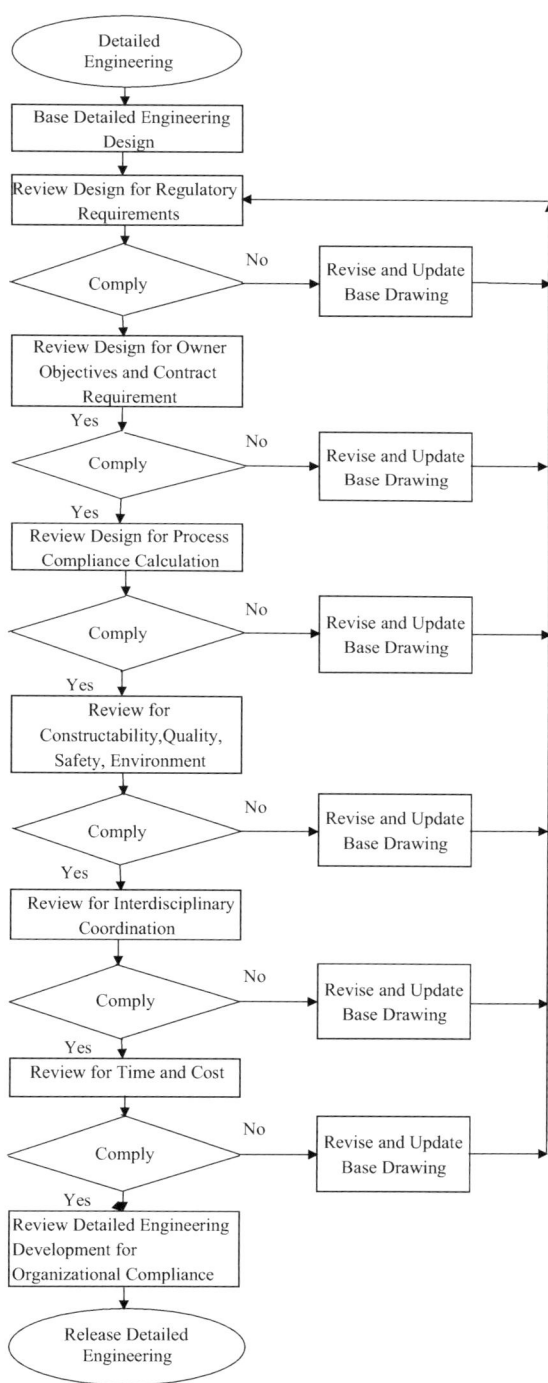

FIGURE 4.42 Detailed engineering review steps. (Abdul Razzak Rumane. (2013). *Quality Tools for Managing Construction Projects*. Reprinted with permission from Taylor & Francis Group.)

Internal Cost	External Cost
• Redesign/Redraw to meet fully coordinated design • Rewrite specifications/documents to meet requirements of all other trades	• Incorporate design review comments by Client/Project Manager • Incorporate specifications/documents review comments by Client/Project Manager • Incorporate comments by Regulatory Authority (ies) • Resolve RFI (Request for Information) during Construction
Appraisal Cost	**Prevention Cost**
• Review of Design/Engineering Drawings • Review of Specifications • Review of Contract Documents to ensure meeting Owner's Needs, Quality Standards, Constructability, and Functionality • Review for Regulatory requirements, Codes	• Conduct technical meetings for proper coordination • Follow quality system • Meeting Submission Schedule • Training of project team members • Update of software used for design

FIGURE 4.43 Cost of quality during design stage.

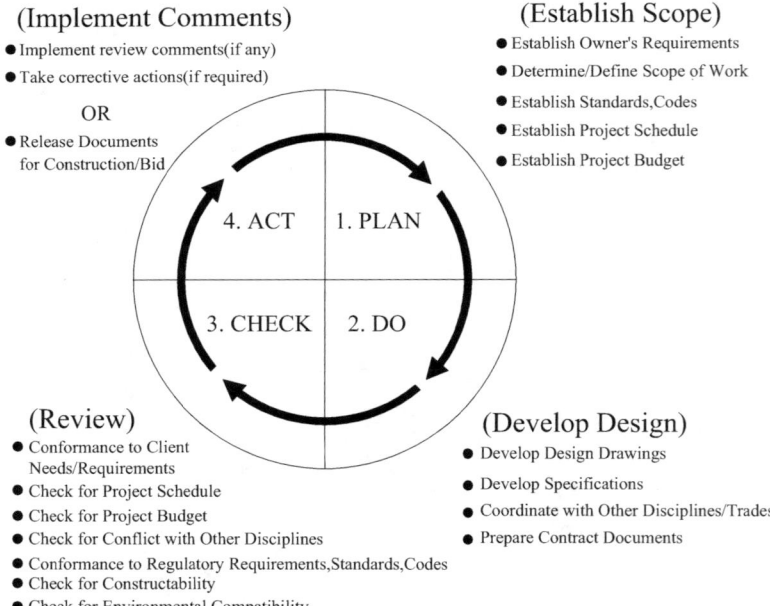

FIGURE 4.44 PDCA cycle for construction projects (design phases). PDCA, plan–do–check–act. (Abdul Razzak Rumane (2010). *Quality Management in Construction Projects*. Reprinted with permission from Taylor & Francis Group.)

TABLE 4.32
Major Elements of Bidding and Tendering Stage

Serial Number	Elements	Description	Related Task
1	Bidding	Bid documents	Prepare tender documents
2	Selection of contractor	Tender announcement	Advertisement for bidders
			Prequalification of contractors
3	Issue tender documents	Request for proposal	Issuing tender documents/request for proposal
4	Receipt of proposal	Receipt of bid documents	
5	Tender analysis	Tabulation of proposals	Analysis of quotation
			Low bid
			Qualification based
6	Selection of contractor		Most competitive bidder (low bid or qualification based as per the company's methodology)
7	Award of contract	Signing of contract	Performance bid
			Insurance
8	Notice to proceed		Contractor to commission

TABLE 4.33
Major Elements of EPC Contracting System Stage

Serial Number	Elements	Description	Related Task
1	Detailed engineering	Contractor to prepare detailed engineering to proceed with construction.	Develop detailed engineering based on contract requirements taking into consideration front-end engineering design
2	Procurement	Procurement of project material, equipment, machinery, and system	Procurement of approved material from approved sources. Inspection of material at site/off site
3	Construction	Contractor to execute contracted works	Mobilization, managing resources, work execution, QA/QC, works approval (Please refer Table 4.34.)
4	Monitoring and control	Monitor and control scope, schedule, cost, quality, risk, and procurement of the project	Scheduling, monitoring, and control (Please refer F**igure 4.45.)
5	Commissioning and handover	Testing, commissioning, and handover.	Project start-up procedures
6	Project closeout	Close the project	Project closeout documents and finalization of claims and payments

EPC, engineering, procurement, and construction.

TABLE 4.34
Major Activities by EPC Contractor during Construction Phase

Serial Number	Activities To Be Performed by Contractor
1	Mobilization
2	Development of detailed engineering
3	Selection of supplier from approved vendor list
4	Procurement of project material
5	Staff approval
6	Selection of subcontractor(s)
7	Selection of resources
8	Preparation of shop drawings
9	Execution of works
10	Project monitoring and control
11	Quality management
12	Health, safety, and environmental compliance
13	Auditing/installation of executed works
14	Approval of works

EPC, engineering, procurement, and construction.

Figure 4.45 illustrates project monitoring and controlling process cycle.

Some of the tools discussed earlier can be applied during construction stage. These are as follows:

- Management and planning tool
  - Networking arrow diagram (Figure 4.12)
- Process analysis tool
  - Root cause analysis (Figure 4.46)
- Process improvement tool
  - Six DMAIC (Table 4.17)
  - PDCA cycle (Figure 4.47)
- Classic quality tool
  - Flowchart for concrete casting (Figure 4.48)

Quality Tools for Oil and Gas Industry

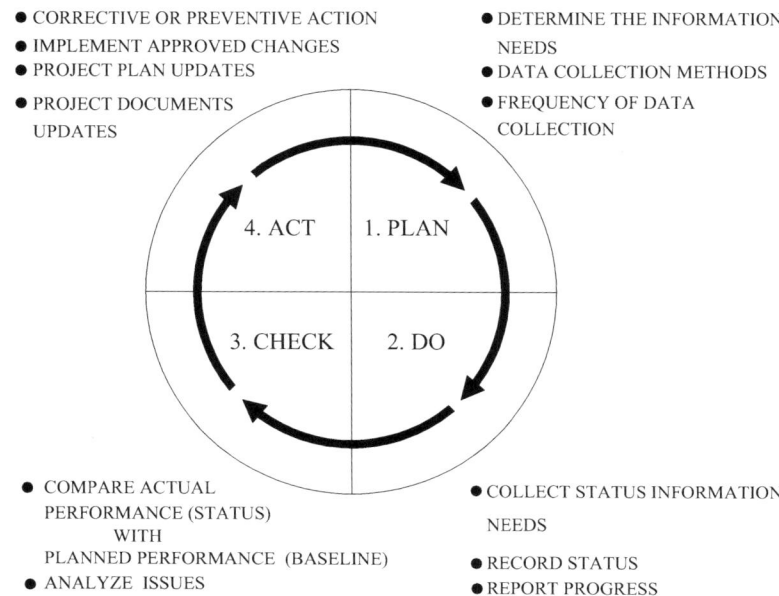

FIGURE 4.45 Project monitoring and controlling process cycle. (Abdul Razzak Rumane. (2013). *Quality Tools for Managing Construction Projects*. Reprinted with permission from Taylor & Francis Group.)

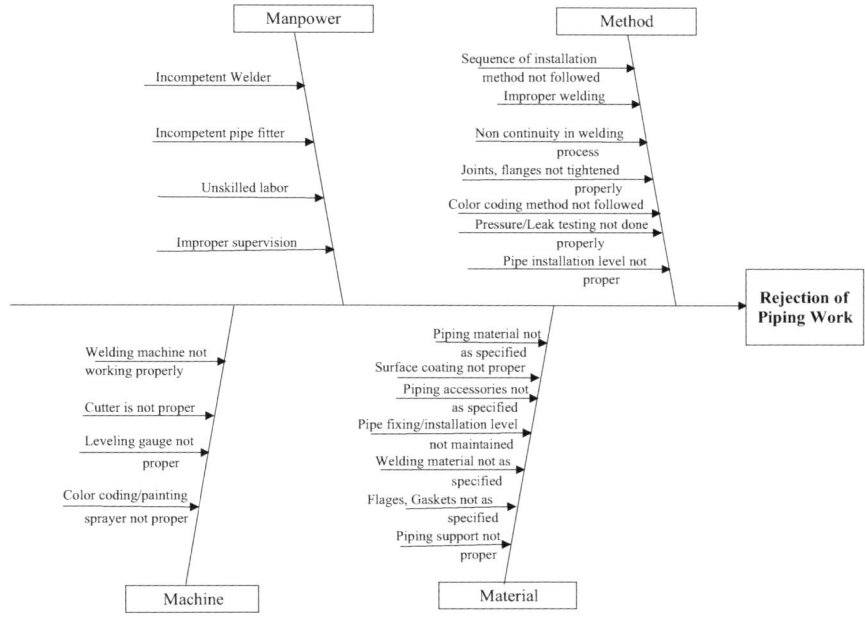

FIGURE 4.46 Root cause analysis for rejection of piping work.

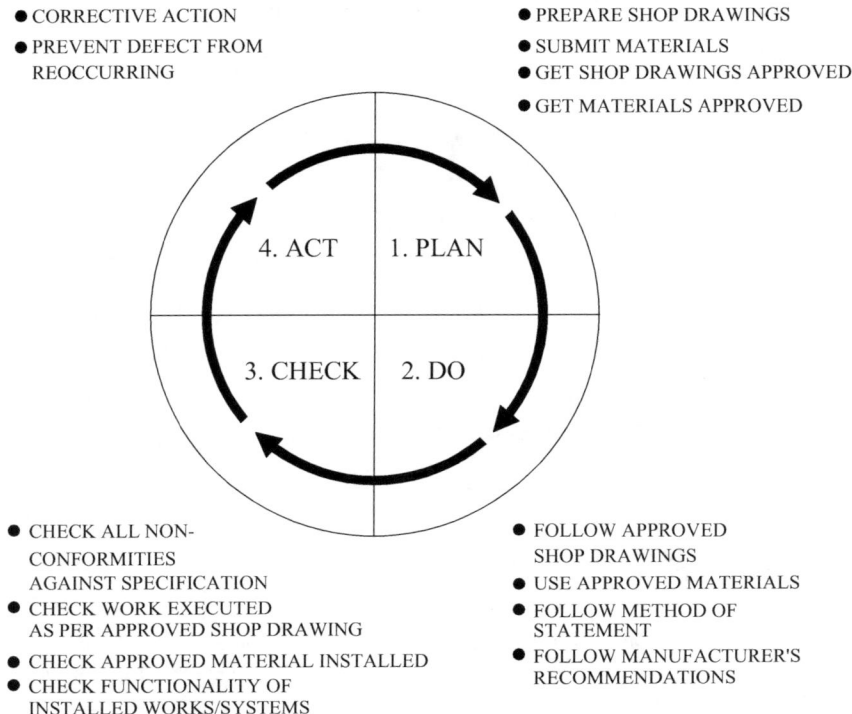

FIGURE 4.47 PDCA cycle (Deming Wheel) for execution of construction works. PDCA, plan–do–check–act. (Abdul Razzak Rumane (2013). *Quality Tools for Managing Construction Projects*. Reprinted with permission from Taylor & Francis Group.)

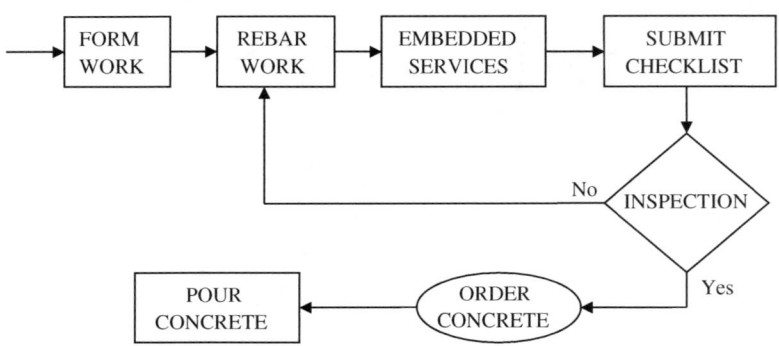

FIGURE 4.48 Flowchart for concrete casting. (Abdul Razzak Rumane. (2010). *Quality Management in Construction Projects*. Reprinted with permission from Taylor & Francis Group.)

5 HSE in Oil and Gas Projects

5.1 HEALTH, SAFETY, AND ENVIRONMENTAL MANAGEMENT

HSE (health, safety, and environmental) management system plays an important role in oil and gas industry. Oil and gas industry is one of the most hazardous industries that need implementation of HSE programs at all the levels of operations and in all sectors (upstream, midstream, and downstream). HSE management system plays an important role in any oil and gas projects right from project inception through substantial completion of the project and also during the operation.

Worker's health and safety and the safety of worksite/work place are important factors and are prime responsibilities of everyone in HSE management in oil and gas industry. Continuous improvement in oil and gas operations is essential in promoting personnel and public safety.

In the upstream sector stages, HSE programs are required to enhance the safety of its exploration and drilling operations.

In the midstream sector, transportation safety including pipelines is essential.

In the downstream sector, protecting the occupational health and safety (OH&S) of the workers, installing and maintaining fire protection system, selecting equipment that are suitable for operations in hazardous area, process safety, environmental protection, focusing on major hazard impacting the environment and surrounding areas, and public health are essential.

Safety is a cornerstone of a successful project. Safety is important to everyone and is a core value for the oil and gas industry. The occurrence of various incidents and hazards in oil and gas industry is quite frequent, thereby necessitating effective HSE management programs. Accidents cause needless loss of human and physical capital. Logically, these losses can only be detrimental to the project schedule and cost. Safety, long neglected, is increasingly being recognized for its value, as contractors and owners strive to avoid the increasing costs of injuries and fatalities.

The construction industry has been considered to be dangerous for a long time. The nature of work at site always presents some dangers and hazards. There are a relatively high number of injuries and accidents at construction sites. Safety represents an important aspect of construction projects. Every project/construction manager tries to ensure that a project is completed without major accidents on the site. Oil and gas construction projects have witnessed many catastrophes that have eventually necessitated for implementation HSE programs.

The trend toward improved safety seems gradual but recognizable. As in other fields, technology advances continually provide better materials, methods, and equipment. Increasingly prefabricated materials and modular construction reduce worker exposure to more hazardous construction activities found with older conventional methods and materials. In recent years, researchers have found that incorporating safety in the design phase has huge potential to impact exposures to hazardous situations. A concept called prevention through design (PtD) is well established in Europe and gaining a foothold in the United States. With PtD, safety is "designed in" a project. For instance, steel columns can receive shop-drilled holes for the easy attachment of safety lanyards during steel erection. This small and insignificant planning effort can save a life. Innovative concepts such as PtD require strong advocates to demonstrate their safety value and make the business case to the industry.

Though safety gains have been made, the situation remains critical. The U.S. annual construction fatality rate hovers around 10 per 100,000. In contrast, the U.S. fatality rate for texting driver incidents is roughly 0.5 per 100,000. It is well established that texting and driving is risky. Sadly, construction workers incur risks 20 times greater solely as a result of their chosen occupation. How can construction be made safer? The process is detailed, but the solution can be simply summarized as in Figure 5.1

Strive to incorporate safety into the project as early as possible. A safety conscious contractor can provide valuable input in the early stages of design. As seen in Figure 5.2, early processes offer the most potential to affect safety.

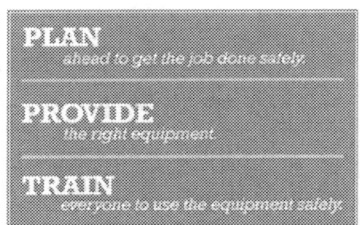

FIGURE 5.1 Safety framework.

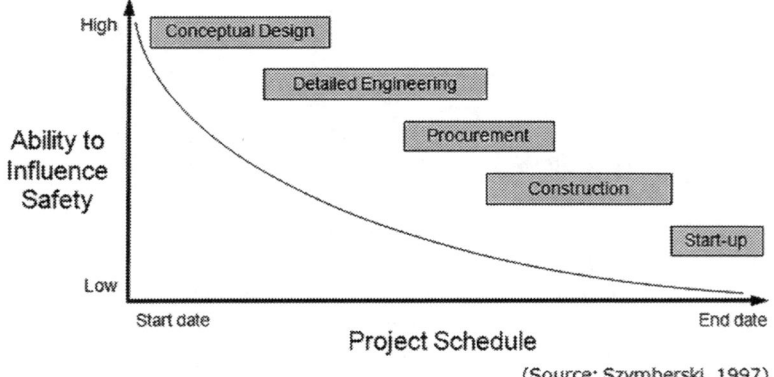

(Source: Szymberski, 1997)

FIGURE 5.2 The ability to influence safety.

HSE in Oil and Gas Projects 215

Prior to construction, the contractor needs to compile a coordinated, comprehensive site plan. Although the details of a comprehensive safety management plan are beyond the scope of this text, we can certainly define its critical elements using our *Plan, Provide, Train* framework.

Plan

Each site is unique and requires a specific plan. In creating a plan, here are some important items.

- Secure emergency contact information from
  - hospitals
  - fire and rescue squad
  - first-aid medical treatment facilities
- Locate buried and overhead utilities
  - notify utilities requiring relocation or deenergizing
  - coordinate the project schedule with relocations
- Provide safe access and egress from the site for
  - typical material and equipment deliveries
  - oversize loads
- Other miscellaneous
  - conduct regular safety inspections and site training
  - plan for traffic control
  - plan for emergency response scenarios

The aforementioned contains only critical items. A compete and comprehensive safety plan should incorporate sufficient detail to address all hazards including those that are site specific.

Provide

Workers have the right to expect that their employer will provide them with the proper equipment, tools, and training to complete all tasks safely. At a minimum, employers should

- provide well-maintained equipment (e.g. cranes, backhoes, etc.),
  - require periodic safety inspections,
  - keep maintenance logs,
  - furnish operating manuals and load charts,
  - utilize certified crane operators,
- require safety assurance checks,
  - inspect chains, slings, and hoists daily
  - replace hand tools with damaged cords or missing guards
- provide the basic safety gear: hardhats, safety glasses, gloves, and hearing protection
- provide fall protection for workers at heights of 6 feet or more, and
- designate a company safety officer.

Providing workers with well-maintained equipment not only enhances safety and improves productivity but also boosts morale. Poorly functioning tools and equipment create hazards, diminish productivity, and damage worker morale.

Train

Even the best plan and top-notch equipment require well-trained employees. In developed countries, construction training is commonly available through unions, trade associations, individual company programs, and even the Internet. In 2012, approximately 350,000 workers in the U.S. construction workforce received a standardized OSHA 10h training course in construction safety. Some localities require such training for all workers on public projects. This is an excellent example of informed owners realizing that increased safety pays long-term benefits by reducing costs. More extensive training courses appropriate for supervisors are also widely available.

Typical basic construction safety training might include the following:

- Four most common fatal hazards (i.e., falls, electrocutions, struck by, and caught in\between)
- Material handling
- Hazard communication
- Lockout/tagout procedures
- Vehicular safety
- Personal protective equipment

Employee turnover, job reassignments, and changing technology necessitate that training be ongoing and frequently updated.

5.1.1 Total Safety Management

Safety management system is a systematic and coordinated approach to managing safety, including the necessary organizational structures, procedures, and accountabilities to optimally manage safety.

Safety management in construction projects refers to a set of measures concerning safety and health during construction to be implemented and operated by the contractor on continuous basis. This system is conducted in conjunction with other management systems.

Safety management system is established as per OHSAS 18001 requirements. (ISO 45000 is a newly established standard). Construction projects are of complex and non-standardized nature. They are exposed to dynamic site conditions and involve coordination among various parties and participants.

Safety management process for construction projects is required to ensure that the construction project is executed with appropriate care to prevent accidents that could cause personal injury or property damage, and safety and health of neighbors to the construction site and general public who may be exposed to the project. Safety management also includes human resource programs, such as drug and alcohol treatment programs, which contribute to reducing accidents on construction jobsites.

Quality has been of great concern throughout the recorded history of human beings. Examples of specification and inspection can be found in Bible dating at least 500 BCE. The desire for product that do as well as, better than, the customer's needs and requirements is constant in human history matched only by the determination of builders and makers to meet the desire.

Quality has different meanings for different people. In technical usage, quality can have two meanings:

1. The characteristics of a product or service that bear on its ability to satisfy stated or implied needs.
2. A product or service free of deficiencies.

The quality in its historical forms has moved through distinct quality eras such as:

1. Quality inspection
2. Quality control
3. Quality assurance
4. Total quality

The definition of quality for construction project is different to that of manufacturing or services industries. The quality management of manufactured products is performed by the manufacturer's own team and has control over all the activities of the product life cycle, whereas construction projects have diversity of interaction and relationship between owners, architects/engineers, and contractors.

Quality for construction projects is different from that of manufacturing or services industries. Quality in construction projects is fulfillment of the owner's needs as per defined scope of work within specified schedule and budget.

The quality actually emerged as a dominant thinking only since World War II, becoming an integral part of overall business system focused on customer satisfaction and becoming known in recent times as "total quality management (TQM)", with its three constitutive elements:

- Total: organization-wide
- Quality: customer satisfaction
- Management: systems of managing

TQM is a philosophy that makes quality values the driving force behind leadership, design, planning, and improvement in activities. TQM is considered a fundamental requirement for any organization to compete, let alone lead, in its market. It is a way of planning, organizing, and understanding each activity of the process and removing all the unnecessary steps routinely followed in the organization. TQM focuses on customers, both internal (within the organization, the next party in the work process) and external (end users, stakeholders, regulatory agencies). TQM focuses on participative management and strong operational accountability at the individual contributor level. Total quality involves not just managers but also everyone in the organization in a complete transformation of the prevailing culture.

The concept of safety management system can be considered like TQM. There are three core aspects of safety management system. These are as follows:

1. Systematic—Safety management activities are in accordance with predetermined safety policies, objectives, and responsibilities to prevent accidents, which cause, or have potential to cause, personal injury, fatalities, or property damage.
2. Proactive—An approach that emphasizes safety hazard identification, investigation and analysis, and risk control and mitigation, before events that affect safety occur.
3. Explicit—All safety management activities are documented, communicated, and implemented.

Like TQM, safety is considered as responsibility of everyone involved at project site. All the main participants (client, designer/consultant, and contractor) should be made responsible for safety at project site to achieve zero accident like quality principle "zero defect". Thus, like TQM, safety can be termed as total safety management (TSM).

Table 5.1 illustrates application of TQM principles to that of TSM.

TABLE 5.1
TQM Approach to TSM

Serial Number	Basic Requirements for Total Quality Management	Total Safety Management Approach
1	Set organization vision	Safety policy
2	Identify customers and their needs	Everyone at site is vulnerable to safety risks and to be protected
3	Develop team approach by involving everyone in the organization	Safety is responsibility of all individuals involved
4	Define critical processes and measures	Hazard identification, risk assessment and control
5	Develop strategic plan	Documenting and detailing safety requirements and regulatory requirements
6	Review planning	Review site communication plan
7	Education and training	System education and training
8	Defect prevention not detection	Accident investigation to determine corrective and preventive actions
9	Improvement of process	System feedback and continuous improvement
10	Identify major products and process	Identify measures for emergency situations
11	Systematic approach to TQM	Systematic and coordinate approach to manage safety

TSM, total safety management; TQM, total quality management.

5.1.2 HEALTH, SAFETY, AND ENVIRONMENTAL STANDARDS

Standards are documents used to define acceptable conditions or behaviors and to provide a base line for assuring that conditions or behaviors meet the acceptable criteria. In most cases, standards define minimum criteria; world class quality is, by definition, beyond the standard level of performance. Standard can be written or unwritten voluntary or mandatory. Unwritten quality standards are generally not acceptable.

The following International Organization of Standardization (ISO) standards are mainly used for managing and monitoring HSE activities in oil and gas projects. These are as follows:

1. ISO 9000
2. ISO 14000
3. ISO 18000
4. ISO 45000

ISO 9000 is already discussed in Chapter 2. It is the most widely recognized quality management standard.

5.1.2.1 ISO 14000 Environmental Management System

ISO 14000 is a series of international standards that have been developed to incorporate environmental aspects into business operations and product standards. ISO 14001 is a specific standard in the series for a management system that incorporates a set of interrelated elements designed to minimize harmful effects on the environment due to the activities performed by an organization and to achieve continual improvement of its environmental performance. ISO 14000 outlines how to put an effective environment system in place. ISO 14001 incorporates quality management system philosophy, terminology, and requirement structure similar to that of ISO 9001 and provides system compatibility.

5.1.2.1.1 Benefits of ISO 14000

The following are the benefits of implementing an environmental management system:

- Pollution prevention and waste reduction opportunities
- Compliance with regulatory requirements on environmental considerations
- Reduction in consumption of energy
- Reduction in use of natural resources
- Minimization of environmental liability and risk
- Reduction in waste generation
- Commitment to social responsibility
- Utilization of recoverable resources
- Cost reduction
- Customer satisfaction

5.1.2.2 Occupational Health and Safety Assessment Series 18000

The Occupational Health and Safety Assessment Series (OHSAS) 18000 has been developed to help organizations control and minimize OH&S risks. OHSAS 18001 is a specific standard for OH&S management systems designed to eliminate or minimize the risk to employees and other related parties who may be exposed to OH&S risks associated with business activities. OHSAS 18000 is compatible with ISO 9001 and ISO 14001 management systems. OHSAS 18001 represents a progression of a management philosophy from quality management to environmental management to OH&S management.

5.1.2.2.1 Benefits of OHSAS Management System

The benefits of implementing an OHSAS management system are as follows:

- Reduced accidents and injuries to the employees
- Reduced insurance liability and risk
- Decreased costs due to personal injury and production downtime
- Reduced worker compensation insurance costs
- Ease of managing safety risks
- Enhanced employee safety awareness

5.1.2.3 ISO 45000

ISO 45000 is a newly established standard. It is a framework for OH&S management system. It has been developed to improve employee safety, reduce workplace risks, and create safer working condition. The new ISO 45000 standard implements a high-level structure, which is compatible with ISO 9001:2015 and ISO 14001:2015. The integration of these management systems adds values to the organizations and raises the health and safety bar.

ISO 45000 enables an organization to integrate worker wellness/well-being under the health and safety system. ISO 45001:2018 can be used in whole or in part to systematically improve OH&S management. However, claims to conformity to this document are not acceptable unless all the requirements are incorporated into an organization's OH&S management system and fulfilled without exclusion. In ISO 45001, the management commitment is central to the standard's effectiveness and integration.

5.1.2.3.1 Benefits of ISO 45000
- Systematic determination and monitoring of business context
- Enhance compliance due to an increased awareness of risks and consequences
- Promote workplace health and safety
- Reduce workplace accidents
- Create safer working condition
- Worker wellness/well-being

Table 5.2 lists differences between OHSAS 18001 and ISO 45001. Table 5.3 illustrates correlation between ISO 45001 and OHSAS 18001, whereas Table 5.4 illustrates correlation between OHSAS 18001 and ISO 45001.

Development and implementation of an effective HSE management system is a strategic and operational decision of an organization that helps to improve its overall

TABLE 5.2
Difference between OHSAS 18001 and ISO 45001

OHSAS 18001	ISO 45001
• British Standard	• International Organization of Standardization Standard
• No such clause	• Concept of context
• Organizational management reviews the process after development	• Leadership and management commitment. Leadership takes leading role to ensure it fits within organization's processes
• Employee participation and consultation	• Company-wide, including leadership involvement in safety. Participation, consultation and participation of workers
• Reactive planning	• Proactive planning
• Procedures are prepared	• Documented results are required
• Communication and information procedures are prepared	• Communication and information documents are required including determination of what, when, and how to communicate
• Safety and health is the responsibility of safety management personnel	• Safety and health is the responsibility of company leadership

TABLE 5.3
Correlation between ISO 45001:2018 and OHSAS 18001

Clause	ISO 45001	Clause	OHSAS 18001
1.0	Scope	1.0	Scope
	1.1 General		1.1 General
	1.2 All exclusions from ISO 9001:2008 Clause 1.2 removed		1.2 Application
2.0	Not valid	2.0	Reference publications
3.0	Included into primary standard, ISO 9000	3.0	Terms and definitions
4.0	Context of the organization/OH&S management system	4.0	New requirements (see 4.6h in management review)
	4.1 Understanding the organization and its context		4.1 New requirements (see 4.6h in Management review)
	4.2 Understanding the needs and expectations of workers and other interested parties		4.4.3.2 Participation and consultation (in part) (see also 4.6b and c in management review)
	4.3 Determine the scope of OH&S safety management system		4.1 General requirements
	4.4 OH&S management system and its processes		4.1 Management system general requirements

(Continued)

TABLE 5.3 (*Continued*)
Correlation between ISO 45001:2018 and OHSAS 18001

Clause	ISO 45001	Clause	OHSAS 18001
5.0	Leadership and worker participation	4.4.3	Communication, participation, and consultation
5.1	Leadership and commitment	4.4.1	Resources roles, responsibility, accountability, and authority
5.2	OH&S policy	4.2	OH&S policy
5.3	Organizational roles, responsibilities, and authorities	4.4.1	Resources roles, responsibility, accountability, and authority
5.4	Consultation and participation of workers	4.4.3.2	Participation and consultation
6.0	Planning for OH&S management system	4.3	Planning
6.1	Actions to address risks and opportunities	4.1	General requirement
		4.3.1	Hazard identification, risk assessment, and determining control
6.1.1	General	4.4.6	Operational control
6.1.2	Hazard identification and assessment of risks and opportunities	4.3.1	Hazard identification, risk assessment, and determining controls
6.1.2.1	Hazard identification	4.3.1	Hazard identification, risk assessment, and determining controls
6.1.2.2	Assessment of OH&S risks and other risks to the OH&S management system	4.3.1	Hazard identification, risk assessment, and determining controls
6.1.2.3	Identification of OH&S and other opportunities to the OH&S management system		New requirement
6.1.3	Determination of legal requirements and other requirements	4.3.2	Legal and other requirements
6.1.4	Planning action	4.3.6	Operational control
6.2	OH&S objectives and planning to achieve them	4.4.6	Objectives and program(s)
6.2.1	OH&S objectives	4.3.3	Objectives and program(s)
6.2.2	Planning to achieve objectives	4.3.3	Objectives and program(s)

(*Continued*)

TABLE 5.3 (*Continued*)
Correlation between ISO 45001:2018 and OHSAS 18001

Clause		ISO 45001	Clause	OHSAS 18001
7.0	Support		4.4	Implementation and operation
	7.1	Resources	4.4.1	Resources roles, responsibility, accountability, and authority
	7.2	Competence	4.4.2	Competence, training, and awareness
	7.3	Awareness	4.4.2	Competence, training, and awareness
	7.4	Communication		4.4.3.1 Communication
		7.4.1 General		4.4.3.1 Communication
		7.4.2 Internal communication		4.4.3.1 Communication
		7.4.3 External communication		4.4.3.1 Communication
	7.5	Documented information		Documentation control of records
		7.5.1 General	4.4.4	Documentation
			4.5.4	Control of records
		7.5.2 Creating and updating	4.4.5	Control of documents
			4.5.4	Control of records
		7.5.3 Control of documented information	4.4.5	Control of documents
			4.5.4	Control of records
8.0	Operation		4.4	Implementation and operation
	8.1	Operational planning and Control	4.4.6	Operational control
		8.1.1 General	4.4.6	Operational control
		8.1.2 Eliminating hazards and reducing OH&S risks	4.3.1	Hazard identification, risk assessment, and determining controls
			4.4.6	Operational control
		8.1.3 Management of change	4.3.1	Hazard identification, risk assessment, and determining controls
			4.4.6	Operational control
		8.1.4 Procurement	4.4.6	Operational control
		8.1.4.1 General	4.4.6	Operational control
		8.1.4.2 Contractors	4.3.1	Hazard identification, risk assessment, and determining controls
				4.4.3.1 Communication

(*Continued*)

TABLE 5.3 (Continued)
Correlation between ISO 45001:2018 and OHSAS 18001

Clause		ISO 45001	Clause	OHSAS 18001
			4.4.3.2	Participation and consultation
			4.4.6	Operational control
		8.1.4.3 Outsourcing	4.3.2	Legal and other requirements
			4.4.3.1	Communication
			4.4.6	Operational control
	8.2	Emergency preparedness and resources	4.4.7	Emergency preparedness and resources
9.0		Performance evaluation	4.5	Checking
	9.1	Monitoring, measurement, analysis, and performance evaluation	4.5.1	Performance measurement and monitoring
		9.1.1 General	4.5.1	Performance measurement and monitoring
		9.1.2 Evaluation of Compliance	4.5.2	Evaluation of compliance
	9.2	Internal audit	4.5.5	Internal audit
		9.2.1 General	4.5.5	Internal audit
		9.2.2 Internal audit program	4.5.5	Internal audit
	9.3	Management review	4.6	Management review
10		Improvement	4.6	Management review
	10.1	General	4.6	Management review
	10.2	Incident, nonconformity, and corrective action	4.5.3	Incident investigation, nonconformity, corrective action, and preventive action
			4.5.3.1	Incident investigation
			4.5.3.2	Nonconformity, corrective action, and preventive action
	10.3	Continual improvement	4.2	OH&S policy
			4.3.3	Objectives and program(s)
			4.6	Management review

ISO, International Organization of Standardization; OH&S, occupational health and safety.

performance by providing safe and healthy workplaces. Implementing OH&S management system conforming to its document enables an organization to manage its OH&S risks and improve its performance. The success of the HSE management system depends on leadership, commitment, and participation from all levels and functions of the organization.

Figure 5.3 illustrates how 4–10 clauses of ISO 45001 can be grouped under HSE management process model in relation to PDCA (plan–do–check–act) cycle.

TABLE 5.4
Correlation between OHSAS 18001 and ISO 45001:2018

Clause	OHSAS 18001	Clause	ISO 45001:2018
1.0	Scope	1.0	Scope
1.1	General	1.1	General
1.2	Application	1.2	All exclusions from ISO 9001:2008 Clause 1.2 removed
2.0	Reference publications	2.0	Not valid
3.0	Terms and definitions	3.0	Included into primary standard, ISO 9000
4.0	OH&S management system requirements	4.0	Context of the organization/OH&S management system
4.1	General requirements	4.3	Determine the scope of OH&S safety management system
		4.4	OH&S management system and its processes
4.2	OH&S policy	5.2	OH&S policy
		10.3	Continual improvement
4.3	Planning	6.0	Planning for OH&S management system
		6.1	Actions to address risks and opportunities
4.3.1	Hazard identification, risk assessment, and determining controls	6.1.2	Hazard identification and assessment of risks and opportunities
		6.1.2.1	Hazard identification
		6.1.2.2	Assessment of OH&S risks and other risks to the OH&S management system
		8.0	8.1.2 Eliminating hazards and reducing OH&S risks
			8.1.3 Management of change

(*Continued*)

TABLE 5.4 (Continued)
Correlation between OHSAS 18001 and ISO 45001:2018

Clause	OHSAS 18001	Clause	ISO 45001:2018
			8.1.4.2 Contractors
4.3.2	Legal and other requirements	6.0	6.1.3 Determination of legal requirements and other requirements
			8.1.4.3 Outsourcing
4.3.3	Objectives and program(s)	6.2	OH&S objectives and planning to achieve them
			6.2.1 OH&S objectives
			6.2.2 Planning to achieve objectives
		10.0	Continual improvement
4.4	Implementation and operation	7.0, 8.0	Support, operation
4.4.1	Resources roles, responsibility, accountability, and authority	5.1	Leadership and commitment
		5.3	Organizational roles, responsibilities, and authorities
		7.1	Resources
4.4.2	Competence, training, and awareness	7.2	Competence
		7.3	Awareness
4.4.3	Communication, participation, and consultation	5.0	Leadership and worker participation
4.4.3.1	Communication	7.4	Communication
			7.4.1 General
			7.4.2 Internal communication
		8.0	7.4.3 External communication
			8.1.4.2 Contractors

(Continued)

TABLE 5.4 (Continued)
Correlation between OHSAS 18001 and ISO 45001:2018

Clause	OHSAS 18001	Clause	ISO 45001:2018
			8.1.4.3 Outsourcing
4.4.3.2	Participation and consultation	4.2	Understanding the needs and expectations of workers and other interested parties
		5.4	Consultation and participation of workers
			8.1.4.2 Contractors
4.4.4	Documentation	7.5	Documented information
		7.5.1	General
4.4.5	Control of documents	7.5.2	Creating and updating
		7.5.3	Control of documented information
4.4.6	Operational control	6.1.1	General
		6.1.4	Planning action
		8.1	Operational planning and control
		8.1.1	General
		8.1.2	Eliminating hazards and reducing OH&S risks
		8.1.3	Management of change
		8.1.4	Procurement
			8.1.4.1 General
			8.1.4.2 Contractors
			8.1.4.3 Outsourcing
4.4.7	Emergency preparedness and resources	8.2	Emergency preparedness and resources
4.5	Checking	9.0	Performance evaluation

(*Continued*)

TABLE 5.4 (Continued)
Correlation between OHSAS 18001 and ISO 45001:2018

Clause	OHSAS 18001		Clause	ISO 45001:2018
4.5.1	Performance measurement and monitoring		9.1	Monitoring, measurement, analysis, and performance evaluation
				9.1.1 General
				9.1.2 Evaluation of compliance
4.5.2	Evaluation of compliance			
4.5.3	Incident investigation, nonconformity, corrective action, and preventive action		10.0	Incident, nonconformity, and corrective action
	4.5.3.1	Incident investigation	10.2	Incident, nonconformity, and corrective action
	4.5.3.2	Nonconformity, corrective action, and preventive action	10.2	Incident, nonconformity, and corrective action
4.5.4	Control of records		7.5	Documented information
				7.5.1 General
				7.5.2 Creating and updating
				7.5.3 Control of documented information
4.5.5	Internal audit		9.2	Internal audit
				9.2.1 General
				9.2.2 Internal audit program
4.6	Management review		4.0	Context of the organization/OH&S management system
			4.1	Understanding the organization and its context

(Continued)

TABLE 5.4 (*Continued*)
Correlation between OHSAS 18001 and ISO 45001:2018

Clause	OHSAS 18001	Clause	ISO 45001:2018
		4.2	Understanding the needs and expectations of workers and other interested parties
9.0		9.3	Management review
10			Improvement
		10.1	General
		10.2	Incident, nonconformity, and corrective action

ISO, International Organization of Standardization; OH&S, occupational health and safety.

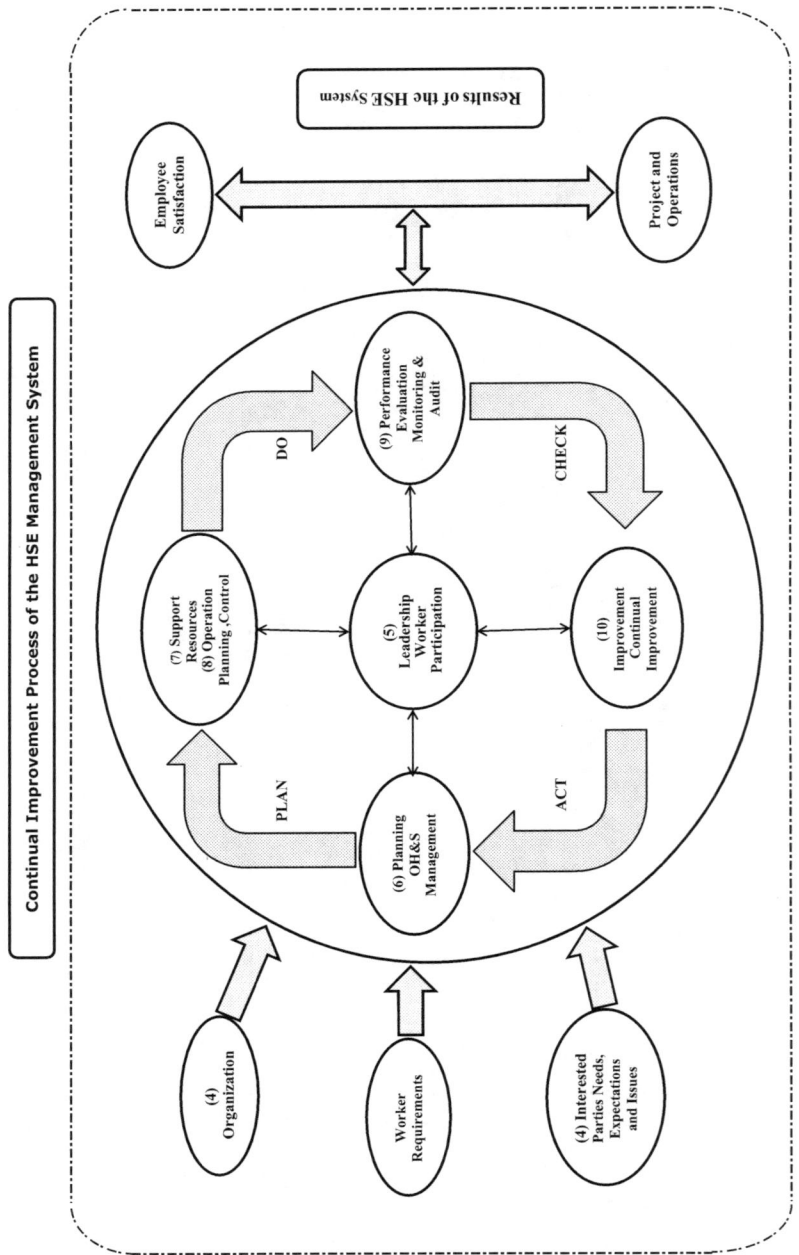

FIGURE 5.3 HSE management process model. HSE, health, safety, and environmental.

5.2 HEALTH, SAFETY, AND ENVIRONMENTAL MANAGEMENT SYSTEM

Safety management system is a systematic and coordinated approach to managing safety, including the necessary organizational structures, procedures, and accountabilities to optimally manage safety. In the early 2000s, there were international standards for environmental management system with ISO 14001, but not for health and safety. Thus, a British Standard was developed, called Occupational Health and Safety Assessment Series OHSAS 18000. The importance of OH&S management system is intended for an organization to improve its OH&S performance and enhance health and safety in the workplace by enabling the management of OH&S risks. It follows an outline and is managed like any other facet of a business, such as with marketing and engineering functions.

In oil and gas projects, OH&S system refers as HSE. Oil and gas industry apply HSE policies in all sectors and to all levels of operations. Health, safety, and environment are separate issues, each with its own technology, but are often combined in the same functional groups within the oil companies.

HSE system is a set of measures concerning OH&S during the construction and environmental issues to be implemented and operated by the designers and contractors on continuous basis. There are three subjects of paramount importance to oil and gas industry. These are as follows:

1. Occupational health
   - The health function typically deals with well-being of the employees as they live and work in their environment. It also focuses on the effects of petroleum chemicals and surrounding environment on the employees.
2. Safety
   - The safety function focuses on protecting the employee from risk involved during any type of operation and duties. It is related to the principle that all injuries should be prevented.
3. Environment
   - The environmental function focuses on the effects that petroleum activities have on the natural resources. The environmental issues relate to reduction/elimination of emissions, effluents, and discharge of waste and hazardous materials that are known to have negative impact on environment.

The adherence to HSE guidelines is a requirement for operators worldwide and is also included under the internal policies of most organizations. HSE system is conducted in conjunction with other management systems.

Construction projects are of complex and non-standardized nature. They are exposed to dynamic site conditions and involve coordination among various parties and participants.

Safety management process for construction projects is required to ensure that the construction project is executed with appropriate care to prevent accidents that could cause personal injury or property damage, and safety and health of neighbors to the construction site and general public who may be exposed to the project. HSE management system also includes human resource programs, such as drug and alcohol treatment programs, which contribute to reducing accidents on construction jobsites.

There are three core aspects of safety management system. These are as follows:

4. Systematic—HSE management activities are in accordance with predetermined HSE policies, strategic objectives, regulatory compliance, and organizational responsibilities to prevent accidents and safety hazards, which cause, or have potential to cause, personal injury, fatalities, or property damage, and protect natural resources due to emission of petroleum hazards.
5. Proactive—An approach that emphasizes safety hazard identification, investigation and analysis, and risk control and mitigation, before events that affect safety occur.
6. Explicit—All safety management activities are documented, communicated, and implemented.

In oil and gas projects, it is fundamental to have and implement HSEMS (health, safety, and environmental management system), which defines the principles by which operations are conducted and control the risks in the whole industry cycle.

The basic components of HSE management system are as follows:

- Leadership and management commitment
- Policies and objectives
- Organizational involvement (resources, documentations)
- Risk evaluation and risk management
- Planning
- Implementation
- Monitoring
- Measuring and recording performance
- Performance/compliance evaluation
- Audit
- Management review
- Continual improvement

Figure 5.4 illustrates PDCA cycle for basic elements of HSE system.

The purpose of HSE management system is to provide a framework for managing OH&S risks and opportunities. The aim and intended outcomes of the system are to prevent work-related injury and ill health to workers and to provide safe and health workplaces; consequently, it is critically important for the organization to eliminate hazards and minimize OH&S risks by taking effective preventive and protective measures. Figure 5.5 illustrates HSE management System pyramid.

5.2.1 Health, Safety, and Environmental Management Plan

In construction projects, the requirements to prepare safety management plan by the contractor are specified under contract documents. In oil and gas project,

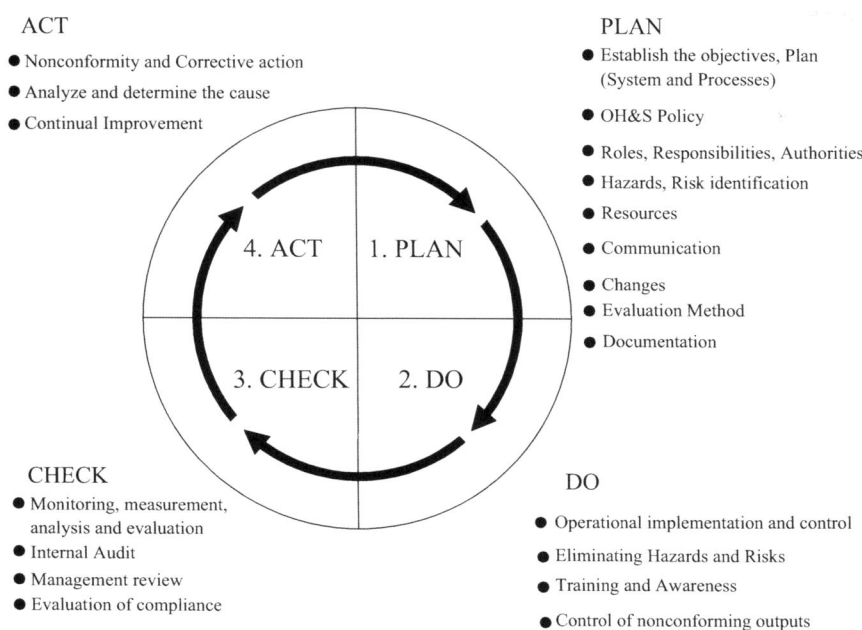

FIGURE 5.4 PDCA cycle for basic elements of HSE system. HSE, health, safety, and environmental; PDCA, plan–do–check–act.

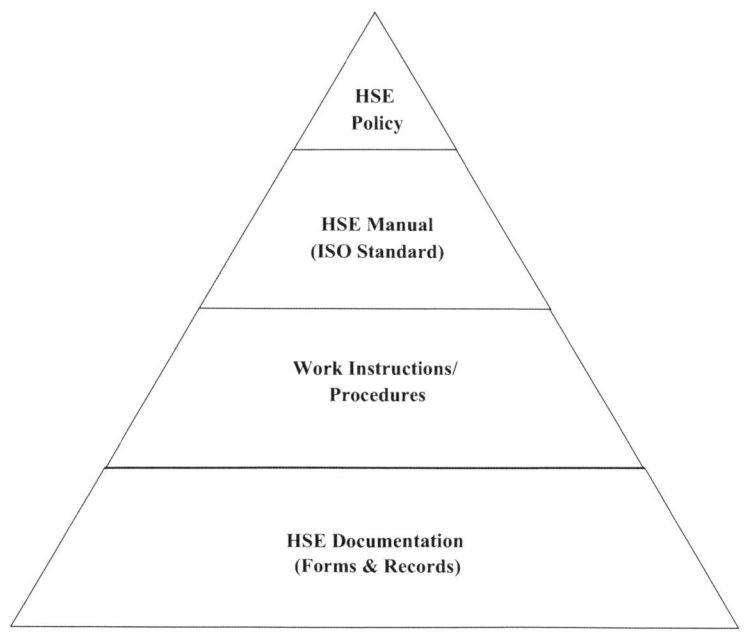

FIGURE 5.5 HSE management system. HSE, health, safety, and environmental.

HSE plan is developed by EPC contractor (other types of contractor, depending on contract requirements). The contractor has to submit the plan for review and approval by supervisor/consultant during mobilization stage of construction phase. The following are outlines to be specified by the designer and to be considered by the contractor to establish safety management system for construction project sites to raise the level of safety and health in the construction sites and prevent accidents:

1. Project scope detailing description of project and safety requirements
2. Safety policy statement documenting the contractor's/subcontractor's commitment and emphasis on safety
3. Regulatory requirements about safety
4. Roles and responsibilities of all individuals involved
5. Site communication plan detailing how safety information will be shared
6. Hazard identification, risk assessment, and control
7. Accident investigation to document root causes and determine corrective and preventive actions
8. Accident reporting system
9. Plant, equipment maintenance, and licensing
10. System education and training
11. Measures for emergency situations
12. Routine inspections
13. Continuous monitoring and regular assessment
14. System feedback and continuous improvements
15. Safety assurance measures
16. Evaluation of subcontractor's safety capabilities
17. Health surveillance
18. Site neighborhood characteristics and constraints
19. Safety audit
20. Documentation
21. Records
22. System update
23. Emergency evacuation plan

Table 5.5 illustrates example contents of contractor's HSE management plan for a construction project detailing hazardous identification, risk assessment, and control for scaffolding work. However, the contractor has to take into consideration all the requirements listed under the contract documents, depending on the complexity and nature of the project.

5.3 MAJOR CAUSES OF SITE ACCIDENTS

Safety statistics for construction indicate high fatality and injury rates. There are significantly more injuries and lost workdays due to injuries or illness in construction than in virtually any other industry. The causes of injuries in construction are numerous. Table 5.6 lists major causes of accidents on construction sites.

TABLE 5.5
Contents of Contractor's HSE Plan

Section	Topic		
1	Introduction		
	1.1	Project description	
2	Project general requirements		
3	HSE policy		
4	Regulatory requirements		
	4.1	Occupational health, health surveillance	
	4.2	Safety	
	4.3	Environmental	
5	Project HSE issues		
6	HSE organization		
	6.1	HSE organization chart	
	6.2	Roles and responsibilities	
	6.3	Communication plan	
7	Safety management plan		
	7.1	Design, detailed engineering	
	7.2	Construction	
	7.3	Process-related activities	
	7.4	Non–process-related activities	
	7.5	Project start-up	
8	Emergency evacuation plan		
	8.1	Accident reporting	
	8.2	Action plan	
	8.3	Measures for emergency situations	
	8.4	Disaster control	
9	Training		
	9.1	Awareness	
	9.2	Meetings	
	9.3	Drills	
10	Monitoring		
	10.1	Routing inspection	
	10.2	Protective equipment	
	10.3	Plant, equipment, machinery	
	10.4	Fire prevention	
	10.5	Gas and chemical leak detection	
	10.6	Safety measures	
		10.6.1	Preparatory activities
		10.6.2	Execution/installation/construction activities
		10.6.3	Start-up activities
	10.7	Hygiene	
	10.8	House keeping	
	10.9	Analysis and evaluation	
	10.10	Sub contractor's compliance	

(*Continued*)

TABLE 5.5 (*Continued*)
Contents of Contractor's HSE Plan

Section	Topic
11	Hazards management
	11.1 Safety hazards
12	Procedure for project HSE review
13	First aid and medical services
14	Environmental protection and control measures
15	Waste management
16	Risk identification, management
17	Material handling
18	Preventive action
19	Documentation
20	Record
21	Internal audit
22	Management review
23	System update/continual improvement

HSE, health, safety, and environmental.

TABLE 5.6
Major Causes of Construction Accidents

Serial Number	Cause of Accident
1	Poor attitude towards safety
2	Lack of safety awareness
3	Lack of proper training
4	Deficient enforcement of training
5	Unsafe site/property conditions
6	Unsafe access
7	Poor lighting
8	Falls and Slip
9	Ladder accidents
10	Struck by object
11	Caught in/between incidents
12	Lack of safety under power lines
13	Electrocution
14	Safe equipment not provided/unsafe equipment
15	Unsafe power tools and machinery
16	Personal protective equipment/safety gear not used
17	Unsafe methods or sequencing
18	Hoisting/lifting machinery not checked for safe operating conditions
19	Unsafe loading/unloading techniques
20	Unskilled operator

(*Continued*)

TABLE 5.6 (*Continued*)
Major Causes of Construction Accidents

Serial Number	Cause of Accident
21	Barriers or guards not provided
22	Lack of protection for working in trenches
23	Scaffolding not properly secured/scaffolding accidents
24	Warning signs not displayed
25	Escape route-site map not displayed and not explained
26	Flammable and combustible material not stored in safe area
27	Fire
28	Isolated, sudden deviation from prescribed behavior

5.4 PREVENTIVE MEASURES

Loss prevention philosophy makes process industries, construction projects, and other industries believe that protection of its resources, including employees and physical assets against human distress and financial loss resulting from accidental occurrences, can be controlled. Disasters have taught us for an improved loss control through a professional management system. There are major elements that may be adopted by any industry and not limited to process industry to loss prevention. These elements are as follows:

1. Policy Declaration
2. Training
3. Work permit procedure
4. Safety inspection
5. Other predictive maintenance inspections
    5.1 On-stream inspection program
    5.2 Relief valve inspection program
    5.3 Rotating equipment monitoring program
    5.4 Crane/heavy equipment inspection
    5.5 Safety meetings
    5.6 Operations safety meetings
    5.7 Maintenance safety meetings
    5.8 Safety talks
    5.9 Safety suggestion program
    5.10 Loss prevention compliance reviews
    5.11 Executive management safety reviews
    5.12 Safe operations committee
    5.13 Contractor safety program
    5.14 Contractor safety orientation program
    5.15 Fire/emergency drills
    5.16 Disaster control plan
    5.17 Incentive programs

Fundamental concepts and methodology of these elements are to identify all loss exposures, evaluate the risk of each exposure, plan how to handle each risk, and manage according to plan. People are the first source of losses. These could be managers, engineers, or workers who plan, design, build, operate, and maintain the plants; also, general public who may be subjected to the hazards! Equipment is considered the second—whether fixed plant, machines, tools, protective gear, or vehicles. Materials such as process substances, supplies, and products that have physical and chemical hazards affecting people, equipment, and environment are considered the third. The fourth is surrounding that includes buildings, surfaces and subsurfaces, atmosphere, lighting, noise, radiation, hot or cold weather, and social or economic conditions, which can affect safe performance of people, equipment, and materials.

The construction site should be a safe place for those working there. Necessary measures are always required to ensure safety of all those working at construction sites. Effective risk control strategies are necessary to reduce and prevent accidents. Table 5.7 lists preventive actions to avoid construction accidents.

TABLE 5.7
Preventive Actions to Avoid Construction Accidents

Serial Number	Cause of Accident	Preventive Action
1	Poor attitude toward safety	Interact with everyone and explain the importance of safety and importance of using safety equipment. Evaluate the response and make the employee understand.
2	Lack of safety awareness	Arrange training and awareness programs.
3	Lack of proper training	Engage trainer/facilitator having expertise in task, able to interview, test or observe employee, and have prior training records. Arrange safety drills.
4	Deficient enforcement of training	Monitor safety program on regular basis. Enforce the plan and penalize for noncompliance.
5	Unsafe site conditions	Know proper site conditions. Study soil conditions and ground reality.
6	Unsafe access	Provide proper ladders and access routes. Keep site clean with proper housekeeping.
7	Poor lighting	Provide temporary lighting on all routes frequently used by the employees which have poor visibility
8	Falls and Slip	Use safety belt. Take precautions while working at high level. Ensure that walking surfaces are not slippery.
9	Ladder accidents	Ensure ladder is strong enough and properly fixed.
10	Struck by object	Take care while walking.
11	Caught in/between incidents	Take precautions and not to be in/between such areas.
12	Lack of safety under power lines	Maintain safe distance. Use protective gears and gloves.

(Continued)

TABLE 5.7 (*Continued*)
Preventive Actions to Avoid Construction Accidents

Serial Number	Cause of Accident	Preventive Action
13	Electrocution	Use insulated gloves. Ensure that earth leakage protection system is installed at the power supply units. Ensure that cable/wires have insulation and are not bare.
14	Safe equipment not provided	Check the equipment before use. Check the license validity if applicable.
15	Unsafe power tools and machinery	Check power tools before use.
16	Personal protective equipment/safety gear not used	Ensure and monitor everyone is using personal protective equipment. Issue warning letters for noncompliance.
17	Unsafe methods or sequencing	Worker to know standard methods and sequence of the task he/she is performing.
18	Hoisting/lifting machinery not checked for safe operating conditions	Inspect and check for maintenance requirement before use. Only authorize employees to operate.
19	Unsafe loading/unloading techniques	Provide proper deck for unloading. Use proper safety gear while lifting the material. Barricade the loading/unloading area. Display signs.
20	Unskilled operator	Engage trained operator.
21	Barriers or guards not provided	Use barriers and guars as necessary. Use warning signs.
22	Lack of protection for working in trenches	Use barriers. Provide protective display signs. Monitor depth of trench. Provide sufficient light.
23	Scaffolding not properly secured	Tighten scaffolding and check the formwork and scaffolding is erected in accordance with applicable standards.
24	Warning signs not displayed	Display signs to keep away employees away from hazardous/unsafe area.
25	Escape route-site map not displayed and not explained	Provide escape route map at various locations. Give training and inform about the escape route and gathering point(s).
25	Flammable and combustible material not stored in safe area	Designate specific area for storage of flammable and combustible material. Display notice as "hazardous area".
27	Fire	Provide temporary firefighting system as per local code. In case of fire, inform local fire brigade immediately.
28	Isolated, sudden deviation from prescribed behavior	Cannot predict or prevent unless employee's emotional or physical condition contributed, and this condition was obvious to others.

The safety program embodies the prevention of accidents, injury, occupational illness, and property damage. Normally the contract documents specify that a safety engineer/officer is engaged by the contractor to monitor safety measures. The safety engineer/officer is normally responsible for

1. conducting safety meetings;
2. monitoring on-the-job safety;
3. inspecting the work and identifying hazardous areas;
4. initiating a safety awareness program;
5. ensuring availability of first aid and emergency medical services per local codes and regulations;
6. ensuring that personnel are using protective equipment such as hard hat, safety shoes, protective clothing, life belt, and protective eye coverings;
7. ensuring that the temporary firefighting system is working;
8. ensuring that work areas are free from trash and hazardous material; and
9. housekeeping.

Construction sites have many hazards that can cause serious injuries and accidents. The contractor has to identify these areas and ensure that all site personnel and subcontractor employees working at the site are aware of unsafe acts, potential and actual hazards, the safety plan and procedures, and the immediate corrective action to be taken if something untoward occurs.

Contract documents normally stipulate that the contractor, upon signing of contract, has to submit safety and accident prevention program. It emphasizes that all the personnel have to put efforts to prevent injuries and accidents. In the program, the contractor has to incorporate requirements of safety and health requirements of local authorities, manuals of accident prevention in construction, and all other local codes and regulations. The contractor has to also prepare emergency evacuation plan (EEP). The EEP is required to protect personnel and to reduce the number of fatalities in case of major accidents at site. The evacuation routes have to be displayed at various locations in a manner. Transfer points and gathering points have to be designated, and sign boards have to be displayed all the time. Evacuation sirens have to be sounded on regular basis in order to ensure smooth functioning of evacuation plan.

A safety violation notice is issued to the contractor/employee if the contractor or any of his employees are not complying with safety requirements. Figure 5.6 illustrates safety violation notice, which is to be actioned by the contractor.

Penalties are also imposed on contractor for noncompliance with the site safety program. Figure 5.7 illustrates sample disciplinary notice form for breach of safety rules. Different colors of card may be issued along with the notice. Figure 5.8 illustrates concepts of issuance of different colors of card.

HSE in Oil and Gas Projects 241

Project Name
Consultant Name

Contract No.:   SVN No.
Contractor: Date :
                                                          Time:

SAFETY VIOLATION REPORT

SAFETY RELATED ITEMS			
Sr.No.	Description	Sr.No.	Description
1	Access Facilities	13	Hygienic
2	Barricade/Railing	14	Poor lighting
3	Construction Equipment	15	Protective Equipment
4	Crane	16	Lifting Gears
5	Earthwork/Excavation	17	Poor lighting
6	Electrical	18	Protective Equipment
7	Firefighting/Protection	19	Safety Gears
8	First Aid	19	Scaffolding
9	Formwork	20	Site Fencing
10	Hand and Power Tools	21	Storage Facilities
11	Hazads/Inflammable Material	22	Vehicles
12	Hoist	23	Welding/Hot Work
12	Housekeeping	24	Others

VIOLATION DESCRIPTION	Action code:	
Item No.	Location	Description

ORIGINATOR:	RESIDENT ENGINEER:

CONTRACTOR'S ACTION				
Item No.	Location	Action	Date	Time

SAFETY OFFICER:	CONTRACTOR'S PROJECT MANAGER:

Action Code: [A]   For immediate action /( )hours   [B]   Within ( ) days

FIGURE 5.6   Safety violation notice.

PROJECT NAME	
CONSULTANT NAME	
Safety Disciplinary Notice	
Notice No.:	Date:

Name of Employee:
Contractor Name:
Area/Floor:

SAMPLE FORM

Date & Time of Observance:

Type of Notice

☐	**First/Verbal Warning** (White Card)
☐	**Second/Written Warning** (Yellow Card)
☐	**Suspension from Site** (Red Card)

Reason for Issuance of Notice:

Action Required by Recipient:

Date by Which Action is Required:

Issued by:

Signature: Date:

Reveived by: Date:

CC: Owner ☐   Resident Engineer ☐   Project Manager ☐

FIGURE 5.7 Safety disciplinary notice.

Serial Number	Card Color	Type of Disciplinary Action	Warning Validity	Reasons for Disciplinary Action
1	White	Verbal followed by Safety Discipline Notice	1 - 3 months	1. Failure to use personal protective equipment 2. Failure to use define access 3. Working on plant, crane, vehicle without license 4. Working with unsafe scaffolding 5. Working on unsafe platform 6. Using unsafe sling or ropes for lifting 7. Working on unsafe ladders
2	Yellow	Issuance of Safety Discipline Notice and suspension from the work for rest of the day	6 months	1. Repetition of activities listed under "White Card" within 1 month of issuance of first notice 2. Failure to observe HSE related instructions 3. Failure to work as per instructed method of work, as per given task 4. Failure to follow storage principles about hazardous materials
3	Red	Issuance of Safety Disciplinary Notice and suspension from site for 1 month	1 Year	1. Breach of safety rules where there is risk to life 2. Removal of safety devices, interlocks, guard rails, barriers without any authority 3. Deliberately exposing public to danger by not complying with agreed safe methods of work 4. Disposal of hazardous material in unsafe area

FIGURE 5.8 Concept of safety disciplinary action.

Section 2

Quality Management Principles, Procedures, Concepts, and Methods in Oil and Gas Projects

6 Overview of Quality in Oil and Gas Projects

6.1 HISTORY OF QUALITY

Quality is a universal phenomenon, which has been a matter of great concern throughout the recorded history of human kind. It was always the determination of the builders and makers of the products to ensure that their products meet the customer desires. Those were the days where products were of "customized nature".

With the advent of globalization and competitive market, the emphasis on quality management has increased. Quality has become the most important single factor for the survival and success of any company. Customer demands for better products and services at the lowest possible costs have put tremendous pressure on developing and manufacturing organizations to improve the quality of products, services, and processes, in order to compete in the market and improve business results. It became important that construction projects be more qualitative, competitive, and economical to meet owner's expectations.

Quality issues have been of great concern throughout the recorded history of human being. During New Stone Age, several civilizations emerged, and some 4000–5000 years ago, considerable skills in construction had been acquired. The Pyramid in Egypt was built approximately 2589–2566 BC. The King of Babylonia (1792–1750 BC) had codified the law, and according to that during Mesopotamian era, builders were responsible to maintain the quality of buildings and were given death penalty if any of their constructed building collapsed and its occupants were killed. The extension of Greek settlement around the Mediterranean after 200 BC had record to show that temples and theaters were built using marble. India had strict standards for working in gold in the 4th century BCE.

During the Middle Ages, guilds took the responsibilities of quality control upon themselves. Guilds and governments carried out quality control while consumers carried out informal quality inspection during every age of humanity.

The guilds' involvement in quality was extensive. All craftsmen living in a particular area were required to join the corresponding guild and were responsible to control the quality of their own products. If any of the items was found defective, then the craftsman was discarding the faulty items. The guilds also initiated punishments for members who turned out shoddy products. Guilds were maintaining inspections and audits to ensure that artisans followed the quality specifications. Guild hierarchy was consisting of three categories of workers: apprentice, the journeyman, and the master. Guilds had established specifications for input materials, manufacturing processes, and finished products, as well as methods of inspection and test. Guilds were active in managing the quality during Middle Ages until the Industrial Revolution reduced their influence.

The Industrial Revolution began in Europe in the mid-18th century. Industrial Revolution gave birth to factories. The goals of the factories were to increase productivity and reduce costs. Prior to Industrial Revolution, items were produced by individual craftsman for individual customers, and it was possible for workers to control the quality of their own products. Working conditions then were more conducive for professional pride. Under factory system, the tasks needed to produce a product were divided up among several or many factory workers. In this system, large group of workmen were performing similar type of work, and each group was working under the supervision of a foreman who also took on the responsibility to control the quality of the work performed. Quality in factory system was ensured through skilled workers, and the quality audit was done by inspectors.

The broad economic result of factory system was mass production at low costs. The Industrial Revolution changed the situation dramatically with the introduction of new approach to manufacturing.

In the early 19th century, the approach to manufacturing in the United States tended to follow the craftsmanship model used in the European countries.

In late 19th century, Frederick Taylor's system of "scientific management" was born. Taylor's goal was to increase production. He achieved this by assigning planning to specialized engineers, and execution of job was left to the supervisors and workers. Taylor's emphasis to increase production had negative effect on the quality. With this change in production method, inspection of finished goods became norm rather inspection at every stage. To remedy the quality decline, factory managers created inspection departments having its own functional boss. These departments were known as quality control departments.

The beginning of 20th century marked inclusion of process in the quality practices. During World War I, the manufacturing process became more complex. The production quality was the responsibility of quality control departments. The introduction of mass production and piecework created quality problems as workmen were interested in earning more money by the production of extra products, which in turn led to bad workmanship. This made factories to introduce full-time quality inspectors and the real beginning of inspection quality control, and thus the beginning of quality control department headed by superintendents. Walter Shewhart introduced statistical quality control (SQC) in process. His concept was that quality is not relevant for the finished product, but for the process that created the product. Shewhart's approach to quality was based on continuous monitoring of process variation. The SQC concept freed manufacturer from time-consuming 100% quality control system because it accepted that variation is tolerable up to certain control limits. Thus, quality control focus shifted from the end of line to the process.

The systematic approach to quality in industrial manufacturing started during the 1930s when some attention was given to the cost of scrap and rework. With the impact of mass production, which was required during World War II, it became necessary to introduce a more stringent form of quality control. Manufacturing units introduced a more stringent form of quality control, which is identified as SQC. SQC had a significant contribution in that it provided a sampling inspection rather than 100% type. This type of inspection, however, did lead to a lack of realization to the importance of the engineering of product quality.

The concept and techniques of modern quality control were introduced in Japan immediately after World War II. Introduction of statistical and mathematical techniques, sampling tables, and process control charts emerged during this period.

From the early 1950s to late 1960s, quality control evolved into quality assurance, with its emphasis on problem avoidance rather than problem detection. The quality assurance perspective suffered from a number of shortcomings as its focus was internal. Quality assurance was generally limited to those activities which were directly under the control of the organization; important activities such as transportation, storage, installation, and service were typically either ignored or give little attention. Quality assurance concept was given little or no attention to the competition's offerings. This resulted in integration of the quality actions on a company-wide scale and application of quality principles in all the areas of business from design to delivery instead of confining the quality to production activities. This concept was called "total quality control".

The concept of "total quality control" was popularized Armand V. Feigenbaum, a quality guru, from the United States.

Quality is a distinguishing characteristic of product or services, which satisfy the customer. Most production or services systems are of repetitive nature and are designed for mass production or batch (lot) production.

Thus, it is evident that quality system, in its different forms, has moved through distinct quality eras such as:

1. Quality inspection
2. Quality control
3. Quality assurance
4. Quality engineering
5. Total quality control
6. Quality management

Introduction and promotion of company-wide quality control led to a revolution in management philosophy. To help sell their products in international markets, the Japanese took some revolutionary steps to improve quality:

1. Upper-level managers personally took charge of leading the revolution.
2. All levels and functions received training in the quality disciplines.
3. Quality improvement projects were undertaken on a continuing basis—at a revolutionary pace.

Thus, the concept of quality management started after World War II, broadening into the development of initiatives that attempt to engage all employees in the systematic effort for quality. Quality emerged as a dominant thinking becoming an integral part of overall business system focused on customer satisfaction and was known as "total quality management (TQM)" with its three constitutive elements:

- Total: organization-wide
- Quality: customer satisfaction
- Management: systems of managing

6.2 TOTAL QUALITY MANAGEMENT

The TQM approach was developed immediately after World War II. There are prominent researchers and practitioners whose works have dominated quality movement. Their ideas, concepts, and approaches in addressing the specific quality issues have become part of the accepted wisdom in the field of quality, resulting in a major and lasting impact within the business. These persons are known as "quality gurus". They all emphasize upon involvement of organizational management in the quality efforts. These philosophers (gurus) are as follows:

1. Philip B. Crosby
2. W. Edwards Deming
3. Armand V. Feigenbaum
4. Kaoru Ishikawa
5. Joseph M. Juran
6. John Oakland
7. Shigeo Shingo
8. Genichi Taguchi

Their approaches to quality emphasize on customer satisfaction, management leadership, teamwork, continuous improvement, and minimizing defects. The common features of their philosophies can be summarized as follows:

1. Quality is conformance to the customer's defined needs.
2. Senior management is responsible for quality.
3. Take a team approach by involving every member of the organization.
4. Establish leadership to help employees perform a better job.
5. Establish performance measurement standards to avoid defects.
6. Provide training and education for everyone in the organization.
7. Continuous improvement of process, product, and services through application of various tools and procedures to achieve higher level of quality.

The TQM was stimulated by the need to compete in global market where higher quality, lower cost, and more rapid development are essential to market leadership. TQM was/is considered as a fundamental requirement for any organization to compete, let alone lead its market. It is a way of planning, organizing, and understanding each activity of the process and removing all the unnecessary steps routinely followed in the organization. TQM is a philosophy that makes quality values the driving force behind leadership, design, planning, and improvement in activities. It places quality as a strategic objective and focus on continuous improvement of products processes and services and cost, to compete in global market by minimizing rework and maximizing profitability to achieve market leadership and customer satisfaction. It is a way of managing people and business processes to meet customer satisfaction. TQM involves everyone in the organization in the effort to increase customer satisfaction and achieve superior performance of the products or services through continuous quality improvement. TQM helps in the following:

Overview of Quality in Oil and Gas Projects

- Achieving customer satisfaction
- Developing corporate culture
- Building motivation within organization
- Developing team work
- Establishing vision for the employees
- Setting standards and goals for the employees
- Achieving continuous improvement

TQM was considered a fundamental requirement for any industry to compete, let alone lead, in its market. TQM concept was there till the end of the 20th century.

Pursuing TQM in construction projects results in the following strategic benefits:

- Survive in an increasingly competitive world.
- Improve market share and profitability.
- Better serve the needs of customers.
- Improve quality and safety of the project/facility.
- Reduce duration project execution.
- Reduce project cost.
- Fully utilize the talents of employees.

6.3 CONSTRUCTION PROJECTS

Construction is the translation of owner's goals and objectives, into a facility built by the contractor/builder as stipulated in the contract documents, plans, specifications on schedule, and within budget. Construction project is a custom rather than a routine, repetitive business and differs from manufacturing. Construction projects work against defined scope, schedule, and budget to achieve the specified result.

Construction projects have involvement of many participants comprising owner, designer, contractor, and many other professionals from construction-related industries. Each of these participants is involved to implement quality in construction projects. These participants are influenced by and depend on each other in addition to "other players" involved in the construction process. Therefore, the construction projects have become more complex and technical, and extensive efforts are required to reduce rework and costs associated with time, materials, and engineering.

Construction projects are mainly capital investment projects. They are customized and nonrepetitive in nature. Construction projects have become more complex and technical, and the relationships and the contractual grouping of those who are involved are also more complex and contractually varied. The products used in construction projects are expensive, complex, immovable, and long lived. Generally, a construction project is composed of building materials (civil), electro-mechanical items, finishing items, and equipment. For oil and gas projects, there are other items such as processing equipment/machines, pumps, piping material, values, storage tanks, and instrumentation. These are normally produced by other construction-related industries/manufacturers. These industries produce products as per their own quality management practices complying with certain quality standards or against specific requirements for a particular project. The owner of the construction project or his representative has no direct control over these companies unless he/his representative/appointed contractor commit to buy their product for use in their facility. These organizations may have their own quality management program. In manufacturing or service industries, the quality management of all in-house manufactured products is performed by manufacturer's own team or under the control of same organization having jurisdiction over their manufacturing plants at different locations. Quality management of vendor-supplied items/products is carried out as stipulated in the purchasing contract as per the quality control specifications of the buyer.

There are numerous types of construction projects.

6.3.1 Process-Type Projects

1. Liquid chemical plants
2. Liquid/solid plants
3. Solid process plants
4. Petrochemical plants
5. Petroleum refineries

6.3.2 Non–Process-Type Projects

1. Power plants
2. Manufacturing plants
3. Support facilities
4. Miscellaneous (R&D) projects
5. Civil construction projects**
6. Commercial/A&E projects**

** Civil construction projects and commercial/A&E projects can further be categorized into four somewhat arbitrary but generally accepted major types of construction. These are as follows:

- Residential construction
- Building construction (institutional and commercial)
- Industrial construction
- Heavy engineering construction

Oil and gas sector projects are considered as the biggest sectors in terms of investment as well as in employment generation. Oil and gas projects are categorized as process type of projects. These types of projects are undertaken by large contraction companies on behalf of large corporations/governmental authorities. These companies are required to have a wide array of expertise and knowledge in such construction. While safety is a prime concern in construction projects, the health, safety, and environmental requirements in process industry projects are to be taken care right from the design and generally need more planning and teamwork coordination. The oil and gas industry has mainly three sectors. These are as follows:

1. Upstream
   - Upstream is exploration and production of crude oil and gas
   - Production and stabilization of oil and gas
2. Midstream
   - Transportation, storage, and processing of oil and gas
   - Gas treatment, LNG (liquefied natural gas) production and regasification plants
   - Oil and gas pipeline system
3. Downstream
   - Filtering, refining of crude oil, and purifying natural gas
   - Processing of oil and condensates into marketable products with defined specifications

Construction projects are custom-oriented and custom design having specific requirements set by the customer/owner to be completed within finite duration and budget. Every project has elements that are unique. No two projects are alike. It is always the owner's desire that his project should be unique and better. To a great extent, each project has to be designed and built to serve a specified need. Construction project is custom than a routine and repetitive business. Construction projects differ from manufacturing or production. Construction projects involve many participants including owners, designers, contractors, and many other professionals from construction-related industries. These participants are influenced by and depend on each other in addition to "other players" who are part of the construction process. Therefore, construction projects have become more complex and technical, and extensive efforts are required to reduce rework and costs associated with time, materials, and engineering.

6.4 LIFE CYCLE OF CONSTRUCTION PROJECT

Systems are as pervasive as the universe in which we live. The world in which we live may be divided into the natural world and the human-made world. Systems appeared first in natural forms, and subsequently with the appearance of human beings, human-made system systems came into existence. Natural systems are those that came into being by natural process. Human-made systems are those in which human beings have intervened through components, attributes, or relationships.

The systems approach is a technique, which represents a broad-based systematic approach to problem that may be interdisciplinary. It is particularly useful when problems are affected by many factors, and it entails the creation of a problem model that corresponds as closely as possible in some sense to reality. The systems approach stresses the need for the engineer to look for all the relevant factors, influences, or components of the environment that surround the problem. The systems approach corresponds to a comprehensive attack on a problem and to an interest in, and commitment to, formulating a problem in the widest and fullest manner that can be professionally handled.

Systems engineering and analysis when coupled with new emerging technologies reveal unexpected opportunities for bringing new improved systems and products into being that will be more competitive in the world economy. Product competitiveness is desired by both commercial and public sector producers worldwide. It is the product or consumer good that must meet customer expectations.

These technologies and processes can be applied to construction projects. Systems engineering approach to construction projects helps understand the entire process of project management in order to understand and manage its activities at different levels of various phases to achieve economical and competitive results. The cost-effectiveness of the resulting technical activities can be enhanced by giving more alteration to what they are to do, before addressing what they are composed of. To ensure economic competitiveness regarding the product, engineering must become more closely associated with economics and economic facilities. This is best accomplished through life cycle approach to engineering. Every system is made up of components, and components can be broken down into similar components. If two hierarchical levels are involved in a given system, the lower is conveniently called a subsystem. The designation of system, subsystem, and components are relatives because the system at one level in the hierarchy is the component at another level.

Most construction projects are custom oriented having a specific need and a customized design. It is always the owner's desire that his project should be unique and better. Furthermore, it is the owner's goal and objective that the facility is completed on time. Expected time schedule is important from both financial and acquisition of the facility by the owner/end user.

The system life cycle is fundamental to the application of systems engineering. Systems engineering approach to construction projects helps to understand the entire process of project management and to manage and control its activities at different levels of various phases to ensure timely completion of the project with economical use of resources to make the construction project most qualitative, competitive, and economical.

However, it is difficult to generalize project life cycle to system life cycle. However, considering that there are innumerable processes that make up the construction process, the technologies and processes, as applied to systems engineering, can also be applied to oil and gas construction projects. The number of phases shall depend on the complexity of the project. Duration of each phase may vary from project to project. Generally, oil and gas construction projects can be divided into six most common phases. These are as follows:

1. Study stage
2. Concept design
3. Front-end engineering design
4. Bidding and tendering
5. Construction
   5.1 Detailed engineering
   5.2 Procurement
   5.3 Construction
6. Testing, commissioning, and handover

Each phase can further be subdivided on work breakdown structure (WBS) principle to reach a level of complexity where each element/activity can be treated as a single unit, which can be conveniently managed. WBS represents a systematic and logical breakdown of the project phase into its components (activities). It is constructed by dividing the project into major elements with each of these being divided into sub-elements. This is done till a breakdown is done in terms of manageable units of work for which responsibility can be defined. WBS involves envisioning the project as a hierarchy of goal, objectives, activities, sub-activities, and work packages. The hierarchical decomposition of activities continues until the entire project is displayed as a network of separately identified and nonoverlapping activities. Each activity will be single purposed, of a specific time duration, and manageable; its time and cost estimates will be easily derived, deliverables will be clearly understood, and responsibility for its completion will be clearly assigned.

WBS facilitates the planning, budgeting, scheduling, and control activities for the project manager and its team. By application of WBS phenomenon, the construction phases are further divided into various activities. Division of these phases will improve the control and planning of the construction project at every stage before a new phase starts. Table 6.1 illustrates construction project life cycle (downstream) phases with subdivided activities/components.

These activities may not be strictly sequential; however, the breakdown allows implementation of project management functions more effectively at different stages

6.5 QUALITY DEFINITION OF CONSTRUCTION PROJECTS

Quality has different meanings to different people. The definition of quality relating to manufacturing, processes, and service industries is as follows:

- Meeting the customer's need
- Customer satisfaction
- Fitness for use
- Conforming to requirements
- Degree of excellence at an acceptable price

The International Organization for Standardization (ISO) defines quality as "the totality of characteristics of an entity that bears on its ability to satisfy stated or implied needs".

However, the definition of quality for construction projects is different to that of manufacturing or services industries as the product is not repetitive, but unique piece of work with specific requirements. In case of mass production and batch-oriented production systems, quality can be achieved by getting the feedback from the process by observing the actual performance and regulating the process to meet the established standards, whereas, because of nonrepetitive nature of construction projects, it is not possible to compare actual performance of the project as past experience may be of limited value. The quality management of manufactured products is performed by the manufacturer's own team and has control over all the activities of the product life cycle, whereas construction projects have diversity of interaction and relationship between owners, architects/engineers, and contractors. Construction projects differ from manufacturing or production. Construction projects are custom-oriented and

TABLE 6.1
Oil and Gas Project (Downstream) Life Cycle Phases

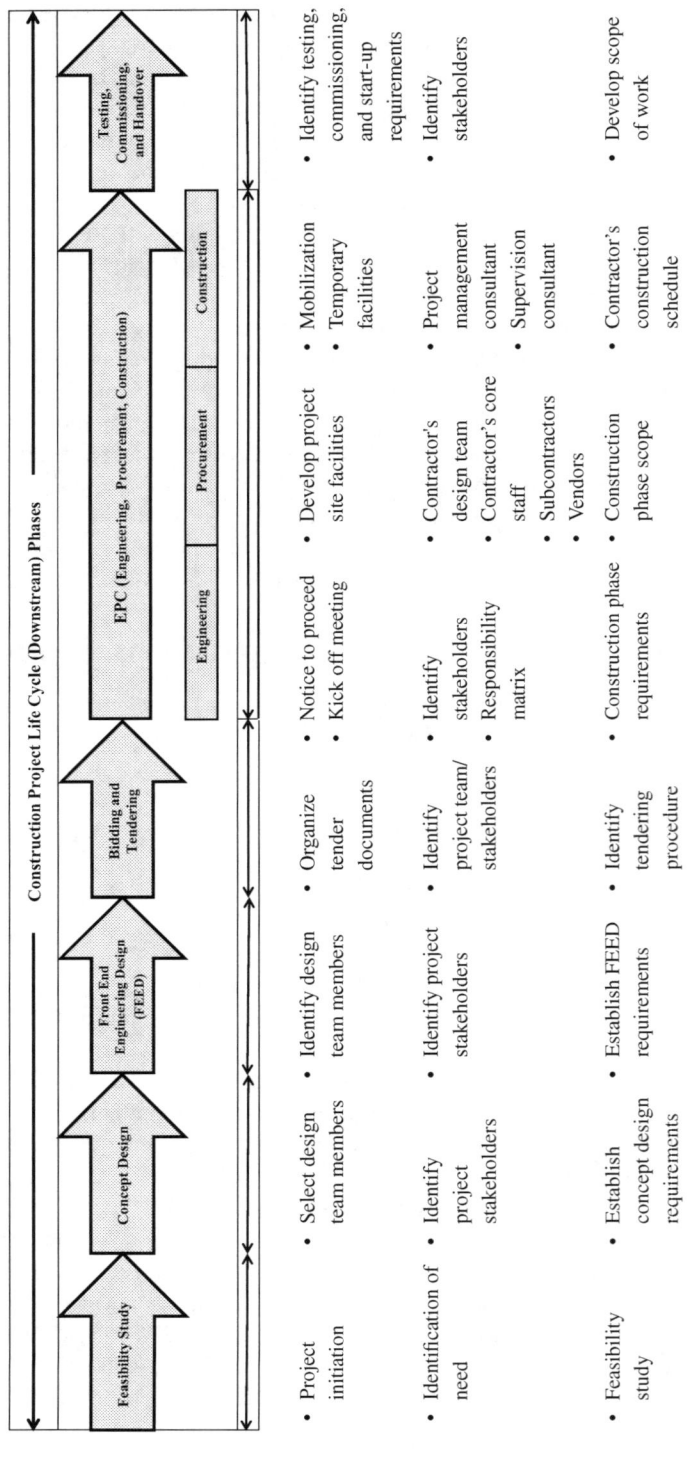

(*Continued*)

Overview of Quality in Oil and Gas Projects

TABLE 6.1 (Continued)
Oil and Gas Project (Downstream) Life Cycle Phases

Phase							
• Establish project goals and objectives	• Identify concept design deliverables	• Develop FEED (technical study to select process)	• Identify bidders	• Project S-curve • Contractor's quality control plan	• Stakeholder management plan • Resource management plan • Communication management plan	• Risk management plan • Contract management plan • HSE management plan	• Develop testing and commissioning plan
• Identification of alternatives/options	• Develop concept design	• Prepare FEED	• Manage tendering process	• Identify detailed engineering deliverables	• Identify procurement deliverables	• Identify construction works deliverables	• Execute testing and commissioning works
• Analyze and evaluate alternatives/options	• Identify/collect regulatory requirement	• Prepare project schedule/plan	• Distribute tender documents	• Develop detailed engineering	• Develop procurement documents	• Manage execution/installation of works	• Manage testing and commissioning quality procedure
• Select preferred alternative	• Prepare preliminary schedule	• Evaluate and estimate project cost	• Conduct prebid meetings	• Perform interdisciplinary coordination	• Prepare material procurement list	• Manage scope change	• As-built drawing
• Finalize project delivery and contracting system	• Prepare preliminary schedule	• Manage FEED quality	• Submit/receive bids	• Develop project schedule	• Identify vendors	• Manage construction quality	• Technical manuals and documents
• Project charter	• Estimate conceptual project cost	• Estimate resources	• Manage bidding and tendering quality	• Estimate project cost	• Manage procurement quality	• Manage construction resources	• Record books (PRB, CRB, MRB)

(Continued)

TABLE 6.1 (Continued)
Oil and Gas Project (Downstream) Life Cycle Phases

• Establish project quality requirements	• Develop communication plan	• Manage risks	• Manage detailed engineering quality	• Finalize procurement	• Train owner's/end user's personnel
• Manage concept design quality	• Manage project risks	• Review bid documents	• Estimate project resources		• Regulatory/authority approval
• Estimate project resources	• Manage HSE requirements	• Evaluate bids	• Manage project risk		• Handover of project to the owner/end user
• Identify project risks	• Develop FEED documents	• Select contractor	• Manage project HSE requirements		• Issue substantial completion
• Identify project HSE issues and requirements	• Perform value engineering study	• Award contract	• Review detailed engineering		• Settle payments
• Review concept design	• Review FEED		• Finalize detailed engineering		
• Finalize concept design	• Finalize FEED				• Manage communication
	• Develop tender documents				• Manage construction risk
					• Manage contract
					• Manage HSE requirements
					• Manage project finances
					• Manage claims
					• Monitor and control project works
					• Validate executed works
					• Settle claims

FEED, front-end engineering design; HSE, health, safety, and environmental.

custom design having specific requirements set by the customer/owner to be completed within finite duration and assigned budget. Every project has elements that are unique. No two projects are alike. It is always the owner's desire that his project should be unique and better. To a great extent, each project has to be designed and built to serve a specified need. Construction project is custom than a routine and repetitive business.

Quality in construction projects encompasses not only the quality of products and equipment used in the construction but also the total management approach to complete the facility as per the scope of works to customer/owner satisfaction within the budget and in accordance with the specified schedule to meet the owner's defined purpose. The nature of the contracts between the parties plays a dominant part in the quality system required from the project, and the responsibility for fulfilling them must therefore be specified in the project documents. The documents include plans, specifications, schedules, bill of quantities, and so on. Quality control in construction typically involves ensuring compliance with minimum standards of material and workmanship in order to ensure the performance of the facility according to the design. These minimum standards are contained in the specification documents. For the purpose of ensuring compliance, random samples and statistical methods are commonly used as the basis for accepting or rejecting work completed and batches of materials. Rejection of a batch is based on nonconformance or violation of the relevant design specifications.

Quality in construction is achieved through the complex interaction of many participants in the facilities development process. In case of construction projects, an organizational framework is established and implemented mainly by three parties: owner, designer/consultant, and contractor. Project quality is the result of aggressive and systematic application of quality control and quality assurance. Figure 6.1 illustrates Juran's triple concept applied to construction.

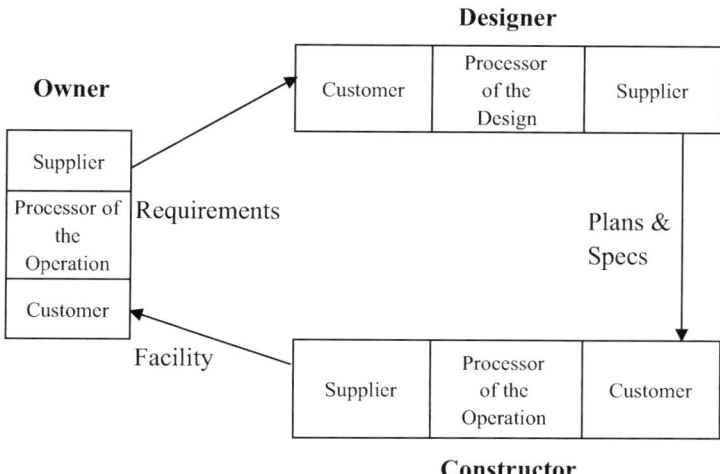

FIGURE 6.1 Juran's triple-role concept applied to construction. (CII Publication 10-4 Reprinted with permission from CII, University of Texas.)

The quality plan for construction projects is part of the overall project documentation consisting of the following:

1. Well-defined specification for all the materials, products, components, and equipment to be used to construct the facility
2. Detailed construction drawings
3. Detailed work procedure
4. Details of the quality standards and codes to be compiled
5. Cost of the project
6. Manpower and other resources to be used for the project
7. Project completion schedule

Figure 6.2 illustrates functional relationships between various participants.

These definitions when applied to construction projects relate to the contract specifications or owner's/end user's requirements to be formulated in such a way that construction of the facility is suitable for the owner's use or meets the owner's/user's requirements.

Based on aforementioned, quality of construction projects can be defined as follows: construction project quality is fulfillment of owner's needs as per defined scope of works within a budget and specified schedule to satisfy owner's/user's requirements. The phenomenon of these three components can be called as "construction project trilogy" and is illustrated in Figure 6.3

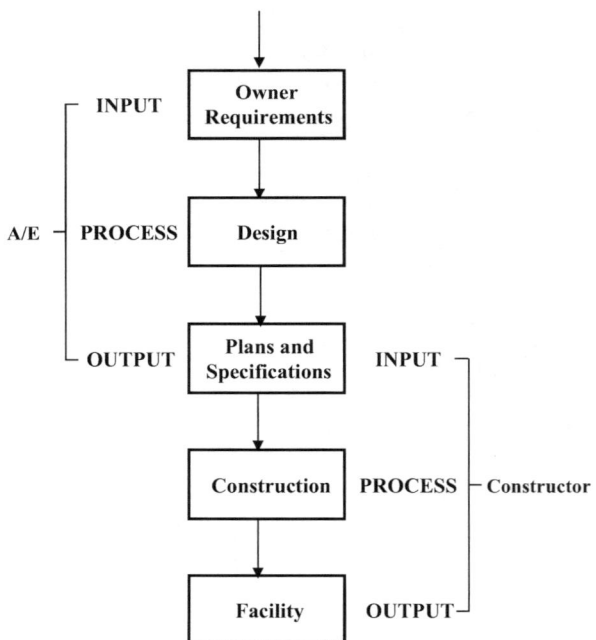

FIGURE 6.2 Juran's triple-role functional relationship. (CII source Document No. 51 Reprinted with permission from CII, University of Texas.)

Overview of Quality in Oil and Gas Projects

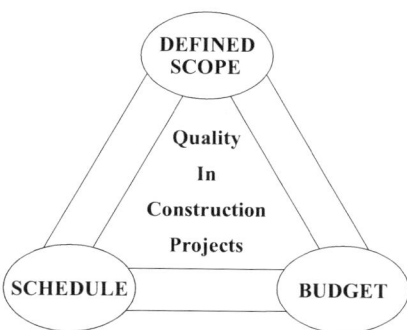

FIGURE 6.3 Construction project trilogy.

6.5.1 CONSTRUCTION QUALITY AND TOTAL QUALITY MANAGEMENT

TQM is an organization-wide effort centered on quality to improve performance that involves everyone and permeates every aspect of an organization to make quality a primary strategic objective. It is a way of managing an organization to ensure the satisfaction at every stage of the needs and expectations of both internal and external customers.

Construction projects being unique and nonrepetitive in nature need specified attention to maintain the quality. Each project has to be designed and built to serve a specific need. TQM in construction projects typically involves ensuring compliance with minimum standards of material and workmanship in order to ensure the performance of the facility according to the design. TQM in a construction project is a cooperative form of doing the business that relies on the talents and capabilities of both labor and management to continually improve quality. The important factor in construction projects is to complete the facility per the scope of works to customer/owner satisfaction within the budget and to complete the work within the specified schedule to meet the owner's defined purpose.

Based on the philosophies and concept of quality of construction projects, the quality of construction projects can be evolved as follows:

1. Properly defined scope of work.
2. Owner, project manager, design team leader, consultant, and constructor's manager are responsible to implement quality.
3. Continuous improvement can be achieved at different levels as follows:
   a. Owner—Specify the latest needs.
   b. Designer—Specification should include the latest quality materials, products, and equipment.
   c. Constructor—Use the latest construction equipment to build the facility.
4. Establishment of performance measures:
   a. Owner:
      i. To review and ensure that designer has prepared the contract documents that satisfy his needs.

ii. To check the progress of work to ensure compliance with the contract documents.
  b. Consultant:
     i. As a consultant designer, include the owner's requirements explicitly and clearly define them in the contract documents.
     ii. As a supervision consultant, supervise contractor's work per contract documents and the specified standards.
     iii. As a project management consultant (PMC), perform project management activities such as planning, monitoring, and managing the project.
  c. Contractor:
     i. To construct the facility as specified and use the materials, products, and equipment that satisfy the specified requirements.
5. Team approach—Every member of the project team should know that TQM is a collaborative effort, and everybody should participate in all the functional areas to improve the quality of the project work. They should know that it is a collective effort by all the participants.
6. Training and education—Both consultant and contractor should have customized training plans for their management, engineers, supervisors, office staff, technicians, and laborers.
7. Establish leadership—Organizational leadership should be established to achieve the specified quality. Encourage and help the staff and laborers to understand the quality to be achieved for the project.

These definitions when applied to construction projects relate to the contract specifications or owner's/end user's requirements to be formulated in such a way that construction of the facility is suitable for the owner's use or meets the owner's requirements. Quality in construction is achieved through the complex interaction of many participants in the facilities development process.

Construction projects are custom-oriented and custom design having specific requirements set by the customer/owner to be completed within finite duration and budget. Construction project is custom than a routine and repetitive business. Construction projects involve many participants including owners, designers, contractors, and many other professionals form construction-related industries. Construction projects being unique and nonrepetitive in nature need specified attention to maintain the quality. Quality in construction is achieved through the complex interaction of many participants in the facilities development process.

Participation involvement of all three parties (owner, designer, and contractor) at different levels of construction phases is required to develop quality system and application of quality tools and techniques. With the application of various quality principles, tools, and methods by all the participants at different stages of construction project, rework can be reduced, resulting in savings in the project cost and making the project qualitative and economical. This will ensure completion of construction and making the project most qualitative, competitive and economical.

TQM is implementation of quality system by organization to satisfy the customer's needs/requirements by developing a system that will make the project qualitative, competitive, and economical.

Overview of Quality in Oil and Gas Projects

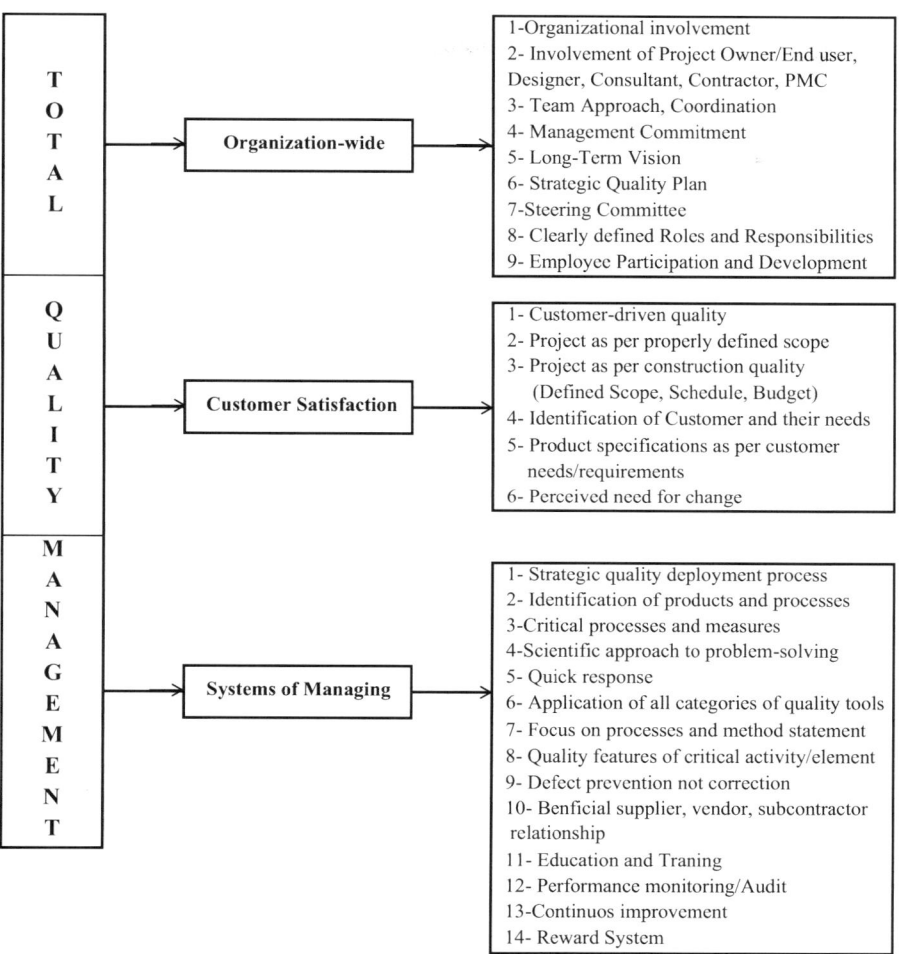

FIGURE 6.4 Concept of total quality management in construction projects.

Figure 6.4 illustrates concept of TQM in construction projects.

Implementation of TQM in construction projects has tangible results in the following:

- Owner satisfaction
- Project execution time and cost
- Owner/contractor relationship
- Simplification in work process
- Supplier/vendor/subcontractor performance
- Employee morale
- Savings in engineering cost
- Savings in material procurement cost

6.6 QUALITY PRINCIPLES IN OIL AND GAS PROJECTS

Construction projects are mainly capital investment projects. They are customized and nonrepetitive in nature. Construction projects have become more complex and technical, and the relationships and the contractual grouping of those who are involved are also more complex and contractually varied. Quality in construction is achieved through complex interaction of many participants in the facilities development process.

Construction projects comprised of a cross section of many different participants. These participants are influenced by and depend on each other in addition to "other players" involved in the construction process. Traditional construction projects have involvement of three main groups. These are as follows:

1. Owner
2. Designer (consultant)
3. Contractor

Participation involvement of all three parties at different levels of construction phases of oil and gas project is required to develop quality system and application of quality tools and techniques. In oil and gas projects, the project owner engages PMC. PMC plays an important role in development of oil and gas project. Construction project quality has mainly three components. These are as follows:

1. Scope
2. Schedule
3. Cost (budget)

In order to achieve a successful project as per owner/end user satisfaction, project documents to be formulated in such a way that construction of project is suitable for owner's use/end user's use or to meet the owner's requirements. An ISO document has listed eight quality management principles (CLIPSCFM) on which the quality management system (QMS) standards of the revised ISO 9000:2000 series are based. These are as follows:

Principle 1—Customer focus
Principle 2—Leadership
Principle 3—Involvement of people
Principle 4—Process approach
Principle 5—System approach to management
Principle 6—Continual improvement
Principle 7—Factual approach to design making
Principle 8—Mutual beneficial supplier relationship

With the application of various quality principles, tools, and methods by all the participants at different stages of construction project, rework can be reduced, resulting in savings in the project cost and making the project qualitative and economical.

Overview of Quality in Oil and Gas Projects

This will ensure completion of construction and making the project most qualitative, competitive, and economical. In order to improve construction project quality and to eliminate/reduce unsatisfactory relations between project owner, designer, and contractor, Table 6.2 summarizes quality principles that are applicable to construction projects.

TABLE 6.2
Principles of Quality in Oil and Gas Projects

Principle	Construction Projects' Quality Principle
Principle 1 (customer focus)	1.1 Designer and consultant are responsible to provide owner's requirements explicitly and clearly defining the standards of the end products and their compliance in the contract documents.
	1.2 Engineering design should include the process, process equipment, and engineering systems' requirements clearly and without any ambiguity for ease of operation.
	1.3 The project and end products should satisfy owner's needs/requirements and be suitable for intended usage
Principle 2 (leadership)	2.1 Owner, designer, consultant, and contractor are fully responsible for application of quality management system to meet customer requirements and strive to exceed customer expectations by complying with defined scope of work in the contract documents.
	2.2 Every member of the project team should exert collaborative efforts in all the functional areas to improve the quality of project
Principle 3 (involvement/engagement of people)	3.1 Each member of the project team should participate and fully involve as per their abilities in all the functional areas by adhering to team approach and coordination to continuously improve quality of the project.
Principle 4 (process approach)	4.1 Contractor has to build the facility as stipulated in the contract documents, plan, and specifications as per the approved schedule and within agreed upon budget to meet owner's objectives.
	4.2 Contractor should study all the documents during tendering/bidding stage and submit his proposal taking into consideration all the requirements specified in the contract documents and identifying, understanding, and managing interrelated processes as a system in achieving the specified product output.
	4.3 Contractor is responsible to provide all the resources, manpower, material, equipment, and so on to build the facility as per specifications to produce the specified products.
	4.4 Contractor has to check executed/installed works to confirm that works have been performed/executed as specified, using specified/approved materials, approved shop drawings, installation methods, and specified references, codes, and standards to meet intended use.
Principle 5 (system approach to management)	5.1 Contractor has to prepare contractor's quality control plan and follow the same to ensure meeting the performance standards specified in the contract documents.

(Continued)

TABLE 6.2 (*Continued*)
Principles of Quality in Oil and Gas Projects

Principle	Construction Projects' Quality Principle
	5.2 Method of payments (work progress, material, equipment, etc.) to be clearly defined in the contract documents. Rate analysis of bill of quantities or bill of materials item to be agreed before signing of contract
	5.3 Contract documents should include a clause to settle the dispute arising during construction stage
Principle 6 (continual improvement)	6.1 Contractor shall follow the submittal procedure specified in the contract documents for detailed design, procurements, checklists, inspection, and testing procedures as per the communication matrix. Review the contents of transmittals and executed works prior to submission for approval.
Principle 7 (factual approach to evidence-based design making)	7.1 Contractor shall follow an agreed-upon quality assurance and quality control plan consultant. PMC shall be responsible to oversee the compliance with contract documents and specified standards and codes.
	7.2 Contractor is responsible to construct the facility to produce the products as specified and use the material, products, systems, equipment, and methods which satisfy the specified requirements (factual approach to design making).
Principle 8 (mutual beneficial relationship management)	8.1 Contractor/all team members should participate and put collective efforts to perform the works as per agreed-upon construction program and handover the project as per contracted schedule to meet the owner's requirements.
	8.2 All team members should focus on participative management and strong operational accountability at the individual contributory level to follow principles of total quality management.

PMC, project management consultant.

6.7 QUALITY MANAGEMENT SYSTEM

A system is an assembly of components or elements having a functional relationship to achieve a common objective for a useful purpose. A system is an assembly of components or elements having a functional relationship to achieve a common objective for useful purpose. A system is composed of components, attributes, and relationships. The simple behavioral approach to systems is generally known as the black box and is represented schematically in Figure 6.5. The black box system phenomenon establishes the functional relationship between system inputs and outputs.

A QMS is a set of coordinated documentation that includes a quality manual, quality procedures, processes as well as work instructions (details of works to be done), quality forms, and records that are developed to meet customer's and regulatory requirements taking into consideration organization's quality policies and business objectives and improve the effectiveness and efficiency on continuous basis. It

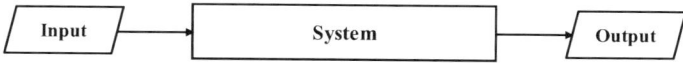

FIGURE 6.5 Black box.

is based on internal and external controls. Internal quality refers to the quality of the product(s) or service and how it is developed. External quality controls refer to customer service and how the organization delivers the finished product(s) or service to the end user. Thus, for QMS, the input is customer's requirement(s) and output is product(s), services as per customer satisfaction.

A QMS covers all the activities, leading to the final product or service. The quality system depends entirely on the scope of operation of the organization and particular circumstances such as number of employees, type of organization, and physical size of the premises of the organization. The quality manual is the document that identifies and describes the QMS.

The ISO is the world's largest developer and publisher of international standards. It is a nongovernmental organization that forms a bridge between the public and private sectors. The ISO has more than 21,000 international standards.

ISO 9000 quality system standards are a tested framework for taking a systematic approach to managing the business process so that organizations turn out products or services conforming to customer's satisfaction. A typical ISO QMS is structured on four levels. These are as follows:

1. Quality policy
2. Quality manual
3. Work instructions/procedures
4. Quality forms and records

The structure of the QMS depends largely on the management structure in the organization. The number of manuals containing work instructions or procedures is determined by the size and complexity of the organization. ISO 9001:2015 identifies certain minimum requirements that all QMSs must meet to ensure customer satisfaction.

In general, QMSs begin with the writing of a quality manual. The quality manual serves as a roadmap for the QMS. It is more practical to build the "road" before preparing the "map". The manual is generic at first, and as the QMS develops, the manual is updated. The quality manual is the document that identifies and describes the QMS. The QMS is based on the guidelines for performance improvement per ISO 9004:2000 and the quality management requirements. ISO 9001:2015 outlines the necessary steps to implement the QMS. The manual is to be developed taking into consideration the following:

1. Eight principles (CLIPSCFM) of QMS as defined by ISO Technical Committee, TC 176.
2. All the related and applicable documents produced taking into consideration ten sections/clauses listed under ISO 9001:2015 to ensure that the manual is in compliant with ISO 9001:2015.

ISO 9001:2015 Sections/Clauses are discussed under chapter 2, *Section 2.4.2.1* Table 2.2.

In the construction industry, a contractor may be working at any time on a number of projects of varied natures. These projects have their own contract documents to implement project quality, which require a contractor to submit a contractor's quality control plan to ensure that specific requirements of the project are considered to meet client's requirements. Therefore, while preparing a QMS at a corporate level, the organization has to take into account tailor-made requirements for the projects, and accordingly, the manual should be prepared. Normally, QMS consists of documents produced taking into consideration relevant sections of ISO 9001:2015 and business-related activities. Development and implementation of an effective QMS is a strategic decision of the organization that helps to improve its overall performance and provide a sound basis for sustainable development initiatives. In construction projects, owner, designer, consultant, and contractor should have their own QMS to meet their business objectives. While developing the QMS, the organizations have to include those documents that are required to perform relevant processes and specific requirements that the organizations have business interest.

QMS approach is a process-based approach that incorporates plan–do–check–act (PDCA) and risk-based thinking. Figure 6.6 illustrates PDCA cycle covering ISO 9001:2015 sections/clauses and elements, activities of each section/clause to develop QMS manual.

Quality management in construction addresses both the management of project and the product of the project and all the components of the product. It also involves incorporation of changes or improvements, if needed. Construction project quality is fulfillment of owner's needs as per defined scope of works, as per specified schedule, and within the budget to satisfy owner's/user's requirements.

QMS in construction projects mainly consists of the following:

- Quality management planning (plan quality)
- Quality assurance (perform quality assurance)
- Quality control (perform quality control)

Chapter 3 discusses about quality management processes (quality planning, quality assurance, and quality control) and activities to be performed while following main stages of oil and gas construction projects:

1. Study stage
2. Design stage (conceptual design, front-end engineering design)
3. Bidding and tendering stage
4. Construction stage (Construction, testing, commissioning, and handover)

Quality management in each phase helps in the following:

- Enhance quality conformance to requirements.
- Produce desired results to satisfy customer's needs.

Overview of Quality in Oil and Gas Projects

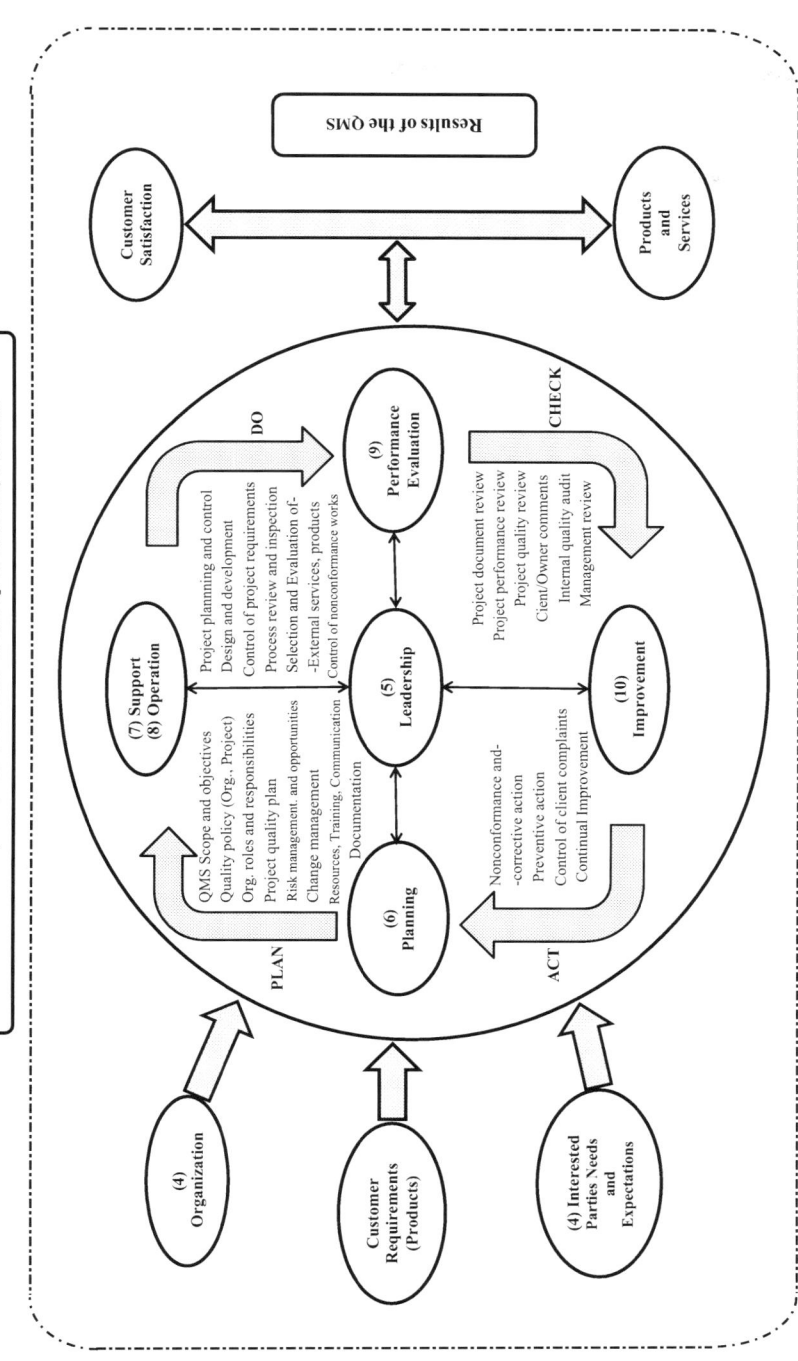

FIGURE 6.6 PDCA cycle for development of quality management system manual. PDCA, plan–do–check–act.

- Achieve quality objectives in the project deliverables.
- Reduce omission/errors in the project activity and element.
- Team approach and coordination.
- Manage and control quality at every stage/phase.
- Engineering design is properly developed to achieve economical and competitive results.
- Construction work is properly executed taking into consideration all the activities to fully meet all the project requirements.
- Timely completion of project with economical use of resources.
- Cost saving in the project.
- Ease of communication.
- Identify risk and take action.
- Understand quality issues and take corrective action for improvement.
- Auditing at each phase.

7 Selection of Project Teams

7.1 PROJECT TEAM

Traditional construction projects involve mainly three groups:

1. Owner
2. Designer(s)
3. Contractor(s)

Figure 7.1 illustrates relationship between these groups.
There are mainly two types of relationships:

1. Contractual
   - Contractual engagement
2. Functional
   - Engagement to fulfill the implied needs

In the case of construction projects, organizational framework is established and performed mainly by three parties: owner, designer/consultant, and contractor.

Participation involvement of all three parties at different levels of construction phases is required to develop the project. Table 7.1 summarizes typical responsibilities of owner, design professional, and contractor.

Construction projects are mainly capital investment projects. They are customized and nonrepetitive in nature. Each project has specific need and customized design. They are executed based on predetermined set of goals and objectives. Oil and gas sector projects are considered as biggest sector in terms of investment as well as in employment generation. Construction projects are constantly increasing

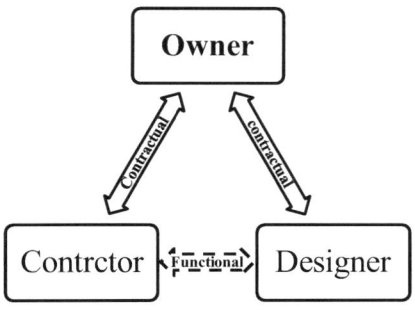

FIGURE 7.1 Traditional construction project team members.

TABLE 7.1
Typical Responsibilities of Construction Project Team Members

Serial Number	Responsibilities
1	Project owner
	1.1 Defining project goals and objectives
	1.2 Project financing (funding)
	1.3 Selection of team members—design professional(s)
	1.4 Selection of team members—contractor(s)
	1.5 Fulfilling of contractual obligations toward other team members
	1.6 Selecting project delivery system
	1.7 Regulatory approvals
	1.8 Provide land/project site
	1.9 Accept project/facility
2	Design professional
	2.1 Design of the project as per TOR, contract to meet owner's goals and objectives
	2.2 Compliance with regulatory requirements
	2.3 Design of project as per applicable codes and standards
	2.4 Consider constructability
	2.5 Environmental and safety considerations
	2.6 Design the project to meet owner's satisfaction and intended use
	2.7 Design the project to be completed within specified schedule
	2.8 Design the project to be completed within agreed-upon budget
	2.9 Attend to request for information during construction
	2.10 Construction supervision, if contracted
3	Contractor
	3.1 Build/construct the project as per contract documents, specification
	3.2 Follow contract documents
	3.3 Complete the project within contracted schedule
	3.4 Complete the project within the agreed-upon budget
	3.5 Manage project quality
	3.6 Provide resources as specified in the contract document
	3.7 Follow communication matrix
	3.8 Manage project risk
	3.9 Testing and commissioning of project as per the contract
	3.10 Handover the project to the owner/end user

TOR, terms of reference.

in technological complexity. Construction projects have become more complex and technical, and the relationships and the contractual grouping of those who are involved are also more complex and contractually varied.

Quality management in construction projects is different to that of manufacturing. Quality in construction projects is not only the quality of products and equipment used in the construction but also the total management approach to complete the facility as per the scope of works to customer/owner satisfaction within specified schedule and within the agreed-upon budget to meet owner's defined purpose. The nature of the contracts between the parties plays a dominant part in the quality system required from the project, and the responsibility for achieving them must therefore be specified in the project documents.

Oil and gas projects are complex construction projects and have many challenges such as delays, changes, disputes, and accidents at site, and therefore, need efficient management of the project from the beginning to the end of construction of the facility/project to meet the intended use and owner's expectations. It is required that these projects are managed properly at different levels of construction phases to ensure completion of construction, making the project most qualitative, competitive, and economical.

There are several types of project delivery systems, contracting systems in which these parties are involved at different levels. All these contract deliverable systems follow generic life cycle phases of construction project; however, the involvement/participation of various parties differs depending on the type of deliverable systems adapted for a particular project. Oil and gas construction projects have become more complex and technical, and the relationships and the contractual grouping of those who are involved are also more complex and contractually varied. Construction projects are facing challenges of ever-changing construction technology, knowledge ideology, management techniques, project delivery system, contracting practices, and dynamic nature of site works. Due to constantly increasing technical complexity in the construction projects, there are many challenges such as delays, changes, disputes, and accidents at site and therefore the construction projects need efficient management of the project from the beginning to the end of construction of the facility/project to meet the intended use and produce products as per owner's expectations. The main area of construction/project management covers planning, organizing, executing, and controlling to ensure that the project is built as per defined scope, maintaining the completion schedule and within agreed-upon budget. The owner/client may not have necessary staff/resources in-house to manage planning, design, and construction of the construction project to achieve the desired results. Therefore, in such cases, the oil and gas project owners engage a professional firm or a person, called project management consultant (PMC), who is trained and has expertise in the management of construction processes, to assist in developing bid documents, overseeing, monitoring, controlling, and coordinating project for the owner. The basic project management concept is that the owner assigns the contract to a firm that is knowledgeable and capable of coordinating all the aspects of the project to meet the intended use of the project by the owner. The professionals and specialists engaged by the owner as PMC brings knowledge and experience that contribute to decisions at every stage of project for successful completion.

PMC mainly involves construction/project management–related activities such as:

- Planning
- Scope management
- Scheduling
- Cost control
- Monitoring and control
- Quality control/quality assurance

- Human resources
- Risk
- Health safety and environment

PMC is responsible to oversee performance of contractor(s) toward construction-related activities (engineering design, construction process, testing, commissioning, and handover).

There are mainly three key attributes in construction projects, which the PMC has to manage effectively and efficiently to achieve successful project. These are as follows:

1. Scope
2. Time (schedule)
3. Cost (budget)

From the quality perspective, these three elements are known as "quality trilogy", whereas when considered with project/construction management perspective, these are known as "triple constraints".

In order for successful management of the project, the PMC should have all the related information about construction/project management principles, tools, processes, techniques, and methods. PMC should also have the professional knowledge about management functions, management processes and project phases (technical processes), and the skills and expertise to manage the project in a systematic manner at every stage of the project.

Project/construction management is a framework for the construction/project manager to evaluate and balance these competing demands. To balance these attributes at each stage of project execution, the project phases and project subdivisions into various elements/activities/subsystems having functional relationship should be developed taking into consideration various management functions, management processes, and interaction and/or combination among some or all of their activities/elements. The major activities of oil and gas project phases are discussed under Chapter 1 and are listed in Table 1.1.

7.2 CONSTRUCTION PROJECT STAKEHOLDERS

Construction projects have direct involvement of the following three stakeholders;

1. Owner
2. Designer
3. Contractor

However, there are many other stakeholders who have significant influence/impact on the outcome of construction project. These stakeholders include members from within organization and people, agencies, and authorities outside the organization. It is important to identify stakeholders who have interest and significant influence in the project. Figure 7.2 illustrates stakeholders having involvement or interest in the construction project.

Selection of Project Teams

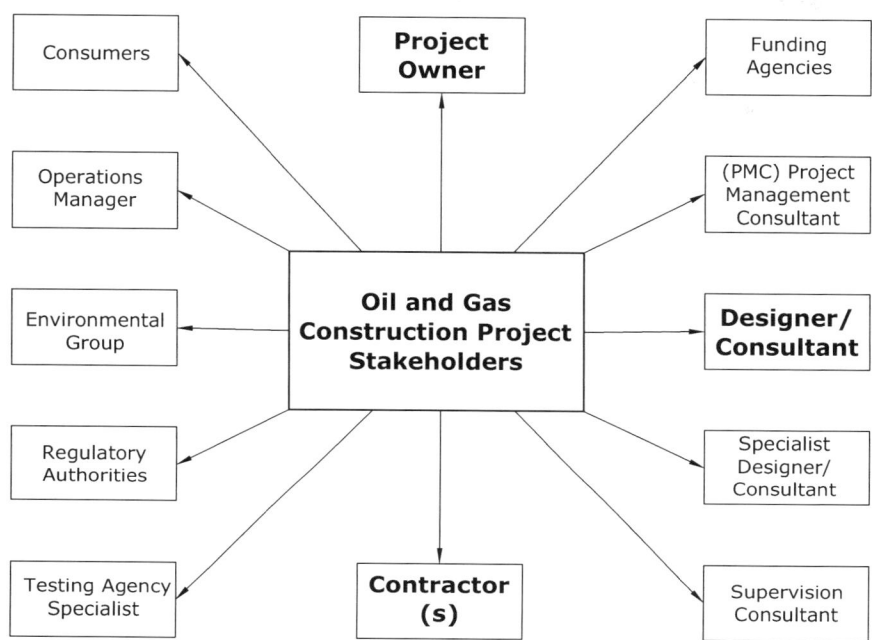

FIGURE 7.2 Construction project stakeholders. (Abdul Razzak Rumane. (2016). *Handbook of Construction. Management: Scope, Schedule, and Cost Control.* Reprinted with permission from Taylor & Francis Group.)

A stakeholder is anyone who has involvement, interest, or impact in the construction project processes in a positive or negative way. Stakeholder plays a vital role in determining, formulation, and successful implementation of project processes. Stakeholders can mainly be classified as follows:

- Direct stakeholders
- Indirect stakeholders
- Positive stakeholders
- Negative stakeholders
- Legitimacy and power

In order to run a successful project, it is important to address the needs of project stakeholders effectively predicting how the project will be affected and how the stakeholders will be affected. Successful completion of construction project is dependent on meeting the expectations of stakeholders; therefore, it is necessary to involve stakeholders through communicating and working together to address the needs/expectations and issues of the stakeholders.

Construction projects involve coordinated actions and input from many professionals and specialists to achieve defined objectives. The involvement and role and responsibilities of construction stakeholders depend on the project delivery system.

There are several types of project delivery systems, contracting systems in which these parties are involved at different levels. These are as follows:

- Traditional system—design–bid–build
- Integrated system—design–build, turnkey–EPC, BOT, etc.
- Management-oriented system
- Integrated project delivery system

All these contract deliverable systems follow generic life cycle phases of construction project; however, the involvement/participation of various parties differs depending on the type of deliverable systems adapted for a particular project. Oil and gas projects mainly follow design–build/turnkey type of project delivery system.

In order to run a successful project, it is important to address the needs of project stakeholders effectively predicting how the project will be affected and how the stakeholders will be affected. Table 7.2 is an example of typical responsibilities matrix for stakeholders of construction project developed depending on the roles and responsibilities of the stakeholders and their needs, expectations, and influence on the project.

7.3 SELECTION OF TEAM MEMBER

Construction is the translation of owner's goals and objectives, by the contractor, to build the facility as stipulated in the contract documents, plans, and specifications on schedule and within budget.

Construction projects are custom-oriented and custom design having specific requirements (defined scope) set by the customer/owner to be completed within finite duration and assigned budget.

Construction projects comprised of a cross section of many different participants. These participants are influenced by and depend on each other in addition to "other players" involved in the construction process. Therefore, efforts are required to ensure completion of construction project within agreed-upon schedule and within approved budget to satisfy owner's/client's/end user's intended need (defined scope).

A project delivery system is defined as the organizational arrangement among various participants comprising owner, designer, contractor, and many other professionals involved in the design and construction of a project/facility to translate/transform the owner's needs/goals/objectives into a finished facility/project to satisfy the owner's/end user's requirements. Generally, design–build/turnkey type of project delivery system is followed in oil and gas project. Accordingly, the owner has to procure services of design professional and contractor. Since oil and gas projects have become more complex and technical, the relationships and the contractual grouping of those who are involved are also more complex and contractually varied. Complex and major construction projects have many challenges such as delays, changes, disputes, and accidents at site and therefore need efficient management of the project from the beginning to the end of construction of the facility/project to meet the intended use and owner's expectations. The owner/client may not have necessary staff/resources in-house to manage planning, design, and construction of the construction project to achieve the desired results. Therefore, in such cases, the owners

TABLE 7.2
Typical Responsibilities Matrix of Stakeholders

Serial Number	Activity	Project Owner	Project Manager Consultant	Designer/ Consultant	Contractor	Supervisor	Regulatory Authority	Funding Agency	End User/ Facility Manager	Notes/ Comments
1	Project initiation	P	-	-	-	-	-	B	B	
2	Selection of project management consultant	P	-	-	-	-	-	-	-	
3	Selection of designer	P	B	-	-	-	-	-	-	
4	Preparation of terms of reference	A	P	-	-	-	-	-	-	
5	Preparation of design	A	B	P	-	-	-	-	-	
6	Value engineering	A	R	P	-	-	R	-	-	
7	Preparation of contract documents	A	B	P	-	-	-	-	-	
8	Project schedule	A	B	P	-	-	-	C	C	
9	Project budget	A	B	P	-	-	-	B	-	
10	Preparation of tendering documents	A	P	B	-	-	-	-	-	
11	Submission of bid	C	C	-	P	-	-	-	-	
12	Evaluation of bid	C	C	P	-	-	-	-	-	
13	Selection of contractor	A	P	B	-	-	-	C	C	
14	Approval of subcontractor	A	B	B	P	-	-	-	-	
15	Approval of contractor's staff	A	B	B	P	-	-	-	-	
16	Execution of works	C	C	R	P	R	-	-	-	
17	Supervision of works	C	C	R	P	P	-	-	-	

(*Continued*)

TABLE 7.2 (Continued)
Typical Responsibilities Matrix of Stakeholders

Serial Number	Activity	Project Owner	Project Manager Consultant	Designer/ Consultant	Contractor	Supervisor	Regulatory Authority	Funding Agency	End User/ Facility Manager	Notes/ Comments
18	Approval of material	C	A	R	P	B	–	–	–	–
19	Approval of shop drawings	C	C	A	P	B	–	–	–	–
20	Construction schedule	C	A	R	P	B	–	–	–	
21	Monitoring progress	C	P	P	P	B	–	–	–	
22	Monitoring cost	C	P	P	B	B	–	–	–	
23	Payments	A	R	R	P	B	–	–	–	
24	quality plan	C	B	R	P	B	–	–	–	
25	Project quality	C	R	R	P	P	–	–	–	
26	Request for information	C	C	R	P	B	–	–	–	
27	Meetings	E	E	P	E	E	–	–	–	
28	Approval of change	A	B	R	P	B	–	–	–	
29	Safety plan	C	B	R	P	B	–	–	–	
30	Site safety	C	C	B	P	P	–	–	–	
31	Testing and commissioning	C	C	R	P	D	–	–	C	
32	Authorities approval	C	C	B	P	B	A	–	–	
33	Snag list	C	C	R	P	P	–	–	C	
34	Substantial completion certificate	A	R	P	C	–	–	–	C	

P, prepare/initiate/responsible; R, review/comment; B, advise/assist; A, approve; E, attend; C, inform.

Selection of Project Teams

engage a professional firm or a person, called project management consultant (PMC), who is trained and has expertise in the management of construction processes, to assist in developing bid documents, overseeing, monitoring, controlling, and coordinating project for the owner. Normally, PMC is engaged as early in the project as possible to guide and assist the owner through all the phases of project.

In construction projects, the involvement of outside companies/parties starts at the early stage of project development process. The owner/client has to decide which works are to be procured and constructed by outside companies/parties. Every organization has their procurement system to procure services, contracts, and product from others. Figure 7.3 illustrates project team selection stages in oil and gas project.

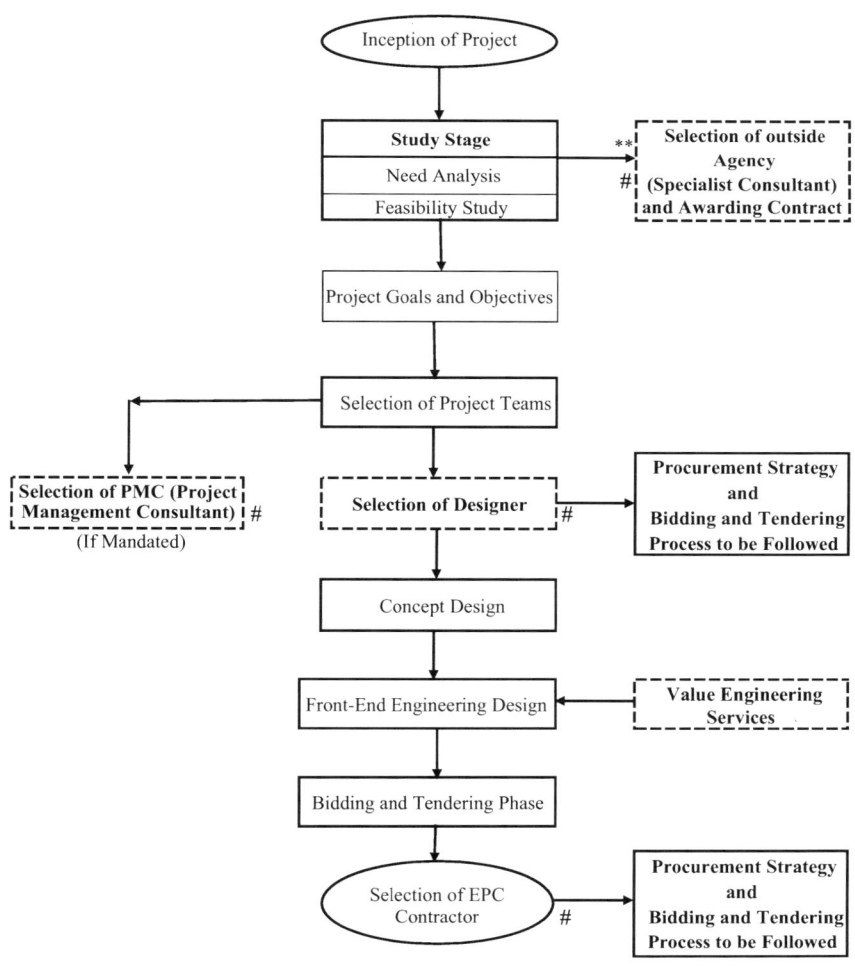

** If in-house facility not available
− − − − − Related activities to engage outside agency (specialist consultant)
\# Procurement Strategy and Bidding and Tendering process to be followed

FIGURE 7.3 Project team selection stages in oil and gas project.

7.3.1 SHORT-LISTING OF TEAM MEMBERS

Procurement in construction projects is an organizational method, process, and procedure to obtain the required services, systems, and products. It includes the processes such as prequalification and bidding and tendering processes to acquire all the related services from designer, contractors, consultants, and companies to the satisfaction of the owner/client/end user. The procurement in construction projects also involves specialist consultant/agencies, commissioning professional services and creating a specific solution. The procurement strategy involves the following:

- Identification of
    - what are the in-house services available;
    - what services to be procured from outside agencies/organizations;
    - how to procure (direct contract, competitive bidding, competitive negotiations, quality-based system (QBS));
    - how much to procure;
    - how to prequalify, select a consultant/designer/contractor/vendor/supplier; and
    - how to arrive at appropriate price, terms, and conditions.
- Signing of contract

Table 7.3 illustrates 5W2H analysis for procurement of project teams, and Figure 7.4 illustrates project team procurement strategy for short-listing/registration of different team members.

While engaging an outside agency (designer, contractor, consultant), the following selection criteria need to be considered as a minimum to prequalify the contractor:

- Available skill level
- Relevant/past performance on similar type of work
- Reputation about their works
- The number of projects (works) successfully completed
- Technical competence
- Knowledge about the type of projects (works) for which likely to be engaged

TABLE 7.3
5W2H Analysis for Procuring Project Teams

Serial Number	Why/How	Related Analyzing Question
1	Why	Why project teams and outside agencies to be procured
2	What	What will be the overall benefits for the project
3	Who	Who will be responsible to manage project teams
4	Where	Where can we get qualified teams having experience and expertise in oil and gas projects
5	When	When the project teams are to be procured and engaged on the project
6	How many	How many project teams are to be procured/selected
7	How much	How much of the project cost to be assigned for procurement of project teams

Selection of Project Teams

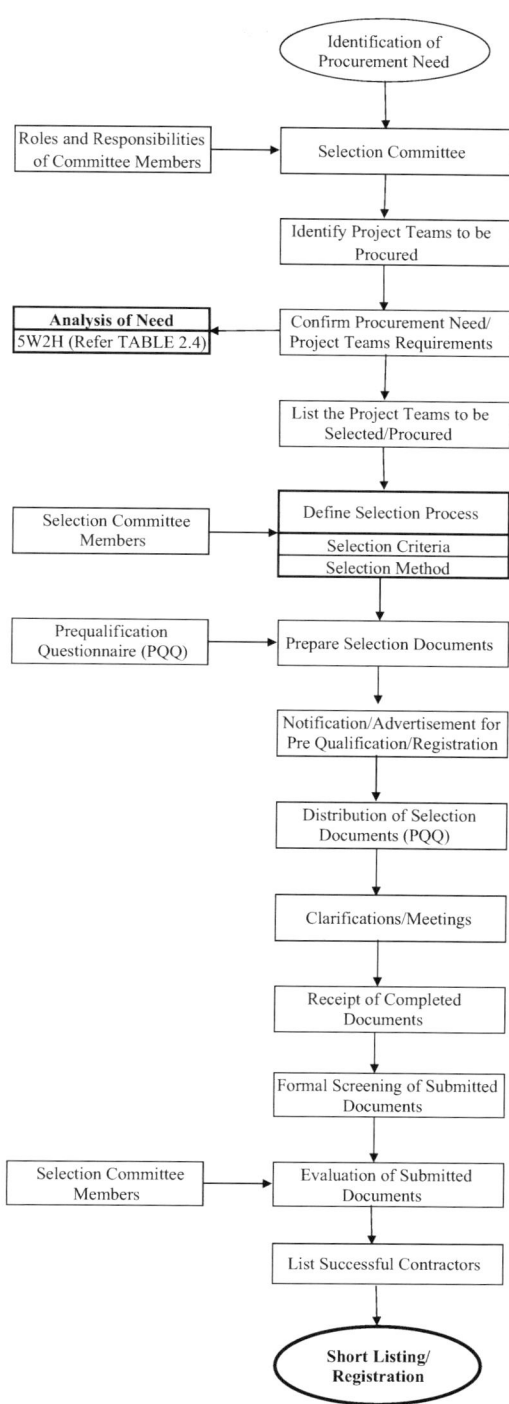

FIGURE 7.4 Project team procurement strategy for short-listing/registration.

- Available resources
- Commitment to creating best value
- Commitment to containing sustainability
- Rapport/behavior
- Communication
- Commitment to health, safety, and environmental requirements

After short-listing/prequalifying/registering of contractors, proposal documents/tender documents are distributed to submit the proposal/quotation. The following are the common procurement methods for selection of project teams:

1. Low bid
   - Selection is based solely on the price.
   - Total construction cost including the cost of work is the sole criterion for the final selection.
2. Best value
   a. Total cost
      - Selection is based on total construction cost and other factors.
      - Both total cost and other factors are criteria in the final selection.
   b. Fees
      - Selection is based on weighted combination of fees and qualification.
      - Both fees and qualifications are factors in the final selection.
3. Qualification-based selection
   - Selection is based solely on qualification.
   - Total costs of the work are **not** a factor in the final selection. Qualification is the sole factor used in the final selection.

Contract is a formalized means of agreement, enforceable by law, between two or more parties to perform certain works or provide certain services against agreed-upon financial incentives to complete the works/services. Regardless of the type of project delivery system selected, the contractual arrangement by which the parties are compensated must also be established. In the construction projects, determining how to procure the product is as important as determining what and when. In most procurement activities, there are several options available for purchase or subcontracts. The basis of compensation type relates to the financial arrangement among the parties as to how the designer or contractor is to be compensated for their services. The following are the most common types of contract/compensation methods:

1. Firm fixed price or lump sum contract
   a. Firm fixed price
   b. Fixed price incentive fee
   c. Fixed price with economic adjustment price
2. Unit price
   - Specific price for a particular task or unit of worked performed by the contractor
   - In this method, each unit of work to be precisely defined

3. Cost reimbursement contract (cost plus)
   a. Cost plus percentage fee
   b. Cost plus fixed fee
   c. Cost plus incentive fee
   d. Cost plus award fee
4. Remeasurement contract
   a. Bill of quantities
   b. Schedule of rates
   c. Bill of materials
5. Target price contract
6. Time and material contract
7. Guaranteed maximum price (GMP)
   a. Cost plus fixed fee GMP contract
   b. Cost plus fixed fee GMP and bonus contract
   c. Cost plus fixed fee GMP with arrangement for sharing any cost saving–type contract

7.3.1.1 Bidding Procedure

Most government, public sector projects follow low bid selection method. There are three international bidding procedures, which may be selected by the project owner to suit the nature of project procurement. These are as follows:

1. Single stage—one envelope: In this procedure, the bidders submit bids in one envelope containing both technical and financial proposals. The bids are evaluated by the selection committee, who then send their recommendation to the owner. Following the review and concurrence by the owner, the contract is awarded to the lowest bidder.
2. Single stage—two envelopes: In this procedure, the bidders submit two envelopes, one containing the technical proposal and the other containing the financial proposal. Initially, technical proposals are evaluated without referring to the price. Bidders whose proposal does not conform to the requirements may be rejected/not accepted. Following the technical proposal evaluation, the financial proposals of technically responsive bidders are reviewed. Upon review and concurrence by the owner, the contract is awarded to the lowest bidder.
3. Two stages: In this procedure, during the first stage, the bidders submit their technical offers on the basis of operating and performance requirements, but without price. The technical offers are evaluated by the selection committee. Any deviations to the specified performance requirements are discussed with the bidders who are allowed to revise or adjust the technical offer and resubmit the same.

During the second stage, the bidders, whose technical offers are accepted, are invited to submit final technical proposal and financial proposal. Both the proposals are reviewed by the selection committee, and following review and concurrence by the owner, the contract is awarded to the lowest bidder. This procedure is mainly applicable for turnkey projects and complex plants.

7.3.2 SELECTION OF DESIGN TEAM

Design team (designer) consists of architects or engineers or consultant. They are the owner's appointed entity accountable to convert owner's conception and need into specific facility/required products with detailed directions through drawings and specifications within the economic objectives. They are responsible for the design of the project and in certain cases supervision of construction process.

Design team is selected based on the procurement strategy set up by the owner. The following is the normal procedure followed to select the design team:

- Short-listing/registration of prequalified teams
- Request for proposal (RFP)
- Prebid meeting(s)
- Submission of proposal/bid
- Selection of design team
- Signing of contract

Procurement strategy for prequalification/short-listing and registration of contractors is already discussed under Section 7.3.1. Please refer Figure 7.4. Table 7.4 lists prequalification questionnaires (PQQ) to select the design team (designer).

7.3.2.1 Request for Proposal

It is a project-based process involving solution, qualifications, and price as the main criteria that define a winning proponent. It is solicitation document requesting proponents requesting proposal in response to required scope of services. The document does not specify in detail how to accomplish or perform the required services. RFP can range from a single-step process for straightforward procurement to a multistage process for complex and significant procurement. Table 7.5 illustrates contents of the requirements of RFP document for construction project design team/consultant, Figure 7.5 illustrates logic flowchart for selection of project design team, and Figure 7.6 illustrates typical proposal submittal procedure by the designer.

Normally, design professionals (consultants) are hired on the basis of **qualifications**. The qualification-based selection can be considered as meeting one of the 14 points of Deming's principles of transformation which states "End the practice of awarding business on the basis of price alone". The basis of selection is solely based on demonstrated competence, professional qualification, and experience for the type of services required. In QBS, the contract price is negotiated after selection of the best qualified firm. Table 7.6 lists the criteria for selection of construction project designer/consultant on QBS basis, and Table 7.7 summarizes criteria for selection of the design team on different weightages basis.

TABLE 7.4
Prequalification Questionnaires for Selecting the Design Team

Serial Number	Question	Answer
1	Name of the organization and address	
2	Organization's registration and license number	
3	ISO certification	
4	LEED or similar certification	
5	Total experience (years) in designing the following type of projects	
	5.1 Process type projects	
	5.1.1 Petroleum plants	
	5.1.2 Petroleum refineries	
	5.1.3 Liquid chemical plants	
	5.1.4 Solid process plants	
	5.1.5 Liquid/solid plants	
	5.2 Non process type projects	
	5.2.1 Power plants	
	5.2.2 Manufacturing plants	
	5.3 Civil construction and commercial A/E	
	5.3.1 Residential	
	5.3.2 Industrial	
	5.3.3 Heavy engineering	
6	Size of project (maximum amount single project)	
	6.1 Process type projects	
	6.1.1 Petroleum plants	
	6.1.2 Petroleum refineries	
	6.1.3 Liquid chemical plants	
	6.1.4 Solid process plants	
	6.1.5 Liquid/solid plants	
	6.2 Non process type projects	
	6.2.1 Power plants	
	6.2.1 Manufacturing plants	
	6.3 Civil construction and commercial A/E	
	6.3.1 Residential	
	6.3.2 Industrial	
	6.3.3 Heavy engineering	
7	List successfully completed projects	
	7.1 Petroleum plants	
	7.2 Petroleum refineries	
	7.3 Liquid chemical plants	
	7.4 Solid process plants	
	7.5 Liquid/solid plants	

(Continued)

TABLE 7.4 (*Continued*)
Prequalification Questionnaires for Selecting the Design Team

Serial Number	Question	Answer
	7.6 Power plants	
	7.7 Manufacturing plants	
	7.8 Residential	
	7.9 Industrial	
	7.10 Heavy engineering	
8	List similar type (type to be mentioned) of projects completed	
	8.1 Project name and contracted amount	
	8.2 Project name and contracted amount	
	8.3 Project name and contracted amount	
	8.4 Project name and contracted amount	
	8.5 Project name and contracted amount	
9	Total experience in oil and gas projects	
10	Joint venture with any international organization	
11	Resources	
	11.1 Management	
	11.2 Engineering	
	11.3 Technical	
	11.4 Design equipment	
	11.5 Latest software	
12	Design production capacity	
13	Design standards	
14	Present work load	
15	Experience in value engineering (list projects)	
16	Financial capability (turnover for the past 5 years)	
17	Financial audited report for the past 3 years	
18	Insurance and bonding capacity	
19	Organization details	
	19.1 Responsibility matrix	
	19.2 CVs of design team members	
20	Design review system (quality management during design)	
21	Experience in preparation of contract documents	
22	Knowledge about regulatory procedures and requirements	
23	Experience in training of owner's personnel	
24	List of professional awards	
25	Litigation (dispute, claims) on earlier projects	

ISO, International Organization for Standardization; LEED, Leadership in Energy and Environmental Design.

TABLE 7.5
Contents of Request for Proposal for Design Team/Consultant

Serial Number	Content
	Project Details (Project Objectives)
1	Introduction
2	Project description
3	Designer's/consultant's scope of work
4	Preliminary project schedule
5	Preliminary cost of project
	Sample Questions (Information for Evaluation)
1	Designer/consultant name
2	Address
3	Quality management system certification
4	HSE considerations in design
5	Organization details
6	Type of firm such as partnership or limited company
7	Is the firm listed in stock exchange?
8	List of awards, if any
9	Design production capacity
10	Current workload
11	Insurance and bonding
12	Experience and expertise
13	Project control system
14	Design submission procedure
15	Design review system
16	Design management plan
17	Design methodology
18	Submission of alternate concept
19	Quality management during design phase

20	Design firm's organization chart
	  a. Responsibility matrix
	  b. CVs of design team members

21	Designer's experience with green building standards or highly sustainable projects
22	Conducting value engineering
23	Authorities approval
24	Data collection during design phase
25	Design responsibility/professional indemnity
26	Designer's relationship during construction
27	Preparation of tender documents/contract documents
28	Review of tender documents
29	Evaluation process and criteria
30	Any pending litigation
31	Price schedule

HSE, health, safety, and environmental.

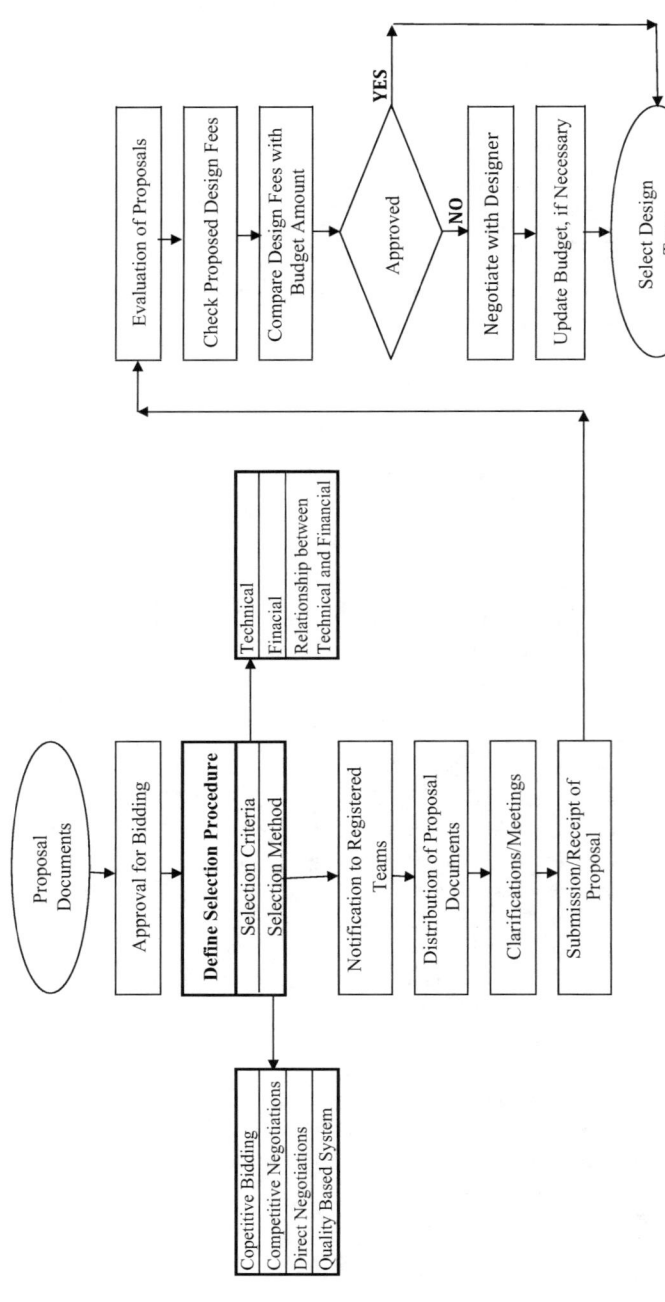

FIGURE 7.5 Logic flowchart for the selection of design team.

Selection of Project Teams

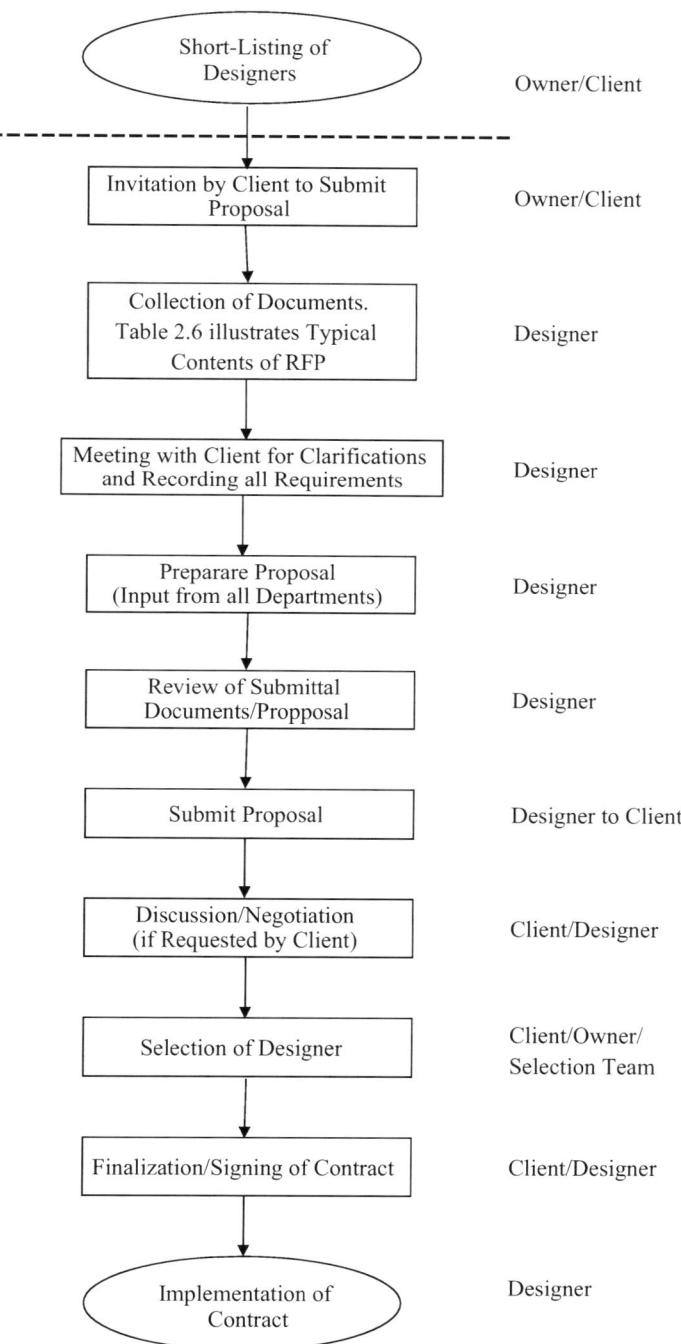

FIGURE 7.6 Typical proposal submittal procedure by designer.

TABLE 7.6
Qualification-Based Selection of Designer (Design Team)

Serial Number	Items to be Evaluated
1	Organization's registration and license
2	Management plan and technical capability
3	Quality certification and quality management system
4	LEED or similar certification
5	Number of awards
6	Design capacity to perform the work
7	Financial strength and bonding capacity
8	Professional indemnity
9	Current load
10	Experience and past performance in similar type of work
11	Experience and past performance in oil and gas projects
12	Experience and past performance in innovative process technology
13	Design of similar value projects in past
14	List of successfully completed projects
15	Proposed design approach in terms of the following: a. Performance b. Effectiveness c. Maintenance d. Logistic support e. Environment f. Green building
16	Design team composition
17	Experience and past performance of proposed individuals in similar projects
18	Professional certification of proposed individuals
19	Record of professionals in timely completion
20	Safety consideration in design
21	Litigation
22	Price schedule

LEED, Leadership in Energy and Environmental Design.
Source: Abdul Razzak Rumane. (2013). *Quality Tools for Managing Construction Projects.* Reprinted with permission from Taylor & Francis Group.

7.3.3 Selection of EPC Contractor

The first stage in the selection of the project team (contractor) is by short-listing the prospective contractors. In construction projects, the selection process for contractor can range from simply deciding to directly award a contract to a multistage process that involves information gathering about the contractor through request for information (RFI), request for qualification, prequalification questionnaire (PQQ) and soliciting activities.

TABLE 7.7
Selection Criteria for Design Team

Serial Number	Evaluation Criteria	Weightage		Notes
1	General Information			
	a. Company information			
2	Business		10%	
	a. LEED or similar certification	5%		
	b. ISO certification	5%		
3	Financial		20%	
	a. Turnover	5%		
	b. Financial standing	5%		
	c. Insurance and bonding limit	10%		
4	Experience		30%	
	a. Design experience	10%		
	b. Process type of projects	10%		
	c. Oil and gas (petroleum) projects	10%		
5	Design capability		10%	
	a. Design approach	4%		
	b. Process design approach	3%		
	c. Design capacity	3%		
6	Resources		20%	
	a. Design team qualification	10%		
	b. Design team composition	5%		
	c. Professional certification	5%		
7	Design quality		5%	
8	Safety consideration in design		5%	

ISO, International Organization for Standardization; LEED, Leadership in Energy and Environmental Design.

Note: The weightages mentioned in the table are indicative only. The % can be determined as per the owner's strategy.

In case of selection of EPC contractor, the information for prequalification is gathered through the following:

- RFI
- Request for qualification
- PQQ

7.3.3.1 Request for Information

It is a procedure where potential contractors are provided with a general or preliminary description of the problem or need and are requested to provide information or advice about how to better define the problem or need or alternative solutions. It may be used to assist in preparing a solicitation document.

7.3.3.2 Request for Qualification/Prequalification Questionnaire

Request for Qualification/Prequalification Questionnaire is a process that enables to prequalify contractors for a particular requirement and avoid having to struggle with a large number of lengthy proposals. It is a solicitation document requesting proponents to submit the qualifications and special expertise in response to required scope of services. This process is used to short-list qualified proponents for procurement process. Table 7.8 illustrates contents of request for qualification for EPC contractor, and Table 7.9 lists PQQ to select EPC contractor.

TABLE 7.8
Contents of Request for Qualification for EPC Contractor

Serial Number	Element	Description
1	General information	a. Company name
		b. Full address
		c. Registration details/business permit
		d. Management details
		e. Nature of company such as partnership and share holding
		f. Stock exchange listing, if any
		g. Affiliated/group of companies, if any
		h. Membership of professional trade associations, if any
		i. Award winning project, if any
		j. Quality management certification
		k. Risk assessment/management capability
		l. Health and safety management certification
2	Financial information	a. Yearly turnover
		b. Current workload
		c. Audited financial report
		d. Tax clearance details
		e. Bank overdraft/letter of credit capacity
		f. Performance bonding capacity
		g. Insurance limit

(*Continued*)

TABLE 7.8 (*Continued*)
Contents of Request for Qualification for EPC Contractor

Serial Number	Element	Description
3	a. Organization details (general)	a. Core area of business
		b. How long in the same field of operation
		c. Quality control/assurance organization
		d. Grade/classification, if any
	b. Organization details (experience)	a. Number of years in the same business
		b. Technical capability
		i. Engineering–construction
		ii. Process design
		iii. Shop drawing production
		c. List of previous contracts
		i. Name of project
		ii. Value of each contract
		iii. Contract period of each contract
		iv. Oil and gas projects
		v. Contract completion delay
		d. List of failed/uncompleted contracts
		e. Overall tender success rate
		f. Claims/dispute/litigation
4	Resources	a. Human resources
		i. Management
		ii. Engineering staff
		iii. Design engineers
		iv. Cad technicians
		v. List of key project personnel
		vi. Skilled labors—permanent/temporary
		vii. Unskilled labors—permanent/contracted
		b. List of equipment, machinery, plant
		c. Human resource development plan
5	Health, safety, and environmental	a. Medical facility
		b. Safety record
		c. Environmental awareness
6	Project reference	
7	Bank reference	

EPC, engineering, procurement, and construction.

Source: Modified from Abdul Razzak Rumane. (2013). *Quality Tools for Managing Construction Projects.* Reprinted with permission from Taylor & Francis Group.

TABLE 7.9
Prequalification Questionnaires for Selecting EPC Contractor

Serial Number	Question	Answer
1	Name of the organization and address	
2	Organization's registration and license number	
3	ISO certification	
4	Registration/classification status of the organization	
5	Joint venture with any international contractor	
6	Total turnover for the past 5 years	
7	Audited financial report for the past 3 years	
8	Insurance and bonding capacity	
9	Total experience (years) as EPC contractor	
10	Total experience (years) as contractor	
	EPC Type of Contracts Information	
11	Total experience (years) in construction of the following type of projects	
	11.1 Petroleum plants	
	11.2 Petroleum refineries	
	11.3 Liquid chemical plants	
	11.4 Solid process plants	
	11.5 Liquid/solid plants	
12	Size of project (maximum amount single project)	
	12.1 Petroleum plants	
	12.2 Petroleum refineries	
	12.3 Liquid chemical plants	
	12.4 Solid process plants	
	12.5 Liquid/solid plants	
13	List successfully completed projects	
	13.1 Petroleum plants	
	13.2 Petroleum refineries	
	13.3 Liquid chemical plants	
	13.4 Solid process plants	
	13.5 Liquid/solid plants	
14	List similar type (type to be mentioned) of projects completed	
	14.1 Project name and contracted amount	
	14.2 Project name and contracted amount	
	14.3 Project name and contracted amount	
	14.4 Project name and contracted amount	
	14.5 Project name and contracted amount	

(*Continued*)

TABLE 7.9 (Continued)
Prequalification Questionnaires for Selecting EPC Contractor

Serial Number	Question	Answer
15	Resources	
	15.1 Management	
	15.2 Engineering	
	15.3 Technical	
	15.4 Foreman/supervisor	
	15.5 Skilled manpower	
	15.6 Unskilled manpower	
	15.7 Plant and equipment	
16	Quality management policy	
17	Health, safety, and environmental policy	
	17.1 Number of accidents during the past 3 years	
	17.2 Number of fires at site	
18	Current projects	
19	Staff development policy	
20	List of delayed projects	
21	List of failed contract	
	Designer's Information	
22	Total years of experience in EPC type of projects	
23	Size of project (maximum value)	
24	List similar type of successfully completed projects	
25	List successfully design–build projects	
26	Resources	
	26.1 Architect	
	26.2 Structural engineer	
	26.3 Civil engineer	
	26.4 Piping engineer	
	26.5 Process engineer	
	26.6 Mechanical engineer	
	26.7 Electrical engineer	
	26.8 Instrumentation and control engineer	
	26.9 Communication engineer	
	26.10 HVAC engineer	
	26.11 Low-voltage system engineer	
	26.12 Landscape engineer	
	26.13 Specialist system designer(s)	
	26.14 Cad technicians	

(*Continued*)

TABLE 7.9 (*Continued*)
Prequalification Questionnaires for Selecting EPC Contractor

Serial Number	Question		Answer
	26.15	Quantity surveyor	
	26.16	Equipment	
	26.17	Design software	
27	LEED or similar certification		
28	Total experience in oil and gas projects		
29	Total experience in green building design		
30	Design philosophy/methodology		
31	Innovative design and process technology		
32	Quality management system		
33	HSE consideration in design		
34	List of professional awards		
35	Number of projects completed before contracted schedule of completion		
36	Litigation (dispute, claims) on earlier projects		
37	Any major accident in the constructed project		

EPC, engineering, procurement, and construction; HSE, health, safety, and environmental; HVAC, heating, ventilation, air conditioning; ISO, International Organization for Standardization.

Normally, every owner maintains a list/register of prospective and previously qualified contractors. Procurement strategy for prequalification/short-listing and registration of contractors is already discussed under Section 7.3.1. Please refer Figure 7.4. Existing list of potential/registered contractors can be expanded by placing advertisement in general publications such as newspapers or trade publications or professional journals.

In oil and gas projects, the selection of contractor is based on the strategy to procure the contract. The corporate procedure to select the contractor is already defined in the company's quality system manual.

7.3.3.3 Request for Quotation

Request for Quotation is a priced-based bidding process that is used when complete documents consisting of defined project deliverables, solution, specifications, performance standards, and schedules are known. Potential bidders are provided with all the related information (bidding and tender documents)—except price—and are requested to submit the price and the evaluation of the bids are only on **price** subject to fulfilling all the required information. Most construction contractors are selected on **price** basis.

Figure 7.7 illustrates logic flowchart for selection of EPC contractor, and Table 7.10 summarizes selection criteria for EPC contractor.

Selection of Project Teams

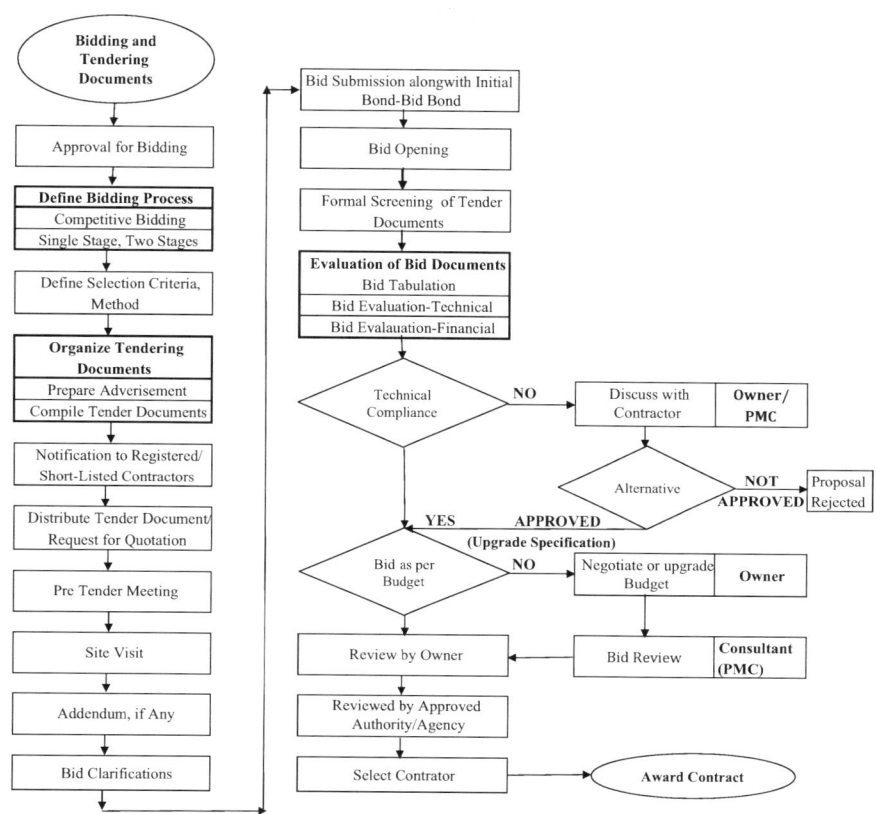

FIGURE 7.7 Logic flowchart for the selection of EPC contractor. EPC, engineering, procurement, and construction.

7.3.4 Selection of Project Management Consultant

Construction/project management is a discipline and management system specially tailored to promote the successful execution of capital and complex projects.

Project management is a professional management practice applied effectively to all projects especially to the construction project from the inception to the completion of the project for the purpose of managing (planning, organizing, executing, and controlling) scope, schedule, cost, and quality. Project management services are generally offered by registered consultant firms/professionals having ability and expertise to manage construction projects.

As per PMBOK® Guide (Project Management Body of Knowledge) published by Project Management Institute, project management is the application of knowledge, skills, tools, and techniques to project activities to meet the project requirements. Project management totally deals with all the processes and activities of construction project.

TABLE 7.10
Selection Criteria for EPC Contractor

		Evaluation Criteria	Weightage	Key Points for Consideration	Review Result
1. General information					
1		Company information		Company's current position—a MUST information	Yes OR No
2. Financial			25%		
1		Total turnover (past 5 years)	25%	Sum of the turnover for the past 5 years	
2		Values of current work-in-hand	25%	Project value/value of current work-in-hand	
3		Audit financial reports	10%	To confirm the ratio given in point 3	
4		Financial standing	30%		
	33%	Assets		Current assets/current liabilities	
	34%	Liabilities		Total liabilities − total equity/total assets	
	33%	Profit/loss		Net profit before tax/total equity	
5		Bonding and insurance limit	10%		
	60%	a. Performance and bonding capacity		Provided or not provided	
	40%	b. Insurance		Provided or not provided	
3. Organization details					
3a. Business			20%		
1		Company's core area of business	30%	Degree of satisfactory answer	
2		Experience of years in business	15%	No. of years	
3		Experience of years in oil and gas projects	15%	Yes or no	
4		ISO certification	15%	Yes or no	
5		Registration/classification status	15%	Grade or classification	

(Continued)

TABLE 7.10 (Continued)
Selection Criteria for EPC Contractor

		Evaluation Criteria		Weightage	Key Points for Consideration	Review Result
6	Organizational chart			10%	Key staff indicated (name/title), balanced resources, departmental (specialization) diversity, lines of communication	
7	Dispute/claims				Degree of satisfactory answer	
3b. Experience				30%		
	50%	a. Projects' value			No. of projects with comparable value	
	25%	b. Projects' type (process type and complexity)			No. of projects with similar complexity	
	25%	c. Projects' type (oil and gas–petroleum projects)			No. of projects with similar complexity	
4. Resources				20%		
1	Personnel			60%		
	30%	Management			No. of Managerial staff	
	30%	Engineers	Design		No. of design engineers	
			Construction		No. of engineers and project staff	
	30%	Technicians	Construction		No. of technicians and foreman	
			Auto-CAD		No. of Auto-CAD technicians	
	10%	Staff development			% turnover spent on training	
2	Technology			10%	% of turnover spent on acquiring latest construction technology	
3	Plant and equipment			30%	List of plant and equipment	

(*Continued*)

TABLE 7.10 (*Continued*)
Selection Criteria for EPC Contractor

		Evaluation Criteria	Weightage	Key Points for Consideration	Review Result
5. General			5%		
1		Bank references	25%	Provided or not provided	
2		Project references	25%	Provided or not provided	
2		Health, safety, and environmental narration	50%	Degree of satisfactory answer	

EPC, engineering, procurement, and construction; ISO, International Organization for Standardization.

Note: The weightages mentioned in the table are indicative only. The % can be determined as per the owner's procurement strategy.

PMC is responsible for administration of construction project on behalf of the owner. They have responsibilities to the following:

- Select project delivering teams.
- Define project baseline.
- Develop project schedule.
- Monitor and control project cost.
- Review project design management/deliverables for compliance.
- Monitor project progress.
- Report project status.
- Update project owner/client about project progress.
- Coordinate with various project teams and stakeholders.
- Manage project risk.
- Administer contracts.
- Manage variation/changes.
- Project safety.
- Manage project payments.
- Claim management.

Normally, the owner/client engages PMC firm from the inception of the project through issuance of substantial completion certificate to manage entire construction project on behalf of the owner.

There are mainly three major activities that are to be performed by the project manager:

1. Selection of project teams
   - Selection of designer/consultant
   - Selection of contractor and subcontractor

2. Design management
   - Review and approval of project design (design phases)
   - Review and approval of construction documents
3. Construction management
   - Construction supervision activities
   - Testing, commissioning, and handover phase

Normally, the PMC firm is hired on the basis of **qualifications**. The basis of selection is solely based on demonstrated competence, professional qualification, and experience for the type of services required. In QBS, the contract price is negotiated after selection of the suitably qualified firm.

Procurement strategy for prequalification/short-listing and registration of contractors is already discussed under Section 7.3.1. Please refer Figure 7.4. Table 7.11 lists PQQ to select the PMC.

Figure 7.8 illustrates logic flowchart for the selection of PMC.

TABLE 7.11
Prequalification Questionnaires for Registration of Project Management Consultant

Serial Number	Element	Question	Response
1	General information	a. Company name	
		b. Full address	
		c. Registration details/business permit	
		d. Management details	
		e. Membership of professional trade associations, if any	
		f. Award winning project, if any	
		g. Quality management certification	
2	Financial information	a. Yearly turnover	
		b. Current workload	
		c. Audited financial report	
		d. Performance bonding capacity	
		e. Insurance limit	
3	a. Organization details (general)	a. Core area of business	
		b. How long in the field of project management consultancy services	
		c. How long in the field of process types of projects	
		d. How long in the field of oil and gas projects	

(*Continued*)

TABLE 7.11 (Continued)
Prequalification Questionnaires for Registration of Project Management Consultant

Serial Number	Element	Question	Response
		e. Quality control/assurance activities	
	b. Organization details (type of services provided)	a. During study stage	
		b. During design stage	
		c. During selection of team members	
		d. During construction	
		e. During start-up	
	c. Organization details (experience)	a. Number of years in the same business	
		b. Technical capability	
		i. Management of construction projects	
		ii. Project team selection	
		iii. Design management	
		iv. Supervision of construction	
		v. Planning, monitoring, and controlling	
		vi. Construction quality audits	
		vii. Risk management	
		viii. Testing and commissioning	
		ix. Conducting value engineering	
		c. List of previous contracts	
		i. Name of project	
		ii. Value of each contract	
		iii. Contract period of each contract	
4	Resources	a. Human resources	
		i. Management	
		ii. Design management	
		iii. Construction supervision	
		iv. Project planning and scheduling	
		v. Project monitoring and controlling	
		vi. Construction quality management/auditing	
		vii. Contract administration	
		viii. Conflict management	
		ix. HSE	
		b. Human resource development plan	
5	Project reference		
6	Bank reference		

HSE, health, safety, and environmental.

Selection of Project Teams

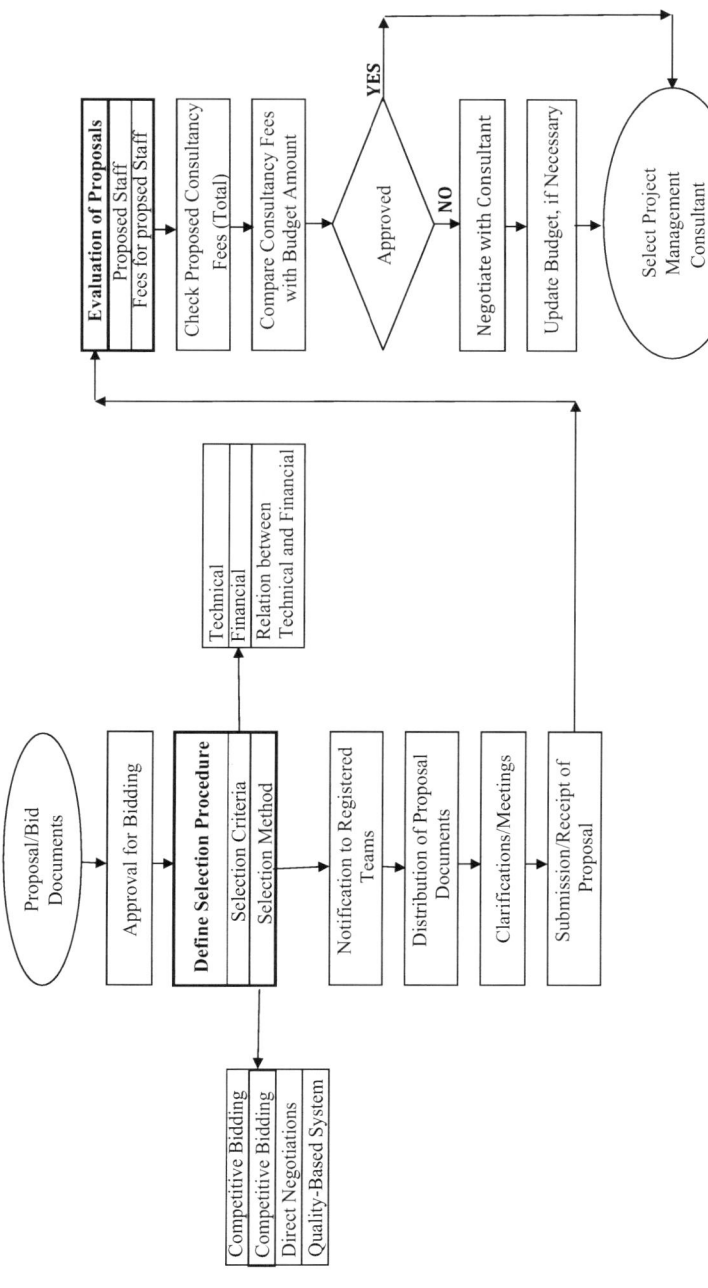

FIGURE 7.8 Logic flowchart for the selection of project management consultant (PMC).

7.3.5 Selection of Construction Supervisor

It is a normal practice that the owner appoints designer/consultant of the project to supervise the construction process. The supervisor is responsible for achieving project quality goals and is also responsible for implementing the procedures specified in the contract documents. Table 7.12 lists the responsibilities the owner delegates to the engineer (consultant).

Table 7.13 lists PQQ to select the construction Supervision team.

Figure 7.9 illustrates typical tender submission procedure by construction supervisor.

7.3.6 Selection of Specialist Consultant

The inception of construction project starts with identification of business case and its needs. A business case typically addresses the business need for the project and the value the project brings to the business (project value proposition). A value proposition is a promise of value to be delivered by the project.

TABLE 7.12
Responsibilities of Supervision Consultant

Sr. No.	Description
1	Achieving the quality goal as specified
2	Review contract drawings and resolve technical discrepancies/errors in the contract documents
3	Review construction methodology
4	Approval of contractor's construction schedule
5	Monitoring and controlling construction time
6	Monitoring and controlling construction expenditure
7	Approval of contractor's quality control plan
8	Regular inspection and checking of executed works
9	Review and approval of construction materials
10	Review and approval of shop drawings
11	Inspection of construction material
12	Conduct progress and technical coordination meetings
13	Coordination of owner's requirements and comments related to site activities
14	Project related communication with contractor
15	Coordination with regulatory authorities
16	Evaluating risk and making decisions related to unforeseen conditions
17	Maintaining project record
18	Processing of site work instruction for owner's action
19	Evaluation and processing of variation order/change order
20	Recommendation of contractor's payment to owner
21	Approval of contractor's safety management plan
22	Monitor safety at site and HSE requirements
23	Supervise testing, commissioning, and handover of the project
24	Issue substantial completion certificate

HSE, health, safety, and environmental.

TABLE 7.13
Prequalification Questionnaires for Registration of Construction Supervisor

Serial Number	Element	Question	Response
1	General information	a. Company name	
		b. Full address	
		c. Registration details/business permit	
		d. Management details	
		e. Membership of professional trade associations, if any	
		f. Award winning project, if any	
		g. Quality management certification	
		h. Safety and environment management certification	
2	Financial information	a. Yearly turnover	
		b. Current workload	
		c. Audited financial report	
		d. Performance bonding capacity	
		e. Insurance limit	
3	a. Organization details (general)	a. Core area of business	
		b. How long in the same field of construction supervision	
		c. How long in the same field of process types of projects	
		d. How long in the same field of oil and gas projects construction supervision	
	b. Organization details (experience)	a. Number of years in the same business	
		b. Technical capability	
		i. Supervision of oil and gas construction projects	
		ii. Construction quality management	
		iii. Construction quality audits	
		iv. Monitoring and control	
		v. Testing and commissioning	
		c. List of previous contracts	
		i. Name of project	
		ii. Value of each contract	
		iii. Contract period of each contract	

(Continued)

TABLE 7.13 (*Continued*)
Prequalification Questionnaires for Registration of Construction Supervisor

Serial Number	Element	Question	Response
4	Resources	a. Human resources	
		i. Management	
		ii. Construction supervision (different trades)	
		iii. Construction quality	
		iv. Construction quality auditing	
		v. Project planning and scheduling	
		vi. Project monitoring and controlling	
		vii. Contract administration	
		viii. Conflict management	
		ix. Information technology	
		x. Safety and environment	
		b. Human resource development plan	
5	Project reference		
6	Bank reference		

Business need assessment and feasibility study is essential to ensure the owner's business case has been properly considered to prepare accurate and comprehensive client brief (terms of reference [TOR]) to achieve a qualitative and competitive project. It is essential that competent consultant is selected to prepare TORs that give the designer a clear understanding for the development of the project. Furthermore, the owner has to select the designer (consultant) to develop project design for construction.

Once the need analysis is carried out, a need statement is prepared, and a feasibility study is conducted to assist project owner/decision-makers in making the decision that will be in the best interest of the owner.

The feasibility study may be conducted in-house if capability exists. However, the services of specialist involved in preparation of economic and financial studies are usually commissioned by the owner/client to perform such study. Since the feasibility study stage is very crucial stage, in which all kinds of professionals and specialists are required to bring many kinds of knowledge and experience into broad-ranging evaluation of feasibility, it is required to engage a firm having expertise in the related fields. The feasibility study establishes the broad objectives for the project and so exerts an influence throughout subsequent stages.

The successful completion of the feasibility study marks the first of several transition milestones and is therefore most important to determine whether or not to implement a particular project or program. The feasibility study decides the possible design approaches that can be pursued to meet the need. After completion of feasibility study and its approval by the owner, project goals and objectives are prepared taking into consideration final recommendations/outcome of the feasibility study.

Selection of Project Teams

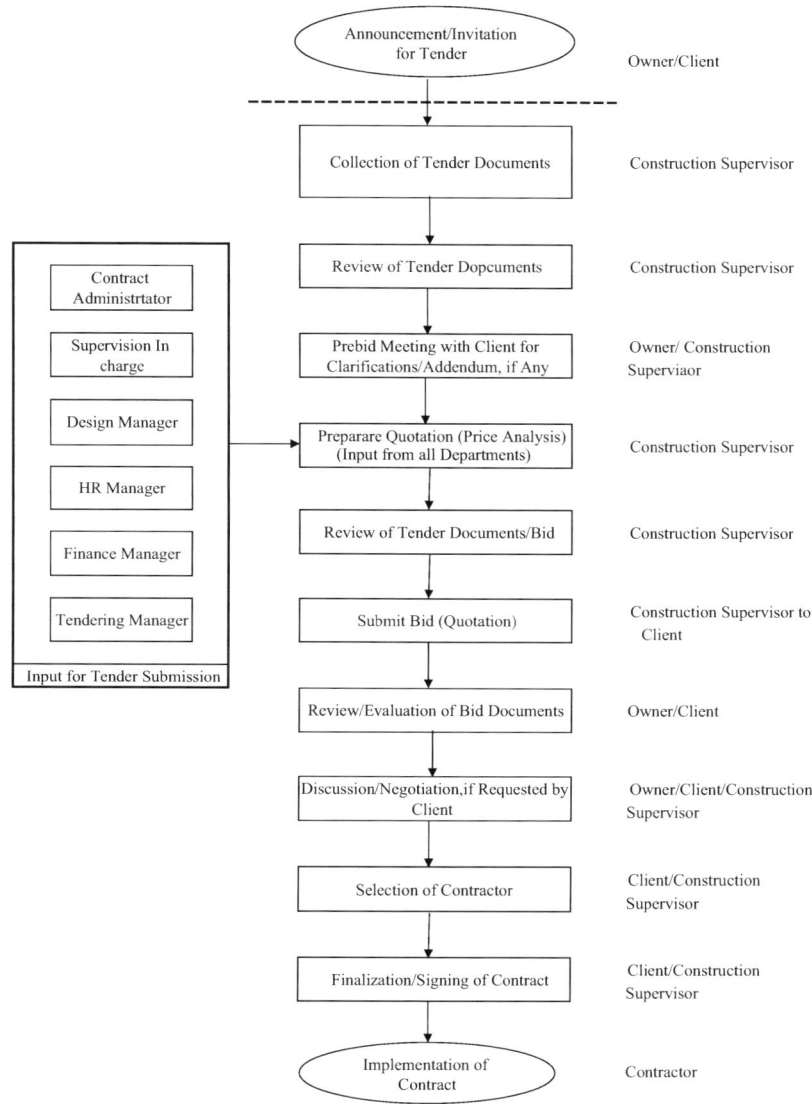

FIGURE 7.9 Typical tender submission procedure by construction supervisor.

Table 7.14 lists the qualification criteria for the selection of specialist consultant to perform study analysis. Normally the consultant to perform need analysis and feasibility study analysis is selected on QBS.

The procurement strategy for short-listing/registration of different team members discussed under Section 7.3.1 (Figure 7.4) can be applied for short-listing of specialist consultant, and team selection procedure discussed under Section 7.3.2 (Figure 7.5) can be applied for the selection of specialist consultant.

TABLE 7.14
Consultant's Qualification for Feasibility Study

Serial Number	Description
1	Experience in conducting feasibility study
2	Experience in conducting feasibility study in similar type and nature of projects
3	Fair and neutral with no prior opinion about what decision should be made
4	Experience in strategic and analytical analysis
5	Knowledge of analytical approach and background
6	Ability to collect a large number of important and necessary data via work sessions, interviews, surveys, and other methods
7	Market knowledge
8	Ability to review and analysis of market information
9	Knowledge of market trend in similar type of projects/facility
10	Multidisciplinary experienced team having proven record in following field: a. Financial analyst b. Engineering/technical expertise c. Policy experts d. Project scheduling
11	Experience in review of demographic and economic data

Source: Abdul Razzak Rumane. (2013). *Quality Tools for Managing Construction Projects.* Reprinted with permission from Taylor & Francis Group.

7.3.7 SELECTION OF SUBCONTRACTOR/SUBCONSULTANT

In most construction projects, the contractor engages special subcontractors to execute certain portion of contracted project works. Areas of subcontracting are generally listed in the particular conditions of the contract document. Generally contractor has to follow the contract conditions while selecting the subcontractor and submit subcontractors/specialist contractors for approval by the owner/PMC/construction supervisor.

Similarly the designer (consultant) has to get approval from the project owner if they want to engage subconsultant for performing a specific activity of the project.

8 Quality Management during Oil and Gas Project Phases

8.1 INTRODUCTION

Construction projects are mainly capital investment projects. They are customized and non-repetitive in nature. Construction projects have become more complex and technical, and the relationships and the contractual grouping of those who are involved are also more complex and contractually varied. The products used in construction projects are expensive, complex, immovable, and long-lived.

Oil and gas sector projects are considered as biggest sector in terms of investment as well as in employment generation. Oil and gas projects are categorized as process type of projects. Oil and gas downstream projects are process type of projects that produce different types of products to satisfy and meet the consumers and end users requirements. The processing of raw materials is very important, and necessary precautions are required to construct the project focusing mainly on sustainability to ensure that the output products meet the required specifications.

Generally, oil and gas construction project comprises building materials (civil), structural material, piping material, processing equipment, rotary machines, petroleum production equipment, process heating equipment, oil and gas handling equipment, oil and gas storage items, pollution and waste water control equipment, pumps, compressors, electromechanical items, instrumentation and control, leak detection, fire and gas detection, security system, finishing items, and many other systems and equipment. These are normally produced by other construction-related industries/manufacturers. These industries produce products per their own quality management practices complying with certain quality standards or against specific requirements for a particular project. Owners of construction projects or their representatives have no direct control over these companies unless they themselves, their representatives, or appointed contractors commit to buying their product for use in their facility. These organizations may have their own quality management program. In manufacturing or service industries, quality management of all in-house manufactured products is performed by the manufacturer's own team or is under the control of the same organization that has jurisdiction over its manufacturing plants at different locations. Quality management of vendor-supplied items/products is carried out as stipulated in the purchasing contract per the quality control specification of the buyer.

Quality management in construction addresses both the management of project and the product of the project and all the components of the product. It also involves incorporation of changes or improvements, if needed. Construction project quality is fulfillment of owner's needs as per defined scope of works within a budget and specified schedule to satisfy owner's/user's requirements.

Construction projects are constantly increasing in technological complexity. In addition, the requirements of construction clients are on the increase, and as a result, construction products (buildings, products in case of oil and gas project) must meet varied performance standards and satisfy the customer. Therefore, to ensure the adequacy of client brief (project charter), which addresses the numerous complex client/user needs, it has become the responsibility of designer to evaluate the requirements in terms of activities and their relationship and follow health, safety, and environmental regulation while designing any building.

There are innumerable processes that make up the construction process. In order to conveniently manage and control the project at each stage, the construction project life cycle is divided into number of phases based on the principle of systems engineering. A systems engineering approach to construction projects helps to understand the entire process of project management and to manage and control its activities at different levels of various phases to ensure timely completion of the project with economical use of resources to make the construction project most qualitative, competitive, and economical.

Each phase is further sub-divided on Work Breakdown Structure (WBS) principle to reach a level of complexity where each element/activity can be treated as a single unit which can be conveniently managed.

The methodology applied in this book for oil and gas project is based on dividing the oil and gas project life cycle into six (6) phases.

Each phase is composed of activities/elements having functional relationship to achieve a common objective for useful purpose. Figure 8.1 illustrates the construction project life cycle (oil and gas project) phases.

Table 1.1 discussed under Chapter 1 summarizes major activities in these phases.

Quality management is an organization-wide approach to understand customer needs and deliver the solutions to fulfill and satisfy the customer. Quality management is managing and implementation of quality system to achieve customer satisfaction at the lowest overall cost to the organization while continuing to improve the process. Quality system is a framework for quality management. It embraces the organization structure, policies, procedures, and processes needed to implement quality management system.

Quality management system in construction projects mainly consists of:

- Quality management planning
- Quality assurance
- Quality control

Each of these processes and activities has to be performed during the following phases of oil and gas project:

1. Feasibility Study
2. Concept Design

Quality Management

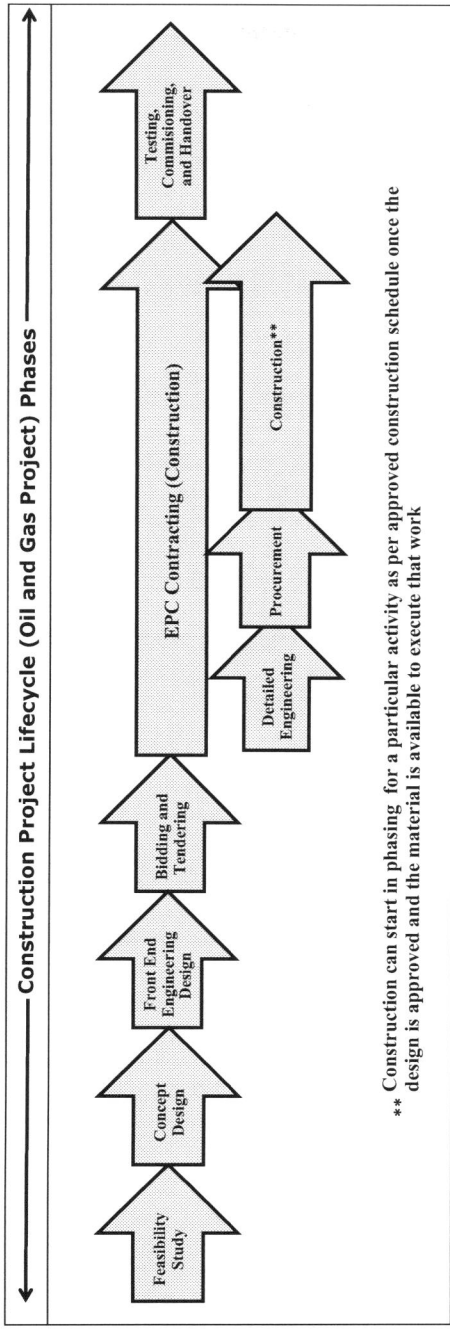

FIGURE 8.1 Construction project life cycle (oil and gas project) phases.

3. Front-End Engineering Design
4. Bidding and Tendering
5. Construction
6. Testing, Commissioning, and Handover

8.2　FEASIBILITY STUDY PHASE

Feasibility study is the first phase of oil and gas construction project life-cycle. This phase is also called/known as

- Programming phase
- Definition phase
- Predesign phase
- Project Inception stage

It is a relatively short period at the start of the project and is about understanding the project goals and objectives and getting enough information to confirm that project should proceed or to convince that it should not proceed and abort. Successful completion of inception stage activities and approval of project by the stakeholder/sponsor marks the initiation of the project.

The primary objectives of this phase are as follows:

- Establish a justification or business case for the project.
- Conduct feasibility study to assess whether the project is technically feasible and economically viable.
- Obtain stakeholders approval to proceed with the project or to stop the project
- Establish project goals and objectives
- Establish scope and boundary conditions
- Estimate overall schedule and cost for the project
- Conduct cost/benefit analysis
- Identify potential risks associated with the project
- Outline key requirements that will drive design trade-offs and identify which requirements are critical.
- Identify and evaluate alternatives/options
- Select preferred alternative
- Develop design basis
- Establish project delivery and contracting system
- Develop project charter

Most important decisions about planning, organization, type of contract take place during this phase.

Figure 8.2 illustrates logic flowchart for feasibility study stage phase.

Quality Management

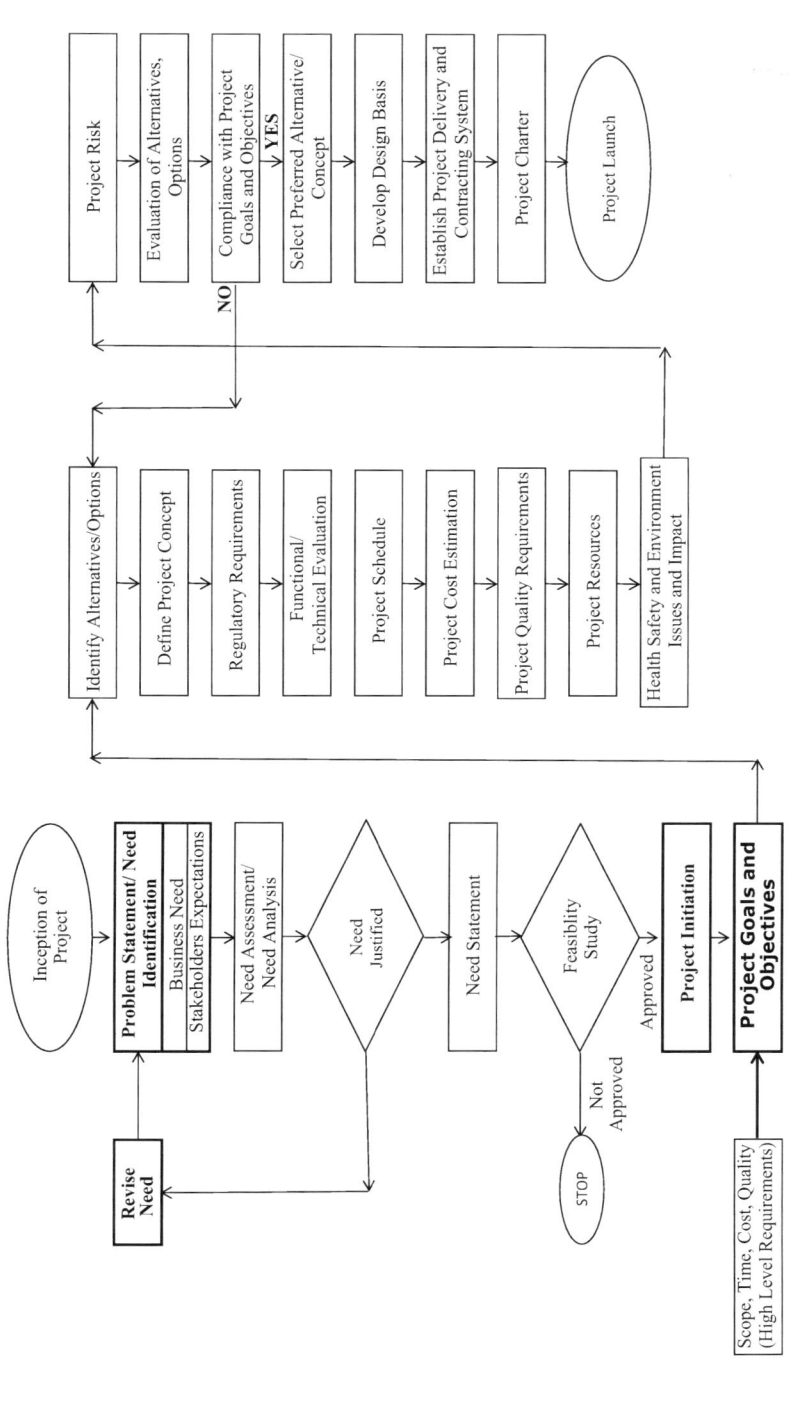

FIGURE 8.2 Logic flowchart diagram for feasibility study stage phase.

8.2.1 Project Initiation

Project development process begins with project initiation and ends with project close-out and finalization of project records after completion of construction of the project. The project initiation starts with identification of need and a business case.

Most construction projects begin with recognition of a new facility and identification of need. The need should be based on real (perceived) requirements or deficiency.

8.2.2 Identification of Need

The need of the project is created by the owner. The owner of the facility could be an individual, a public/private sector company, or a governmental agency. The need could be to develop a new facility/project or renovation/refurbishment of existing facility or to produce new products. The need of the project is created by the owner and is linked to the available financial resources to develop the facility/project. The owner's need must be well defined describing the minimum requirements of quality, performance, project completion date, and approved budget for the project.

Owner's needs are quite simple and are based on following:

- To have best facility/project for the money, i.e., to have maximum profit or services at a reasonable cost
- On time completion, i.e., to meet the owner's/user's schedule
- Completion within budget, i.e., to meet the investment plan for the facility

It is essential to get a clear definition of the identified need or the problem to be solved by the new project. The owner's need must be well defined, indicating the minimum requirements of quality and performance, an approved budget, and required completion date. The need should be based on real (perceived) requirements. The identified need is then assessed and analyzed to develop the need statement. Need assessment is conducted to determine the need. Need assessment is a systematic process for determining and addressing needs or "gaps" between current conditions and desired conditions "want." Need analysis is the process of identifying and evaluating the need. The need statement is written based on the need analysis and is used to perform feasibility study to develop project goals and objectives and subsequently to prepare project scope documents.

The owner's need must be well defined indicating the minimum requirements of quality and performance, an approved main budget and required completion date. Sometimes, the project budget is fixed and therefore the quality of building system, materials, and finishes of the project need to be balanced with the budget. A business case typically addresses the business need for the project, and the value the project brings to the business (project value proposition). A value proposition is a promise of value to be delivered by the project. Following questions address the value proposition:

a. How the project solves the current problems or improves the current situation?

TABLE 8.1
5 W 2 H Analysis for Project Need

Serial Number	Why	Related Analyzing Question
1	Why	Why new project?
2	What	What advantage it will have over other similar projects producing similar items?
3	Who	Who will be the customer for the products produced in this plant?
4	Where	Where from the raw material to feed the project requirements to produce desired products will be available?
5	When	When the project will be ready?
6	How many	How many competitors are in the market for the products produced in this project?
7	How much	How much market share we will have for the products/items produced in this project?

b. What specific benefits the project will deliver?
c. Why the project is the ideal solution for the problem?

Business need assessment is essential to ensure that owner's business case has been properly considered before the initial Project Brief (Need Statement) is developed.

Table 8.1 illustrates 5W2H analysis of the project need.

8.2.3 Feasibility Study

The feasibility study takes its starting point from the output of the project identification need. Feasibility study is conducted to assist owner/decision-makers in making the decision that will be in the best interest of the owner. The main purpose of feasibility study is to evaluate the project need and decide whether to proceed with the project or to stop. Depending on the circumstances, the feasibility study may be short or lengthy, simple or complex. In any case, it is the principle requirement in project development as it gives the owner an early assessment of the viability of the project and the degree of risk involved. Project feasibility study is usually performed by the owner through his own team or by engaging individuals/organizations involved in the preparation of economical and financial studies. However, the feasibility study can be conducted by a specialist consultant in this field.

Feasibility study can be categorized into the following functions:

- Legal
- Marketing
- Technical and engineering
- Financial and economical
- Social
- Environmental

- Risk
- Scheduling of project

The objective of the feasibility study is to review technical/financial viability of the project to give sufficient information to enable the client to proceed or abort the project. Feasibility study is undertaken to analyze the ability to complete a project successfully, taking into account various factors such as economical, technological, scheduling, etc. A feasibility study looks into the positive and negative effects of a project before investing the Company resources, viz., time and money.

Feasibility study is defined as an evaluation or analysis of the potential impact of the identified need of the proposed project. The feasibility study assists decision-makers (investors/owners/clients) in determining whether or not to implement the project. Since the feasibility study stage is a very crucial stage, in which all kinds of professionals and specialists are required to bring many kinds of knowledge and experience into broad-ranging evaluation of feasibility, it is required to engage a firm having expertise in the related fields. The feasibility study establishes the broad objectives for the project and so exerts an influence throughout subsequent stages. The successful completion of the feasibility study marks the first of several transition milestones and is therefore most important to determine whether or not to implement a particular project or program. The feasibility study decides the possible design approaches that can be pursued to meet the need.

Following are the contents of the feasibility study report:

1. Purpose of the feasibility study
2. Project history (project background information)
3. Description of the proposed project.
    a. Project location
    b. Plot area
    c. Interface with adjacent/neighboring area
    d. Expected project deliverables
    e. Key performance indicators
    f. Constraints
    g. Assumptions
4. Business case
    a. Project need
    b. Stakeholders
    c. Project benefits
    d. Estimated time
    e. Financial benefits
    f. Estimated cost
    g. Justification
5. Feasibility study details
    a. Technical
    b. Economical
    c. Time scale

d. Financial
   e. Environmental
   f. Ecological
   g. Sustainability
   h. Political, legal
   i. Social
6. Risk
7. Environmental impact (considerations)
8. Social impact (considerations)
9. Final recommendation

8.2.4 Project Goals and Objectives

The outcome of the feasibility study helps selection of a defined project which meets the stated project objectives, together with a broad plan of implementation. If the feasibility study shows that the objectives of the owner are best met through the ideas generated then the project is moved to further stage to deliver intended objectives of the project passing through different stages of project life cycle.

After completion and approval of feasibility study, it is possible to establish project goals and objectives. Project goals and objectives are prepared taking into consideration final recommendations/outcomes of the feasibility study. Clear goals and objectives provide the project team with appropriate boundaries to make decisions about the project and ensure that the project/facility will satisfy the owner's/end user's requirements fulfilling owner's needs. Establishing properly defined goals and objectives is the most fundamental elements of project planning. Therefore, the project goals and objectives must be

- Specific (Is the goal specific?)
- Measurable (Is the goal measurable?)
- Agreed upon/achievable (Is the goal achievable?)
- Realistic (Is the goal realist or result-oriented?)
- Time (cost) limited (Does the goal have time element?)

The project objective definition usually includes the following information:

1. Project scope and project deliverables
2. Preliminary project schedule
3. Preliminary project budget
4. Specific quality criteria the deliverables must meet
5. Type of contract to be employed
6. Design requirements
7. Regulatory requirements
8. Potential project risks
9. Environmental considerations
10. Logistic requirements

8.2.5 IDENTIFICATION OF ALTERNATIVES/OPTIONS

Once the owner/client defines the project objectives, a project team (in-house or outside agency) is selected to start identification of alternatives. Normally, the owner assigns a specialist consultant, in certain cases it may be the designer of the project, to identify conceptual alternatives, evaluation of conceptual alternatives and selection of preferred conceptual alternative in consultation with the owner, project management consultant (PMC) if it is commissioned, and is included in the Terms of Reference (TOR)/Project Charter.

The project goals and objectives serve as a guide for the development of alternatives. The team develops several alternative schemes and solutions. Each alternative is based on the predetermined set of performance measures to meet the owner's requirements. In case of construction projects, it is mainly the extensive review of development options which are discussed between the owner and the team members. The team provides engineering advice to the owner to enable him asses its feasibility and relative merits of various alternative schemes to meet his requirements.

Quality tools such as Brainstorming, Delphi technique, 5W2H can be used to identify alternatives.

8.2.5.1 Analyze Alternatives

Qualitative and Quantitative comparison, evaluation and analysis of identified alternatives are carried out by considering the advantages and disadvantages of each item systematically. Social, economical (time and cost), sustainability, performance of equipment, environmental impacts, functional capability, safety, and reliability should be considered while development of alternatives. Each alternative is compared by considering the advantages and disadvantages of each element systematically to meet the predetermined set of performance measures and owner's requirements. The evaluation of alternatives requires cooperative efforts between the owner and other team members involved in performing evaluation and analysis, regulatory authorities, and other stakeholders who have involvement, interest, or impact on the processes of construction project. The team makes a brief presentation to the owner, and the project is selected based on preferred conceptual alternatives.

The following elements are considered to evaluate and analyze each of the identified alternatives:

1. Suitability to the purpose and objectives
2. Performance parameters
3. Process technology
4. Equipment and machinery
5. Product output and specifications
6. Economy
7. Sustainable (environmental, social, economical)
8. Cost-efficiency
9. Life cycle costing
10. Environmental impact
11. Environment-preferred material and products

Quality Management

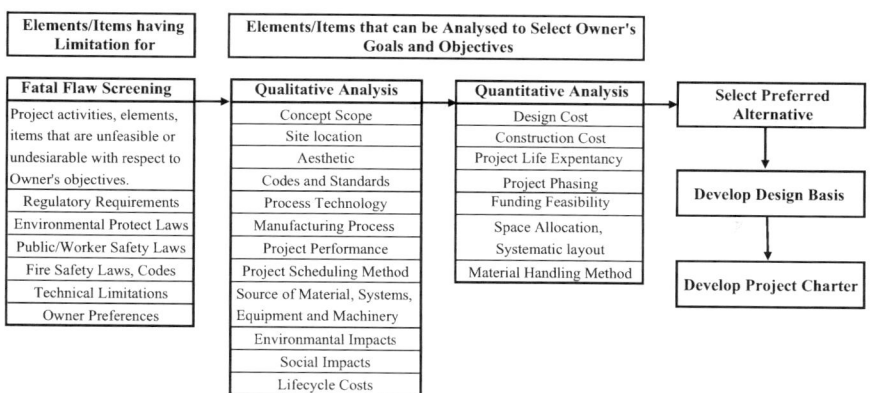

FIGURE 8.3 Evaluation and analysis of alternatives method to select preferred alternative.

12. Physical properties, thermal comfort, insulation, fire resistance
13. Utilization of space
14. Accessibility
15. Power consumption and energy-saving measures—use of renewable energy—alternate energy
16. Green building concept
17. Aesthetic
18. Safety and security
19. Statutory/regulatory requirements
20. Codes and standards
21. Any other critical issues

8.2.5.2 Select Preferred Alternative

Based on the analysis of identified alternatives, the preferred alternative that satisfies the project goals and objectives is selected.

With the approval of Preferred Alternative by the owner, the project proceeds towards the next stage of project development process.

Figure 8.3 illustrates Evaluation and Analysis Method to select Preferred Alternative.

8.2.6 Finalize Project Delivery and Contracting System

During this phase, the owner has to decide the procurement method and contract strategy. The selection of project delivery system mainly depends on the project size, complexity of the project, innovation, uncertainty, urgency, and the degree of involvement of the owner. Oil and gas projects mainly follow Design-Build/Turnkey project delivery system with firm fixed lump sum contracting. In certain cases, the owner may give incentive to the contractor for completing the project ahead of the specified schedule.

8.2.7 Project Charter

Project charter is also known as Terms of Reference (TOR), Design Brief, which is prepared by the owner/client or by PMC on behalf of owner describing the objectives and requirements to develop the project. A client brief (TOR) defines the objectives for the project and agrees a project brief to guide the next stage of the project.

A TOR is a written document stating what will be done by the designer (consultant) to develop the project/facility. It is issued to the designer (consultant) by the owner to develop the project design and construction documents. A well-prepared accurate and comprehensive client brief (TOR) is essential to achieve qualitative and competitive project. In case of oil and gas projects, the TOR generally requires the designer (consultant) to perform the following:

- Development of concept design
- Preparation of Front End Engineering Design (FEED) and Contract Documents for tendering purpose.
- Preparation of project schedule, budget, and obtaining authorities' approvals.

The TOR gives the project team (designer) a clear understanding of the development of project. Further, TOR is used throughout the project as a reference to ensure that the established objectives are achieved. Client brief or TOR describes information such as:

- Project objectives which has triggered the project
- Propose location of the project
- Project/facility to be developed
- Project assumptions and constraints
- Project function and size
- Performance characteristics of the project
- Environmental considerations
- Design approach
- Estimated timescale
- Estimated cost
- Codes, standards, and specifications
- Project control guidelines
- Regulatory requirements
- Initial list of defined risks
- Description of approval requirements
- Drawings
- Report
- Models (if applicable)
- Presentation

8.3 CONCEPT DESIGN PHASE

Concept design is the second phase of oil and gas construction project life cycle. This phase is also called/known as

Quality Management

- Pre-FEED Engineering
- Basic Engineering

The selected preferred alternative is the base for development of the concept design. Project charter (TOR) defines the activities to be performed by the designer during this phase and also the deliverables that have to be prepared for further proceedings, action in the next phase of the design development. The most significant impacts on the quality of a project occur during the conceptual phase. This is the time when specifications, statement of work, contractual agreements, and initial design are developed. Initial planning has the greatest impact on a project because it requires the commitment of processes, resources schedules, and budgets. A small error that is allowed to stay in the plan is magnified several times through subsequent documents that are second or third in the hierarchy. Concept design, or the basic design phase, is often viewed as most critical to achieving outstanding project performance.

The primary objectives of this phase are as follows:

- Develop concept design taking into consideration requirements listed under project charter
- Follow regulatory requirements, codes, and standards

While developing the concept design, the designer must consider the following:

1. Project goals and objectives
2. Usage
3. Process engineering
4. Facilities engineering
5. Technical and functional capability
6. Aesthetics
7. Constructability
8. Sustainability (environmental, social, and economical)
9. Project quality
10. Reliability
11. Availability of resources
12. Health and safety
13. Environmental compatibility
14. Fire protection measures
15. Project automation and control requirements
16. Supportability during maintenance/maintainability
17. Cost-effectiveness over the entire life cycle (economy)

Figure 8.4 illustrates the logic flowchart for the concept design phase

8.3.1 Select Design Team Members

Design team is selected as per the procurement strategy discussed under Section 7.3.2. Upon signing of contract with the client to design the project and offer other services,

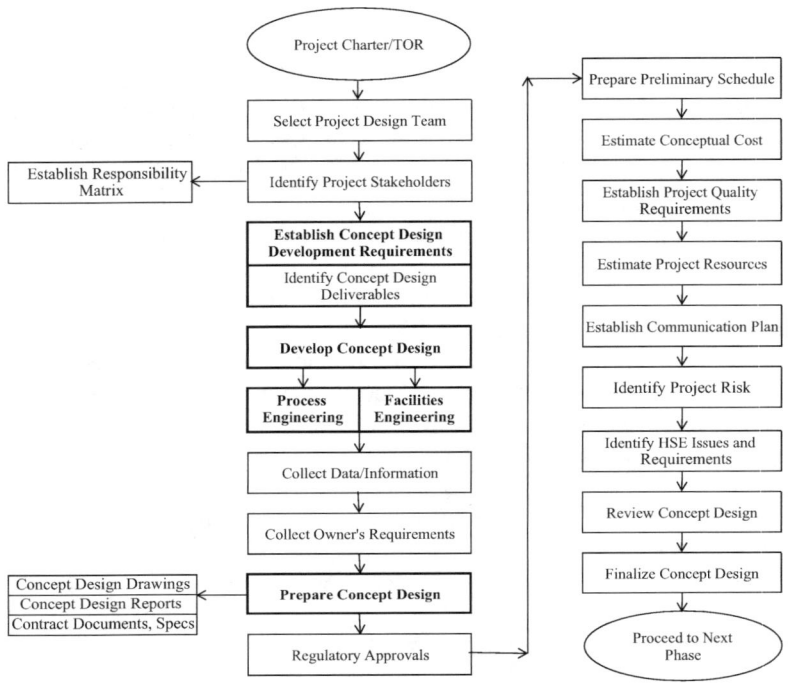

FIGURE 8.4 Logic flowchart for concept design phase.

the designer (consultant) assigns a Project Manager to execute the contract and is responsible to manage the development of design and contract documents, as per the project charter (TOR), to meet the client need and objectives. The Project Manager coordinates with other departments and acquires design team members to develop project design. A project team leader along with the respective design engineer(s), quality engineer, and AutoCAD technician from each trade is assigned to work for the project. The project team is briefed by the project manager about the project objectives and the roles and responsibilities and authorities of each team member. Quality Manager also joins the project design team to ensure compliance to organization's quality management system. Figure 8.5 illustrates the project design organization structure.

8.3.2 IDENTIFY PROJECT STAKEHOLDERS

The following stake holders have direct involvement in the project during the conceptual design:

- Owner
- Designer
- Regulatory authorities
- PMC (as applicable)

Quality Management

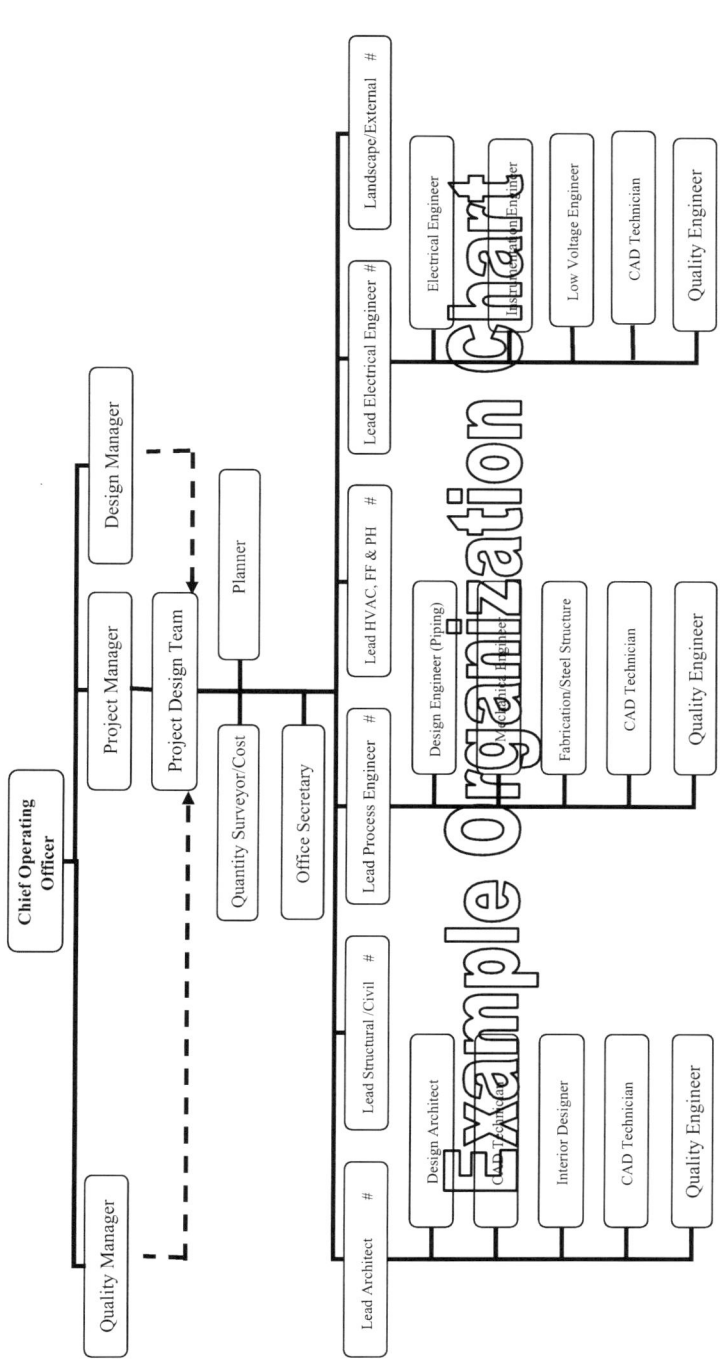

FIGURE 8.5 Project design team organization chart.

TABLE 8.2
Responsibilities of Various Participants during Concept Design Phase

	Responsibilities		
Phase	Owner/PMC	Designer	Regulatory
Concept design	Approval of team members Approval of time schedule Approval of budget Approval of concept design	Data collection Preparation of schedule Estimation of project cost Project quality requirements (concept design quality, project quality) Estimation of project resources Identify project risks Identification of HSE Issues and requirements Regulatory approvals development of concept design	Approval of project submittals

8.3.2.1 Develop Responsibility Matrix

A Responsibility Matrix listing key project activities during the concept design phase and stakeholders who are involved to advice, monitor, and review and/or approve various activities, documents, or deliverables is prepared and distributed to the concerned stakeholders for appropriate action. Table 8.2 illustrates the responsibilities of various participants during the concept design phase of the construction project life cycle.

8.3.3 Establish Concept Design Phase Requirements

In order to establish design phase requirements, the designer has to review all the documents which are part of the contract that designer has signed with the owner of the project. It basically includes Project Charter (TOR). In case of oil and gas project, the TOR generally requires the Designer (Consultant) to perform following:

- Development of Concept Design
- Preparation of FEED, Budget, and Schedule, and obtaining Authorities' Approvals
- Preparation of Contract Documents for bidding and tendering the purpose of selecting the EPC Contractor

8.3.3.1 Identify Concept Design Deliverables

Concept design deliverables are listed in TOR. Following are the concept design deliverables:

1. Concept design report (narrative/descriptive report)
    i. Overall site plan
    ii. Process technology selection

Quality Management 325

    iii. Product characteristics
    iv. Process flow scheme
    v. Piping system
    vi. Pipeline management system
    vii. Supervisory control
    viii. Instrumentation and control
    ix. Process automation
    x. Fire and gas detection
    xi. Leak-detection method
    xii. Corrosion monitoring system
    xiii. Storage method
    xiv. Material handling method
    xv. Facilities
    xvi. Building details (plan, elevation, and finishes)
    xvii. Structural system
    xviii. MEP system (Mechanical, Electrical, Plumbing, or Public Health)
    xix. Conveyance system
    xx. Fire suppression system
    xxi. Information and Communication Technology (ICT)
    xxii. Low-voltage systems
    xxiii. Data security system
    xxiv. Project risks and assessment
    xxv. HSE issues and environmental impact
    xxvi. Landscape
    xxvii. External work
    xxviii. Parking
    xxix. Constructability report
    xxx. Project execution plan
2. Drawings
    i. Overall site plan
    ii. Process flow diagram
    iii. Piping and instrumentation diagram
    iv. Piping layout
    v. Pumping station
    vi. Storage system layout
    vii. Schematic of process unit equipment
    viii. Supervisory control system
    ix. General arrangement drawing
    x. Electrical schematic diagram
    xi. Electrical substation
    xii. Building plan and elevation
    xiii. Low-voltage system
    xiv. Tie-in drawings (for all-utility services)
    xv. Sketches
    xvi. Sections (indicative to illustrate the overall concept)
3. Data collection and study reports

4. Existing site conditions
5. Calculations
6. Utility and infrastructure requirements
7. Concept schedule of material and equipment
8. Equipment and machinery data sheet
9. Tie-in material (for all-utility services)
10. Fire- and gas-detection system
11. Leak-detection and control systems
12. Safety and security system
13. Preliminary project schedule
14. Preliminary cost estimate
15. Project quality requirements
16. Resources for the project
17. Energy conservation
18. Economic analysis
19. Facility management requirements
20. Approvals from regulatory/statutory authorities and other agencies
21. Meeting records
22. Models

8.3.4 Develop Concept Design

Project charter or TOR is the base for the development of concept design. There is a significant impact the design has on product and customer satisfaction. The designer can use quality tools such as benchmarking for the development of product and quality function deployment (QFD) to translate the owner's need into technical specifications.

Benchmarking tool identifies and measures the factors critical to the success of the product comparing with the similar product of other businesses. It is used to determine the performance level of product with other products and achieve functional performance requirement.

QFD is a technique to translate customer needs into technical requirements to develop a product, project. It is also known as "House of Quality" because of its shape. It was developed in Japan by Dr. Yoji Akao in 1960s to transfer the concepts of quality control from the manufacturing process into the new product development process.

Figure 8.6 illustrates the concept of QFD "House of Quality."

It is the designer's responsibility to pay greater attention to improving the environment and to achieve sustainable development. The designer has to address environmental and social issues and comply with local environmental protection codes. During the design stage, the designer must work jointly with the owner to develop details regarding the owner's needs and give due consideration to each part of the requirements. The owner on his part should ensure that the project objectives are

- Specific
- Measurable
- Agreed upon by all the stakeholders
- Realistic

Quality Management

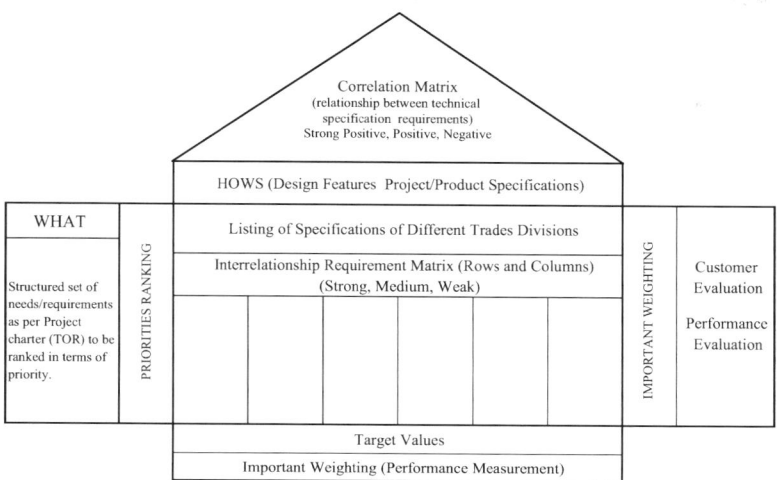

FIGURE 8.6 Concept of QFD house of quality.

- Possible to complete within definite time
- Within the budget

The designer can use Design for Six Sigma, an organizational approach for translating customer requirements into design features and then developments of design to meet customer requirements. DMADV tool can be used to develop concept design. The phases of DMADV are as follows:

D: Define: Requirements listed under Projects Charter (TOR) to develop new project/product

M: Measure: Identify the needs/requirements and then translate into the technical requirements (Critical to Quality—CTQ) and specifications

A: Analyze: Analyze product/process options and prioritize based on capabilities to satisfy customer requirements (TOR)

D: Design: Design product/process to satisfy customer requirements as per TOR

V: Verify: Verify and review the design to ensure that the design meets customer needs

8.3.4.1 Collect Data/Information

The purpose of data collection is to gather all the relevant information on existing conditions, both on project site and surrounding area that will impact the planning and design of the project. The data related to following major elements are required to be collected by the designer:

1. Certificate of title
    a. Site legalization
    b. Historical records

2. Topographical survey
   a. Location plan
   b. Site visits
   c. Site coordinates
   d. Photographs
3. Geotechnical investigations
4. Field and laboratory test of soil and soil profile
5. Existing structures in/under the project site
6. Existing utilities/services passing through the project site
7. Existing roads and structures surrounding the project site
8. Shoring and underpinning requirements with respect to adjacent area/structure
9. Requirements to protect neighboring area/facility
10. Environmental studies
11. Energy conservation requirements
12. Wind load, seismic load, dead load, and live load
13. Site access/traffic studies
14. Applicable codes, standards, and regulatory requirements
15. Usage and space program
16. Material handling area
17. Availability of resources
18. Design protocol
19. Scope of work/client requirements

8.3.4.2 Collect Owner's Requirements

Based on the scope of work/requirement mentioned in the TOR, a detailed list of requirements is prepared by the designer (consultant). The project design is developed taking into consideration owner's requirements and all other design criteria. The designer can prepare checklists of various trades by listing the items to obtain owner-preferential requirements.

Table 8.3 is a sample checklist to collect owner's preferred requirements.

8.3.4.3 Collect Regulatory Requirements

The designer has to collect regulatory requirements to be taken into consideration while preparing concept design.

8.3.5 PREPARE CONCEPT DESIGN

While preparing the concept design, the designer has to consider various available options/systems to ensure project economy, performance, and operation. The concept design should be suitable for further development.

8.3.5.1 Concept Design Drawings

Table 8.4 lists the elements which comprise the concept design drawings of various trades (disciplines).

TABLE 8.3
Checklist for Owner Requirements

Serial Number	Description of Item/Activity	Owner's Preferred Requirements (Trade Name)		
		Yes	No	Notes
1				
2				
3				
4				
5				
6				
7				
8				
9				
10				
11				
12				
13				
14				
15				
16				
17				
18				
19				
N…				

8.3.5.2 Concept Design Report

Concept design reports should be concise covering all the related information about concept design considerations. Reports are prepared for each of the trades mentioned in the TOR. Table 8.5 is example showing contents of concept design report for the oil and gas project.

8.3.6 PREPARE PRELIMINARY SCHEDULE

The duration of construction project is finite. It has a definite beginning and a definite end; therefore, during the conceptual phase, the expected time schedule for the completion of the project/facility is worked out. Expected time schedule is important from both financial and acquisition of the facility by the owner/end user. It is the owner's goal and objective that the facility is completed in time for further usage. During this phase, very limited information and details about the project activities are available and have very wide variance in the schedule.

In order to improve the understanding and the communication among stakeholders involved with preparing, evaluating, and using project schedule, AACE

TABLE 8.4
Elements to be included in Concept Design Drawings

Serial Number	Elements to Be Included in Drawing
1	General arrangement
	a. Site layout
	b. Space allocation layout (area mark-up plan)
	c. Unit plot pan
2	Process
	a. Process flow diagram
	b. Utility flow diagram
	c. Process automation
3	Piping
	a. Piping and Instrumentation diagram
	b. Pipeline
	c. Pumping station
4	Mechanical
	a. Storage tanks
	b. Piping works
	c. Steel structure
5	Instrumentation and Control
6	Supervisory Control System
7	Pipeline Management System
8	Architectural
	a. Overall site plans
	b. Floor plans
	c. Roof plans
9	Structural
	a. Building structure
	b. Stairs
	c. Roof and general sections
10	Elevator
	a. Traffic studies
	b. Elevator/escalator location
	c. Equipment room
11	Plumbing and Fire suppression
	a. Sprinkler layout plan
	b. Development of preliminary system schematics
12	HVAC
	a. Ducting layout plan
	b. Piping layout plan
	c. Estimation of HVAC electrical load
	d. Development of BMS schematics showing interface
	e. Location of plant room, chillers, cooling towers

(*Continued*)

TABLE 8.4 (*Continued*)
Elements to be included in Concept Design Drawings

Serial Number	Elements to Be Included in Drawing
13	Electrical
	a. Lighting layout plan
	b. Power layout plan
	c. System design schematic without any sizing of cables and breakers
	d. Substation layout and location
	e. Total connected load
	f. Location of electrical rooms and closets
	g. Riser requirements
14	Low-voltage system
	a. ICT
	b. Public address, audiovisual system layout
15	Safety and fire protect system
16	Security system
17	Leak detection system
18	Loss prevention system
19	Corrosion monitoring system
20	Tie-in DRAWING
21	Landscape
	a. Green area layout
	b. Selection of plants
	c. Irrigation system
22	External
	a. Street/road layout
	b. Street lighting
	c. Bridges (if any)
	d. Security system
	e. Pedestrian walkways
23	Narrative description, reports

International has published the guideline to classify Guidelines to classify schedules into five classes and five levels. Figure 8.7 illustrates Schedule Classifications versus Schedule Levels that schedule can be developed and/or presented.

Table 8.6 illustrates generic schedule classification matrix and Table 8.7 illustrates characteristics of schedule classifications.

The preliminary project schedule is developed using top-down planning using key events. It is also known as Class 4 schedule or Schedule Level 1

8.3.7 Estimate Conceptual Cost

In construction projects, the cost estimates vary as the project design progresses. At the inception of project, the cost estimate is based on Rough Order of Magnitude. When detailed design is available, the cost estimates become definitive.

TABLE 8.5
Contents of Concept Design Report (Trade Name)

Section	Topic
1	Introduction
	1.1 Description of project
	1.2 Project goals and objectives
2	Owner's schedule requirement
3	Project directory
4	Trade name (process, architecture, MEP, .etc.)
	4.1 Applicable codes and standards
	4.2 International codes
	4.3 Local codes
	4.4 Regulatory requirements
	4.5 Applicable design system/details of the project trade item/element
5	Project requirement
6	Existing site conditions
7	Data collection
8	Design options/strategy/criteria
9	Design software
10	Geographical investigation
11	Risk assessment
12	HSE issues
13	Drawings
14	Models (if applicable)

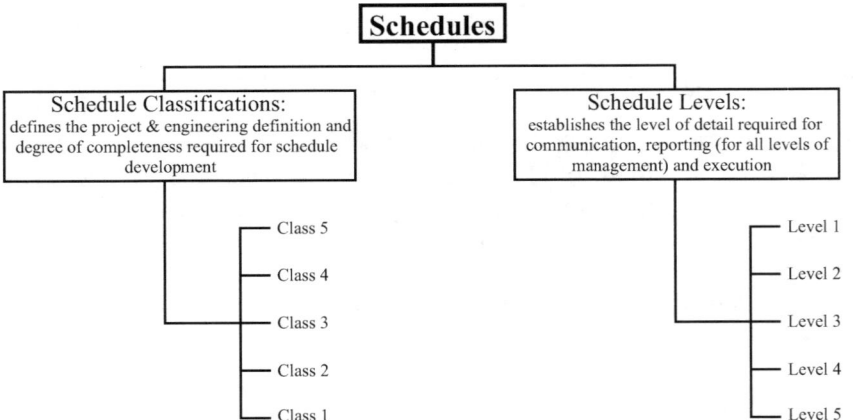

FIGURE 8.7 Schedule: classifications versus levels. (Adapted from AACE International Recommended Practice 27R-03. Copyright © 2010 AACE International; all rights reserved. Reprinted with permission from AACE International.)

TABLE 8.6
Schedule Levels

Schedule Level 1

Schedule Title	**Project Master Schedule (Executive Summary)**
Description	Level 1 schedule is a high-level schedule that reflects key milestones and high-level project activities by major phase, stage, or project being executed. This schedule level may represent summary activities of an execution stage, specifically engineering, procurement, construction, and start-up activities.
	Level 1 schedules provide high-level information that assists in the decision-making process (go/no go prioritization and criticality of projects).
	Level 1 schedule can be used to integrate multiple contractors/multiple schedules into an overall program management process.
	Level 1 audience includes, but are not limited to client, senior executives and general managers.
	Level 1 schedule may be used in the proposal stage of a potential project/contract.
Scheduling method	Graphical representation in the bar chart format.
Usage in construction projects	Study stage, predesign phase, feasibility study

Schedule Level 2

Schedule Title	**Summary Master Schedule (Management Summary)**
Description	Level 2 schedules are generally prepared to communicate the integration of work throughout the life cycle of a project.
	Level 2 schedules may reflect, at a high level, interfaces between key deliverables and project participants (contractors) required to complete the identified deliverables.
	Level 2 schedules provide high-level information that assist in the project decision-making process (re-prioritization and criticality of project deliverables).
	Level 2 schedules assist in identifying project areas and deliverables that require actions and/or course correction. Audiences for this type of schedule include but are not limited to general managers, sponsors, and program or project managers.
Scheduling method	Typically presented in the Gantt (bar chart) format.
Usage in construction projects	Overall design schedule, overall construction schedule.

Schedule Level 3

Schedule Title	**Project Coordination Schedule (Publication Schedule)**
Description	Level 3 schedules are generally prepared to communicate the execution of the deliverables for each of the contracting parties. Level 3 reflects the interfaces between key workgroups, disciplines, or crafts involved in the execution of the stage.
	Level 3 schedule includes all major milestones, major elements of design, engineering, procurement, construction, testing, commissioning, and handover.
	Level 3 schedules provide enough detail to identify critical activities.

(Continued)

TABLE 8.6 (*Continued*)
Schedule Levels

Schedule Level 3	
Schedule Title	**Project Coordination Schedule (Publication Schedule)**
	Level 3 schedules assist the team in identifying activities that could potentially affect the outcome of a stage or phase of work, allowing for mitigation and course correction in short course. Level 3 audiences include, but are not limited to program or project managers, CMs or owner's representatives, superintendents, and general foremen.
Scheduling method	Typically presented in Gantt or CPM network format and is generally the output of CPM scheduling software.
Usage in construction projects	Detail design, detail construction

Schedule Level 4	
Schedule Title	**Project Working Level Schedule (Execution Schedule)**
Description	Level 4 schedules are prepared to communicate the production/execution of work packages at the deliverable level. Level 4 schedule reflects interfaces between key elements that drive completion of activities.
	Level 4 schedules usually provide enough detail to plan and coordinate contractor or multidiscipline/craft activities.
	Level 4 schedule displays the activities to be accomplished by identifying all the required resources.
	Level 4 audiences include but are not limited to project managers, superintendents, and general foremen.
Scheduling method	Typically presented in Gantt or CPM network format.
Usage in construction projects	Monthly schedule and look-ahead schedules

Schedule Level 5	
Schedule Title	**Detail Schedule**
Description	Level 5 schedules are prepared to communicate task requirements for completing activities identified in a detailed schedule.
	Level 5 schedules are usually considered working schedules that reflect hourly, daily or weekly work requirements.
	Level 5 schedules are used to plan and schedule utilization of resources (labor, equipment, and materials) in hourly, daily, or weekly units for each task.
	Level 5 audiences include but are not limited to superintendents, general foremen and foremen.
Scheduling method	Typically presented in an activity listing format without time scaled graphical representation of work to accomplish.
Usage in construction projects	Daily work schedule and daily progress report

Source: Adapted from AACE International Recommended Practice 27R-03 Schedule Classification System and AACE International Recommended Practice 37R-06, Schedule Levels of Detail – As Applied in Engineering, Procurement, and Construction. "Copyright © 2010 by AACE International; all rights reserved." Reprinted with permission from AACE International.

TABLE 8.7
Generic Schedule Classification Matrix

Schedule class	Primary Characteristic		Secondary Characteristic
	Degree of project definition (Expressed as % of Complete definition)	End usage	Scheduling methods used
Class 5	0%–2%	Concept screening	Top-down planning using high-level milestones and key project events
Class 4	1%–15%	Feasibility study	Top-down planning using high-level milestones and key project events. Semi-detailed
Class 3	10%–40%	Budget, authorization, or control	"Package" top-down planning using key events. Semi-detailed
Class 2	30%–75%	Control or bid/tender	Bottom-up planning. Detailed
Class 1	65%–100%	Bid/tender	Bottom-up planning. Detailed

Source: Adapted from AACE International Recommended Practice 18R-97. Recommended Practice 18R-97 provides range of percentages for each class. "Copyright © 2010 by AACE International; all rights reserved." Reprinted with permission from AACE International.

The cost estimated by the designer (consultant) is based on assumptions and historical data available from experience on similar projects.

Table 8.8 illustrates cost estimation levels for construction projects.

A cost estimate during concept design is required by the owner to know how much the capital cost of construction is. This is required by the owner to enable arranging the financial resources. Conceptual cost is also known as budgetary cost. Parametric cost estimation methodology is used to estimate the conceptual cost as discussed in Table 8.8. The designer has to ensure that the conceptual cost does not exceed the cost estimated during feasibility stage. The accuracy and validity of the conceptual cost estimation are related to the level of information available while developing the concept design. The designer has to properly estimate the resources required for the successful completion of project. Table 8.9 illustrates quality check for cost estimation.

8.3.8 Establish Project Quality Requirements

During this phase, the designer has to plan and establish quality criteria for the project. This includes mainly the following:

- Owner's requirements
- Quality standards and codes to be complied
- Regulatory requirements
- Conformance to owner's requirements

TABLE 8.8
Characteristics of Schedule Classification

	Class 5 Schedule Characteristics
Description	Class 5 schedules are generally prepared based on very limited information and subsequently have a wide accuracy range.
	The Class 5 schedule is considered a preliminary document, usually presented in either Gantt (Bar chart) or Table form. The class 5 schedule should have, as a minimum, a single summary per stage with major project milestones identified
Degree of project definition required	0%–2%
End usage	Class 5 schedules are prepared for any number of strategic business planning purposes, such as but not limited to: market studies, assessment of initial viability, evaluation of alternative schemes, project screening, project location studies, evaluation of resource need and budgeting, long range capital planning, etc.
Scheduling methods used	Gantt, bar chart, milestone/activity table. Top-down planning using high-level milestones and key project events.
	Class 4 Schedule Characteristics
Description	Class 4 schedules are generally prepared based on limited information and subsequently have fairly wide accuracy ranges. They are typically used for project screening, determination of "do-ability," concept evaluation, and to support preliminary budget approval. The Class 4 schedule is usually presented in either Gantt (bar chart) or table form. The Class 4 schedule should define the high-level deliverables for each specific stage going forward (since the previous stage has passed). This document should also provide an understanding regarding the timing of key events, such as independent project reviews, committee approvals, as well as determining the timing of funding approvals. A high-level WBS may be established at this time.
Degree of project definition required	1%–15%
End usage	Class 4 schedules are prepared for a number of purposes, such as but not limited to: detailed strategic planning, business development, project screening at more developed stages, alternate scheme analysis, confirmation of "do-ability," economic and/or technical feasibility, and to support preliminary budget approval or approval to proceed to next stage. It is recommended that the Class 4 schedule be reconciled to the Class 5 schedule to reflect the changes or variations identified as a result of more project definition and design. This will provide an understanding of the changes from one schedule to the next.
Scheduling methods used	Gantt, bar chart, milestone/activity table top-down planning using high-level milestones and key project events. Semi-detailed.

(Continued)

TABLE 8.8 (*Continued*)
Characteristics of Schedule Classification

Class 3 Schedule Characteristics

Description	Class 3 schedules are generally prepared to form the basis of execution for budget authorization, appropriation, and/or funding. As such, they typically form the initial control schedule against which all actual dates and resources will be monitored.
	The Class 3 schedule should be a resource-loaded, logic-driven schedule developed using the precedence diagramming method (PDM). The schedule should be developed using relationships that support the overall true representation of the execution of the project (with respect to start to start and finish to finish relationships with lags). The amount of detail should define, as a minimum, the work package (WP) level (or similar deliverable) per process type/unit and any intermediate key steps necessary to determine the execution path. (The WP rolls up into the predefined WBS.) In some circumstances, where there is a high degree of parallel activities, the critical nature of the project, or the extreme complexity and/or size of the project, it may be warranted to provide further detail of the schedule to assist in the control of the project.
Degree of project definition required	10%–40%
End usage	Class 3 schedules are typically prepared to support full project-funding requests and become the first of the project phase "control schedules" against which all start and completion dates and resources will be monitored for variations to the schedule. They are used as the project schedule until replaced by more detailed schedules.
	It is recommended that the Class 3 schedule be reconciled to the Class 4 schedule to reflect the changes or variations identified as a result of more project definition and design.
Scheduling methods used	PDM, PERT, Gantt/bar charts
	"Package" top-down planning using key events.
	Semi-detailed.

Class 2 Schedule Characteristics

Description	Class 2 schedules are generally prepared to form a detailed control baseline against which all project works are monitored in terms of task starts and completions and progress control.
	The Class 2 schedule is a detailed resource loaded, logic-driven schedule that should be developed using the CPM process. The amount of detail should define as a minimum, the required deliverables per contract per WP. The schedule should further define any additional steps necessary to determine the critical path of the project necessary for the appropriate degree of control.

(*Continued*)

TABLE 8.8 (*Continued*)
Characteristics of Schedule Classification

Degree of project definition required	30%–70%
End usage	Class 2 schedules are typically prepared as the detailed control baseline against which all actual start and completion dates, and resources will now be monitored for variations to the schedule and form a part of the change/variation control program. It is recommended that the Class 2 schedule be reconciled to the Class 3 schedule to reflect the changes or variations identified as a result of more project definition and design.
Scheduling methods used	Gantt/bar charts, PDM, PERT Bottom-up planning. Detailed.
Class 1 Schedule Characteristics	
Description	Class 1 schedules are generally prepared for discrete parts or sections of the total project rather than generating this amount of detail for the entire project. The updated schedule is often referred to as the current control schedule and becomes the new baseline for the cost/schedule control of the project. The Class 1 schedule may be a detailed, resource-loaded, logic-driven schedule and is considered a "production schedule" used for establishing daily or weekly work requirements
Degree of project definition required	70%–100%
End usage	Class 1 schedules are typically prepared to form the current control schedule to be used as the final control baseline against which all actual start and completion dates and resources will now be monitored for variations to the schedule, and form a part of the change/variation control program. They may be used to evaluate bid-schedule checking, to support vendor/contractor negotiations, or claim evaluations and dispute resolution. It is recommended that the Class 1 schedule be reconciled to the Class 2 schedule to reflect the changes or variations identified as a result of more project definition and design.
Scheduling methods used	Gantt/bar charts, PDM, PERT Bottom-up planning. Detailed.

Source: Adapted from AACE International Recommended Practice 27R-03. "Copyright © 2010 by AACE International; all rights reserved." Reprinted with permission from AACE International.

- Conformance to requirements listed under project charter or TOR
- Design review procedure
- Drawings review procedure
- Document review procedure
- Quality management during all the phases of project life cycle

Table 8.10 lists contents of designer's quality control plan.

TABLE 8.9
Quality Check for Cost Estimate during Concept Design

Serial Number	Points to be Checked	Yes/No
1	Check for use of historical data	
2	Check if estimate factors used to adjust historical data	
3	Check the estimate is updated with revision or update of concept	
4	Check if the updates/revisions are chronologically listed and cost estimate updated	
5	Whether scope of work is descriptive/narrative, well enough for estimation purpose?	
6	Is the estimate based on area schedule and overall plan for the project provided by the architect?	
7	Is the estimate based on schedule of equipment and machinery for the project provided by the process engineer?	
8	Is the estimate based on schedule of piping material provided by the piping/mechanical engineer?	
9	Whether estimate is updated taking into consideration feedback from each trade?	
10	Is the cost estimate clearly identify the quantities and associated work?	
11	Whether all the assumption are as per current market data?	
12	Is cost estimates from all the relevant trades included in the final sum?	
13	Is the estimate include all the requirements of specialist consultants?	
14	Whether the total estimate is reviewed and verified?	

Source: Modified from Abdul Razzak Rumane. (2013). *Quality Tools for Managing Construction Projects.* Reprinted with permission from Taylor & Francis Group.

8.3.8.1 Manage Concept Design Quality

During this phase, the designer has to manage the quality for development of concept design and also to plan and establish quality criteria for quality management during all the phases of the project life cycle.

Figure 8.8 illustrates the quality management procedure for development of concept design.

8.3.9 ESTIMATE RESOURCES

Designer has to estimate the resources required to complete the project. This includes the estimation of manpower required during the construction phase, testing, commissioning, and handover phase.

The designer has to also prepare equipment and machinery data sheet and schedule of material.

TABLE 8.10
Contents of Designer's Quality Control Plan

Section	Topic
1	Introduction
2	Description of project
3	Quality control organization
4	Qualification of QC staff
5	Responsibilities of QC personnel
6	Procedure for submittals
7	Quality control procedure for concept design
	7.1 Establish concept design requirements
	7.1.1 Review of project charter, terms of requirements (TOR)
	7.1.2 Identify project goals and objectives
	7.1.3 Identify concept design scope
	7.1.4 Review feasibility reports
	7.1.5 Review preferred alternative selection report
	7.1.6 Establish concept design deliverables
	7.1.7 Establish design procedure
	7.2 Identify quality management requirements
	7.2.1 Plan quality
	7.2.1.1 Owner's requirements
	7.2.1.2 TOR requirements
	7.2.1.3 Numbers of drawing, reports, models
	7.2.1.4 Scope of design work
	7.2.1.5 Responsibility matrix
	7.2.1.6 Codes and standards
	7.2.1.7 Regulatory requirements
	7.2.1.8 Submittal plan
	7.2.1.9 Drawings, specifications, documents
	7.2.1.10 Design review plan
	7.2.2 Quality assurance
	7.2.2.1 Information/data collection
	7.2.2.2 Site investigation
	7.2.2.3 Engineering surveys
	7.2.2.4 Preparation of drawings
	7.2.2.5 Interdisciplinary coordination
	7.2.2.6 Project risks
	7.2.2.7 Environmental issues
	7.2.2.8 Preparation of specification, documents
	7.2.2.9 Functional and technical compatibility
	7.2.3 Quality control
	7.2.3.1 Design drawing
	7.2.3.2 Quality of drawings
	7.2.3.3 Project schedule
	7.2.3.4 Project cost

(Continued)

TABLE 8.10 (*Continued*)
Contents of Designer's Quality Control Plan

Section	Topic
	7.2.3.5 Project resources
	7.2.3.6 Specification, contract documents
8	Project-specific design requirements
9	Design software
10	Design development procedure
	10.1 Design criteria
	10.2 Preparation of drawings
	10.3 Interdisciplinary coordination
	10.4 Design review
11	Company's quality manual and procedure.
12	Sub consultant's work
13	Value Engineering (VE)
14	Quality auditing program
15	Quality control record
16	Innovative and latest technology
17	Quality updating program

8.3.10 IDENTIFY PROJECT RISKS

The designer has to identify the risks which will affect the successful completion of project. The following are typical risks which normally occur during the conceptual design phase:

- Lack of input from owner about the project goals and objectives
- Project charter or TOR not defined clearly
- The related project data and information collected are incomplete
- The related project data and information collected are likely to be incorrect and wrongly estimated
- Environmental consideration
- Regulatory requirements
- Errors in estimating the project schedule
- Errors in cost estimation
- Errors in resources estimation
- Environmental issues not correctly identified
- Design basis is not suitable for further development/development of concept design

Designer has to take into account the above-mentioned risk factors while developing the concept design.

Further, the designer has to consider the following risk while planning the duration for completion of the conceptual phase.

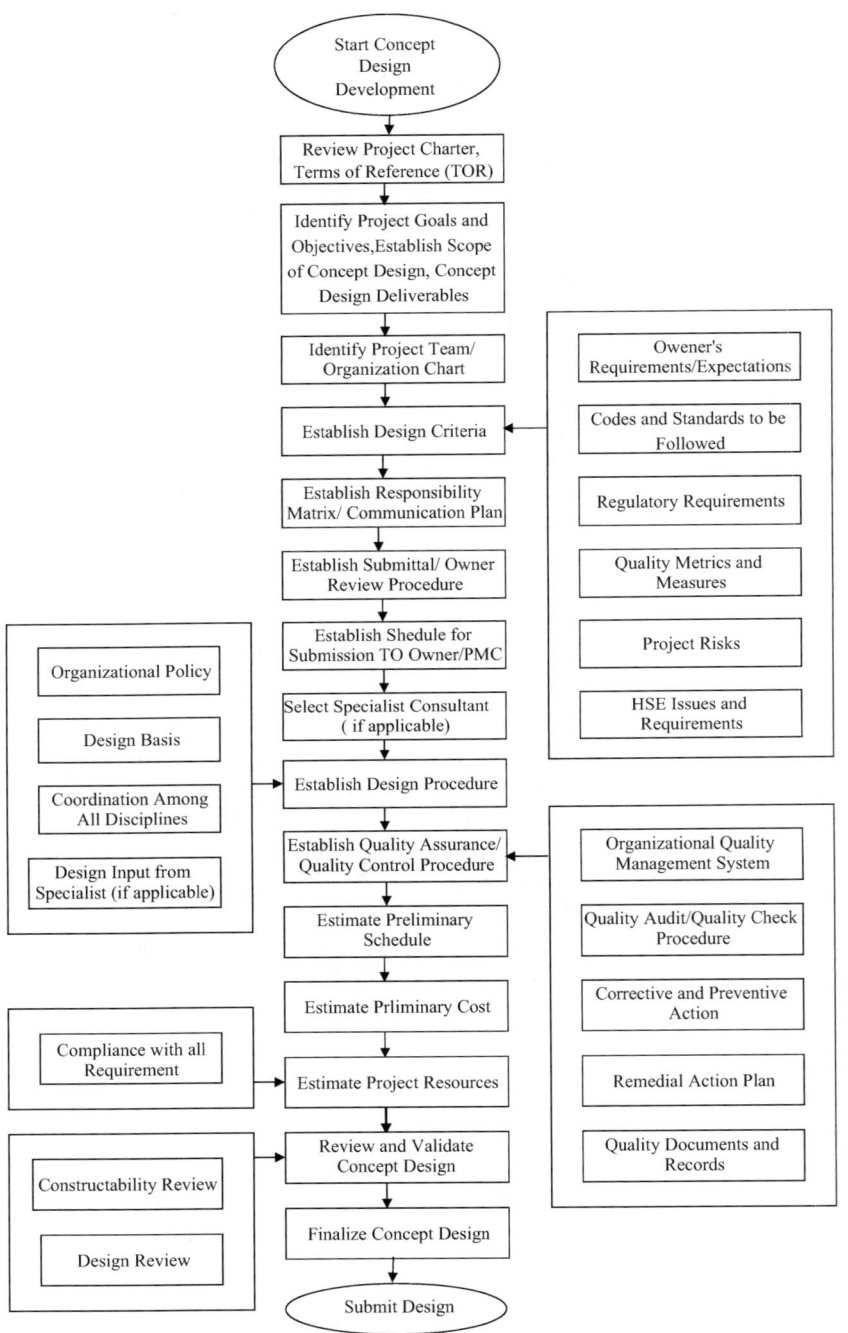

FIGURE 8.8 Quality management procedure for development of concept design.

Quality Management

- Impractical conceptual design preparation schedule
- Delay in data collection
- Delay to obtain authorities' approval
- Delay in environmental approval

The designer has to prepare the Risk Assessment Report for submission to the owner along with other documents while submitting the concept design package. The typical contents of the report are as follows:

1. Purpose of report
2. Description of project (goals and objectives)
3. References (relevant codes, standards, owner references)
4. Scope of work (during all the life cycle phases of the project)
5. Hazard identification and assessment method

8.3.11 Identify HSE Requirements

The designer has to identify HSE issue that could affect the environment due to the project. While developing the design, the designer has to consider the following:

- Hazardous properties of the materials and products used in the project
- Safety in process and operations
- Inherent history of process technology being considered for the project
- Hazardous emissions
- Pollution and its impact
- Impact on health
- Safety in design (safe design)
- Regulatory and other environmental protection agencies' requirements
- Environmental compatibility

The designer has to prepare Environmental Impact Assessment Report for submission to the owner along with other documents while submitting the concept design package. The typical contents of environmental impact assessment report are as follows:

1. Purpose of report
2. Description of project (goals and objectives)
3. References (relevant codes, standards, owner references)
4. Scope of work (during all the life cycle phases of the project)

The designer is also required to submit HSE Management Plan during this phase.

8.3.12 Review Concept Design

Table 8.11 illustrates major points for review of concept design.

TABLE 8.11
Review of Concept Design

Serial Number	Description	Yes	No	Notes
	A: Concept Design			
A.1	Does the design support owner's project goals and objectives?			
A.2	Does the design meets all the elements specified in project charter (TOR)?			
A.3	Does the design meet all the performance requirements?			
A.4	Whether constructability has been taken care?			
A.5	Whether usage of space is optimal			
A.6	Whether technical and functional capability considered?			
A.7	Whether process technology considered is suitable to produce the required products?			
A.8	Whether ease of operations is considered while selecting the process technology?			
A.9	Whether the design meets all the specified codes and standards?			
A.10	Whether the designer has considered into design all the data and information collected?			
A.11	Whether design confirm with fire and egress requirements?			
A.12	Whether design risks been identified, analyzed, and responses planned for mitigation?			
A.13	Whether risk assessment report is prepared			
A.14	Whether health and safety requirements in the design are considered?			
A.15	Whether environmental constraints considered?			
A.16	Whether environmental impact assessment report is prepared and considered?			
A.17	Whether the design meets LEED requirements?			
A.18	Does energy conservation is considered?			
A.19	Does sustainability considered in design?			
A.20	Whether cost-effectiveness over the entire project life cycle is considered?			
A.21	Whether all reasonable design options/systems are considered for project economy?			
A.22	Whether accessibility is considered?			
A.23	Does the design meet ease of maintenance?			
A.24	Does the design have provision for inclusion of facility management requirements?			
A.25	Does the design have provision for interface with all the low-voltage and control systems?			
A.26	Whether the design is coordinated with all trades?			
A.27	Whether all the regulatory/statutory requirements taken care?			
A.28	Does the design support proceeding to next design development stage?			

(Continued)

TABLE 8.11 (*Continued*)
Review of Concept Design

Serial Number	Description	Yes	No	Notes
	B: Schedule			
B.1	Project schedule is practically achievable?			
	C: Financial			
C.1	Project cost is properly estimated?			
	D: Resources			
D.1	Whether availability of resources during construction phase is considered?			
D.2	Whether schedule of equipment, machinery is prepared?			
	E: Reports			
E.1	Whether the reports are complete and include adequate information about the project?			
E.2	Whether the reports are prepared for all the trades mentioned in TOR?			
E.3	Whether the report is properly formatted and has Table of Contents for each report?			
	F: Drawings, Sketches			
F.1	Whether drawings, sketches for all trades prepared as per TOR?			
F.2	Whether number of drawings is as per TOR requirements?			
	G: Models			
G.1	Whether the Models meet the design objectives? (As applicable)			
	H: Submittals			
H.1	Whether numbers of sets prepared as per TOR?			

8.3.13 FINALIZE CONCEPT DESIGN

Final designs are prepared incorporating the comments, if any, found during analysis and review of the drawings and documents for submission to the owner/client.

8.3.14 SUBMIT CONCEPT DESIGN PACKAGE

Normally, the following are submitted to the owner for their review and approval in order to proceed with development of FEED:

1. Outline specifications
2. Concept design drawings
3. Concept design reports

4. Project schedule
5. Cost estimate
6. Model

8.4 FEED PHASE

FEED is the third phase of oil and gas construction project life cycle project. It is generally known as "FEED."

FEED is mainly a refinement of the elements in the conceptual design phase. It is developed based on the output from Concept Design or Basic Engineering phase. FEED is design intent documents which quantify functional performance expectations and parameters for each system to be commissioned. FEED adequately describes information about all proposed project elements in sufficient details for obtaining regulatory approvals, necessary permits, and authorization. At this stage, the project is planned to the level where sufficient details are available for the initial schedule and cost. The FEED is a subjective process suitable for transforming ideas and information into plans, drawings, and specifications of the project to be built. Component/equipment/machinery configurations, material specifications, and functional and operational performance of the process equipment, machinery are decided during this stage. The FEED design focuses the technical requirements and studies to identify technical issues as well as estimate investment cost for the project. During this phase, accurate total investment cost (TIC) is evaluated and decision is made for development of overall project execution planning and preparation of tender documents for selection of EPC contractor. The FEED defines the implementation details needed for a successful project. The construction execution strategy is determined during Front End Engineering Design Phase. A good FEED reflects the client's project-specific requirements with clear justification scope, schedule, and budget to avoid significant changes during the execution phase. The FEED is used as a basis for detailed engineering in the Engineering, Procurement, and Construction (EPC) contracting system.

The main objectives of this phase are as follows:

- Develop FEED taking into consideration the concept design output and requirements listed under project charter
- Follow regulatory requirements, codes, and standards
- Make studies for selection of process technology
- Make studies to identify technical and operational issues of the process
- Evaluate accurate TIC
- Develop project execution plan
- Perform value engineering (VE)
- Tender documents to select EPC contractor

While developing the FEED, the designer must consider the following:

1. Concept design deliverables
2. Project charter requirements (project goals and objectives)

3. Requirements of all stakeholders
4. Usage
5. Process engineering
6. Facilities engineering
7. Technical and functional capability
8. Operational capability of the process
9. Constructability
10. Calculations to support the design and process equipment
11. Collection of data and information
12. Optimization of design
13. Optimization and usage of site layout
14. Energy conservation measure
15. Sustainability (environmental, social, and economical)
16. Project quality
17. Reliability
18. Availability of resources
19. Health and safety
20. Environmental issues and compatibility
21. Fire protection measures
22. Supportability during maintenance/maintainability
23. Cost-effectiveness of the project
24. Data security

Figure 8.9 illustrates the logic flowchart of FEED phase.

8.4.1 Identify Design Team Members

Design team members are selected based on the organizational structure and suitable skills required to perform the job. Normally, the design team consists of

1. Project Manager
2. Design Managers (one for each trade)
3. Quality Manager
4. Team Leader (Principal Engineer) – Each Trade
5. Team Members (Engineers and CAD technicians for each discipline)
6. Quantity surveyor (cost engineer)

Figure 8.10 illustrates an example process engineering design team organization chart.

8.4.2 Identify Project Stakeholders

During FEED phase, the following are the stake holders directly involved in the project:

- Owner
- Designer

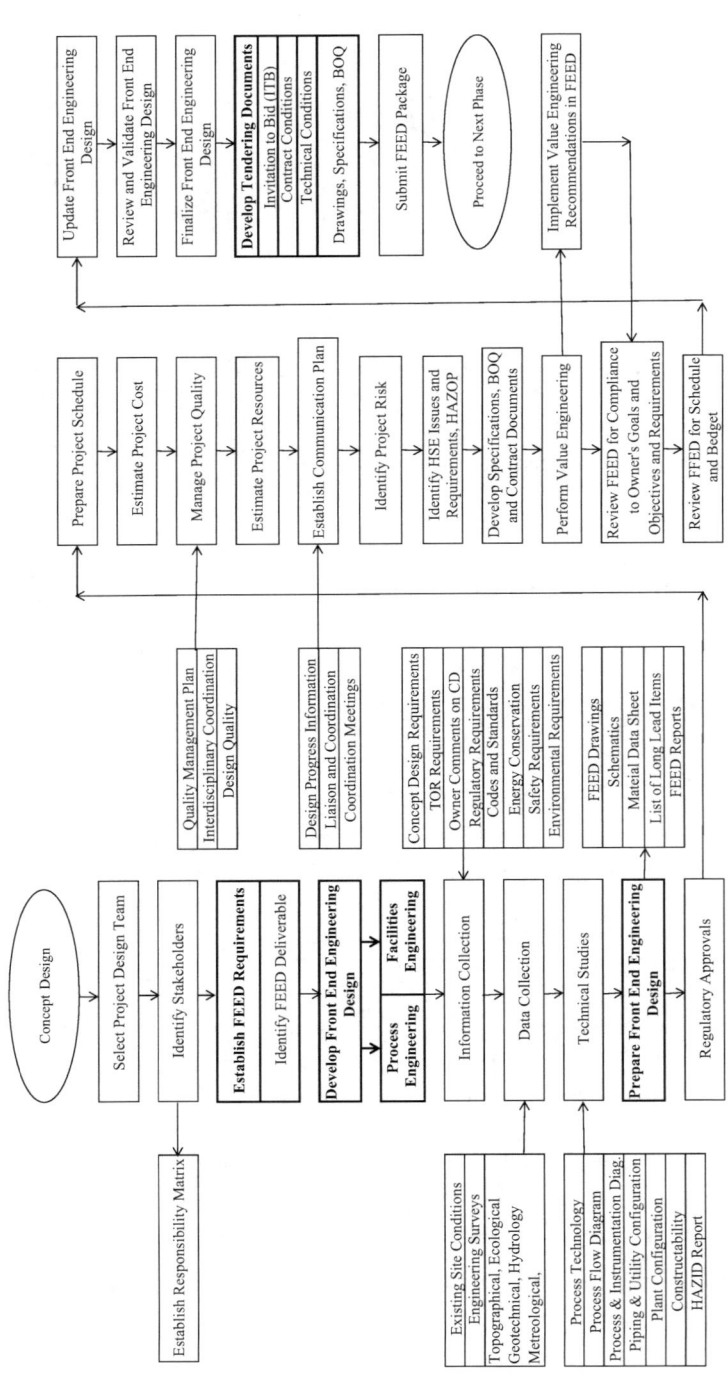

FIGURE 8.9 Logic flowchart for FEED phase.

Quality Management

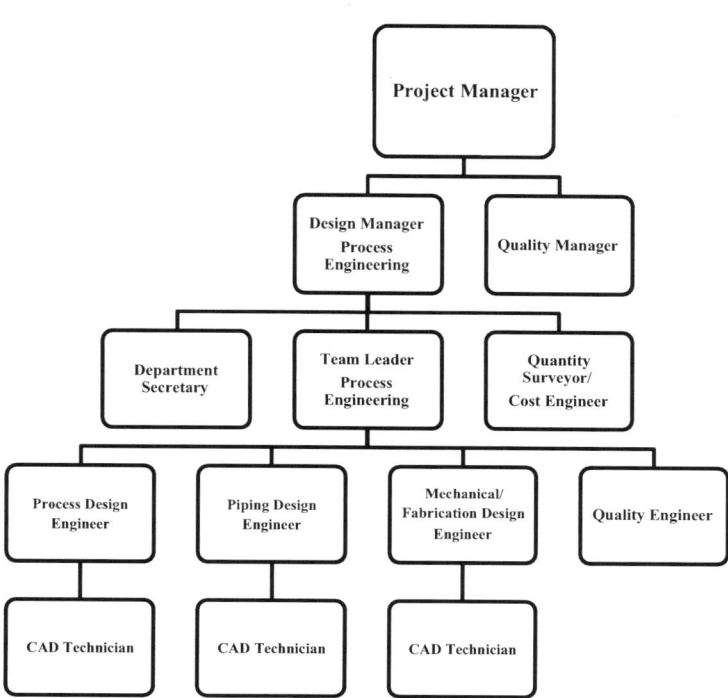

FIGURE 8.10 Process engineering design team organization chart.

- Regulatory authorities
- PMC (as applicable)
- Specialist consultant (as applicable)

8.4.2.1 Develop Responsibility Matrix

Table 8.12 illustrates typical responsibilities of various participants during FEED phase.

8.4.3 ESTABLISH FEED REQUIREMENTS

The central activity of FEED is the designer's concept of owner's goals and objectives which can help EPC contractor to develop detailed engineering for the required project and construct the project to produce the project(s) to meet and satisfy owner's requirements. Before starting FEED, the scope must adequately define and the deliverables should be identified to ensure compliance to TOR requirements.

8.4.3.1 Identify FEED Deliverables

Normally, the TOR lists the requirements guidelines to develop FEED. It mainly consists of the following:

1. Complete FEED drawings for the selected/approved concept design
2. FEED report

TABLE 8.12
Responsibilities of Various Participants during FEED Phase

		Responsibilities	
Phase	Owner	Designer	Regulatory Authorities
FEED	Update Owner's Requirements Approval of Project Schedule Approval of Project Cost Approval of FEED	Data/Information Collection and Surveys Develop general layout Develop scope of work Selection of process technology Develop FEED drawings Regulatory requirements and approvals Estimate project Schedule Estimate project cost Contract terms and conditions VE Develop tendering documents	Approval of project submittals

3. Basic engineering to develop detailed engineering by EPC contractor
4. Invitation to bid (ITB) documents

The designer has to identify deliverables to such a level that there will be not be any major changes in the scope of works included in the ITB documents. Table 8.13 lists FEED deliverables. While developing the design, the designer has to consider these items.

8.4.4 DEVELOP FEED

In order to identify requirements to develop FEED, the designer has to gather comments made by the owner/PMC on the submitted concept design, collect TOR requirements, regulatory requirements, perform technical studies about the process and engineering surveys, and also collect other related data to ensure that the developed design is error-free and has minimum omissions.

8.4.4.1 Information Collection

Prior to the start of FEED, the designer has to identify and collect the following information:

- TOR requirements
- Concept design comments
- Owner's requirements
- Regulatory requirements
- Energy conservation requirements

TABLE 8.13
FEED Deliverables

Serial Number	Deliverables
1	Basic engineering design data
2	General arrangements
	2.1 General plot layout
	2.2 Unit plot plan (equipment layout)
3	Process Engineering
	3.1 Process flow diagrams
	3.2 Piping and instrumentation diagram
	3.3 Material data sheet
	3.4 Long-lead items
	3.5 Utility system design
4	Piping
	4.1 Piping layout
	4.2 Piping routing drawing
	4.3 Pipeline management system
5	Mechanical
	5.1 Storage tank
	5.2 Fabrication drawings
	5.3 material handling
6	Instrumentation and automation
	6.1 instrumentation design
	6.2 Control system
	6.3 Process automation
7	Fire and gas detection system
8	Flaring and venting system
9	Supervisory control system
10	Leak-detection system
11	Corrosion monitoring system
12	Tie-in drawings
13	Buildings
	13.1 Architectural plans, elevation, roof, interiors, external finishes, floor plans, doors, windows, partitions, toilets, wash rooms
	13.2 Structural details, elevations, beams, columns, sections, stairs, calculations
	13.3 Elevators and conveying systems, locations, equipment room, traffic study
	13.4 MEP (mechanical, electrical, and public health)
	13.5 Schematics for electrical
	13.6 Schematics for piping
	13.7 Electrical and mechanical rooms
14	Pumping station
15	Electrical sub station

(*Continued*)

TABLE 8.13 (*Continued*)
FEED Deliverables

Serial Number	Deliverables
16	Fire suppression system
	16.1 Fire alarm system
17	Loss prevention system
18	Cathodic protect system
19	Information and Communication Technology (ICT)
	19.1 Information technology (computer network)
	19.2 IP telephone system (telephone network)
	19.3 Smart building system
	19.4 Public address system
20	Security and access control
21	Data security
22	Landscape
	22.1 Plants
	22.2 Irrigation system
23	External works
	23.1 Street/road layout
	23.2 Pedestrian walkways
	23.3 Street lighting
	23.4 Parking
	23.5 Parking control system
24	Narrative reports
	24.1 Technology selection
	24.2 Design criteria
	24.3 Process selection
	24.4 Process automation
	24.5 Material selection
	24.6 H&M balance
	24.7 Risk assessment
	24.8 HAZID
	24.9 HAZOP
	24.10 Environmental, health, and safety requirements, and environment impact assessment
	24.11 Energy code requirements
	24.12 Construction methodology
	24.13 Site survey
	24.14 Site investigation
25	Catalyst and chemical summary
26	Start-up and shut-down philosophies
27	Operational philosophies
28	Calculations
29	Cause and effect diagrams

(*Continued*)

Quality Management

TABLE 8.13 (*Continued*)
FEED Deliverables

Serial Number	Deliverables
30	Material
	30.1 Material data sheet
	30.2 Long lead items/material
	30.3 Tie-in material
	30.4 MTO list
31	Permits and regulatory approvals
32	Estimate construction period (project execution plan)
33	Estimated cost
34	Specifications
35	BOQ
36	VE suggestions and resolutions
37	Bidding and tendering documents
	37.1 ITB
	37.2 Contract documents
	37.3 Technical documents
	37.3.1 Specifications
	37.3.2 BOQ

8.4.4.1.1 Identify TOR Requirements

The TOR lists the requirements to develop FEED. It mainly consists of the following:

1. Complete FEED
2. Drawings
   - 2.1 Site layout
   - 2.2 Process engineering,
   - 2.3 Process automation
   - 2.4 Process diagram
   - 2.5 Mechanical
   - 2.6 Piping
   - 2.7 Instrumentation and control
   - 2.8 Fire- and gas-detection system
   - 2.9 Supervisory control system
   - 2.10 Leak detection system
   - 2.11 Corrosion monitoring system
   - 2.12 Buildings (architectural, interiors, structural, and MEP)
   - 2.13 Schematics
   - 2.14 Special systems
   - 2.15 Low-voltage systems
   - 2.16 Data security

2.17 Utilities
   2.18 Landscape
   2.19 External works
   2.20 Tie-in drawings
3. Reports
   3.1 Technology selection
   3.2 Narrative reports
   3.3 Site surveys
   3.4 Calculations
   3.5 Assumptions and constraints
   3.6 Design criteria
   3.7 HAZOP
4. Specifications
5. MTO list
6. Project schedule
7. Project cost
8. Approval
9. VE
10. Basic engineering to develop detailed engineering by EPC contractor
11. ITB documents

8.4.4.1.2 Gather Concept Design Comments

The objective of FEED is to refine and develop a clearly defined design based on the client's requirements. While developing the FEED, the designer to review the submitted concept design drawings and reports should take into consideration the comments, if any, made on the concept design. The designer can discuss in detail with the owner/PMC and incorporate all their requirements to develop FEED.

8.4.4.1.3 Collect Owner's Requirements

The owner requirements discussed earlier under Section 8.3.4.2 are further developed in detail and are updated in order to be considered in the FEED.

8.4.4.1.4 Collect Regulatory Requirements

The designer has to collect regulatory/statutory requirements to incorporate the same while developing the design. These include

- Fire department
- Environmental protection agency
- Utilities department/agencies
- Traffic department
- Local codes and standards

In certain countries, the FEED drawings are submitted to the regulatory bodies for their review and approval for compliance with the regulations, codes, and licensing procedures. Any comments on the drawings are incorporate on the drawings and are resubmitted, if required.

8.4.4.1.5 Collect Energy Conservation Requirements

The designer has to collect applicable energy conservation requirements and Leadership in Energy and Environmental Design (LEED) requirements, and consider the same while developing the design.

8.4.4.2 Data Collection

The following data have to be collected to develop FEED for the oil and gas project:

- Project usage (product output)
- Space program
- Site investigations
- Site surveys
- No. of drawings to be produced
- Disabled (special needs) access requirements

8.4.4.2.1 Perform Site Investigations

The site investigation covers mainly the following activities:

- Soil profile and laboratory test of soil
- Topography of the project site
- Hydrological information
- Wind load, seismic load, dead load, and live load
- Existing services passing through the project site
- Existing roads and structure surrounding the project site
- Shoring and underpinning requirements with respect to adjacent area/ structure

8.4.4.2.2 Engineering Surveys

The following Engineering Surveys have to be conducted:

- Topographical
- Ecological
- Geotechnical
- Hydrology
- Metrological

8.4.4.3 Technical Studies

The designer has to study and evaluate following:

- Process technology
- Technical and functional capability
- Operational capability
- Project constraints
- Constructability

Figure 8.11 illustrates the logic flowchart for selection of process technology.

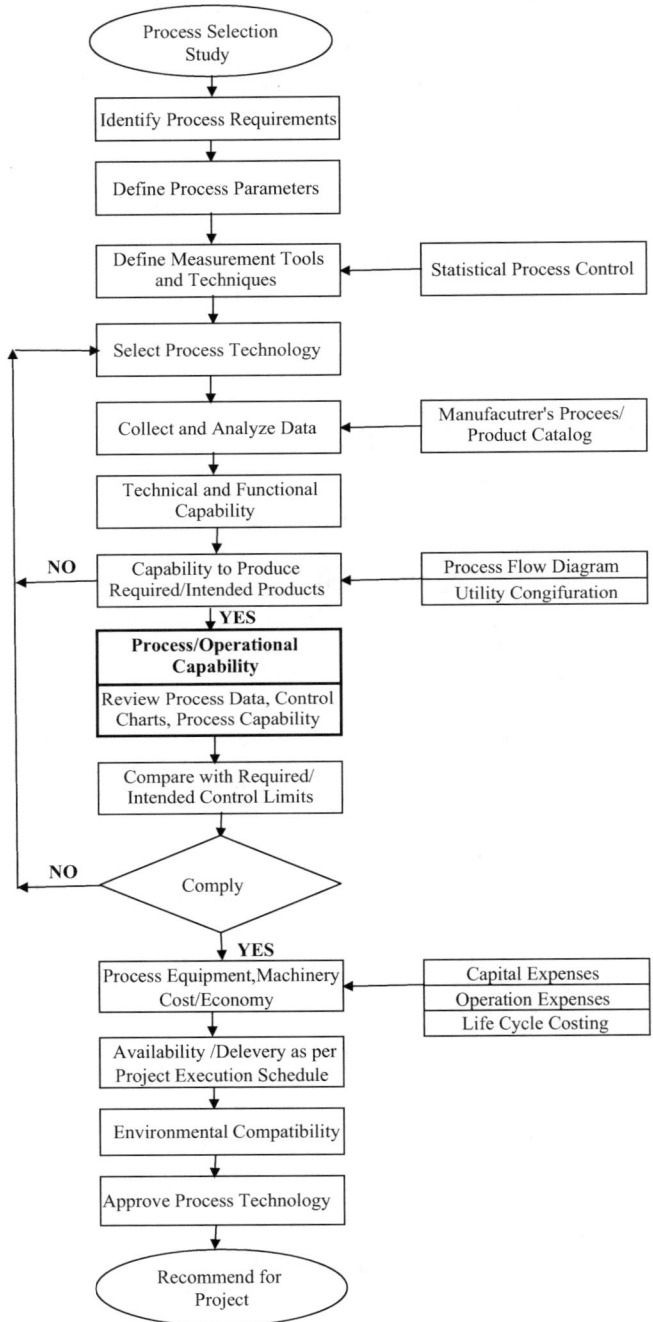

FIGURE 8.11 Logic flowchart for selection of process technology.

Quality Management

8.4.4.3.1 Process Selection

The designer has to consider the following in order to select the process:

- Selection of process technology
- Selection of process

8.4.4.3.2 Process Diagrams and Configurations

The following diagrams and reports are to reviewed by the designer:

- Process flow diagrams
- Piping and instrumentation diagram
- Piping and utility configuration
- Plant configuration
- HAZID

8.4.5 Prepare FEED

During this phase, several options to select the process, equipment, machinery, and systems are reviewed, one scheme which meets the owner objectives is selected, and the design/drawings are prepared.

8.4.5.1 FEED Drawings

The following drawings for the oil and gas project are generally prepared during this phase:

1. Process engineering
2. Piping system
3. Pumping station
4. Instrumentation and control
5. Process automation
6. Storage thanks
7. Fabrication works
8. Supervisory control
9. Fire- and gas-detection system
10. Flaring and venting system
11. Leak-detection system
12. Corrosion monitoring system
13. Loss prevention system
14. Tie-in drawings
15. Buildings
16. Structural
17. MEP
18. Electrical substation
19. Conveyance/elevator system
20. Fire suppression system

21. ICT
22. Security and access control
23. Low-voltage systems
24. Landscape
25. External works
26. Specialist systems
27. Data security
28. Schematics and flow diagrams

8.4.5.2 FEED Reports

The FEED reports mainly consists of the following:

1. Design criteria
2. Design feature describing the selection option
3. Process selection
4. process automation
5. System selection
6. Cause and effect diagrams
7. Material data sheets
8. Construction methodology
9. Project risks
10. Environmental impact assessment
11. HAZID
12. Site surveys and investigation reports

8.4.6 Prepare Project Schedule/Plans

After preparation of FEED, the logic of the construction program is set and the project execution plan/project schedule is developed. On the basis of logic, a critical path method (CPM) schedule (bar chart) is prepared to determine the critical path and set the contract milestones.

The project schedule is developed using top-down planning using key events. It is also known as Class 3 schedule (Please refer Tables 8.6 and 8.7).

8.4.7 Evaluate and Estimate Project Cost

FEED is the phase of a project during which a percentage of the overall engineering required for construction of a project is completed, which help in developing the TIC. Oil and gas projects are capital-intensive projects. Capital cost of equipment and machinery has a major bearing on return on investment, so the sanction estimate is an important part of the investment decision. Based on the FEED, project cost estimate is prepared by estimating the cost of activities and resources. The evaluate of accurate cost estimate is essential because of its impact on overall success of the project. The preparation of the budget is an important activity that results in a timed phased plan summarizing the expected expenses toward the construction cost

Quality Management

and also the income or the generation of funds necessary to achieve the milestone. The budget for a construction project is the maximum amount the owner is willing to spend for design and construction of the facility that meets the owner's need. The budget is determined by estimating the cost of activities and resources and is related to the schedule/execution plan of the project.

The cost estimate during this phase is based on elemental parametric methodology. It is also known as budgetary (preliminary estimate) (Please refer Table 8.8).

8.4.8 Manage FEED Quality

In order to minimize design errors, minimize design omissions, and reduce rework during FEED, the designer has to plan quality (planning of design work), perform quality assurance, and control quality for preparing FEED. This will mainly consist of the following:

1. Plan quality:
    - Establish owner's requirements
    - Identify requirements listed under TOR
    - Determine number of drawings to be produced
    - Establish scope of work for developing FEED
    - Identify quality standards and codes to be complied
    - Establish design criteria
    - Identify regulatory requirements
    - Establish quality organization with responsibility matrix
    - Develop design (drawings and documents) review procedure
    - Establish submittal plan
    - Establish design review procedure
2. Quality assurance:
    - Collect information
    - Collect data
    - Investigate site conditions
    - Prepare FEED drawings
    - Prepare specifications
    - Coordinate with all disciplines
    - Ensure functional and technical compatibility
    - Ensure ease of operation of the selected process
    - Constructability
    - Select material to meet project specifications, codes, and standards
3. Control quality:
    - Check design drawings
    - Check specifications/contract documents
    - Check for regulatory compliance
    - Check project schedule
    - Check estimated cost of the project

The designer has to follow the designer's quality control plan to manage quality in this phase.

The designer has to include contractor's quality control requirements in the contract documents. Figure 8.12 illustrates quality management procedure for FEED phase.

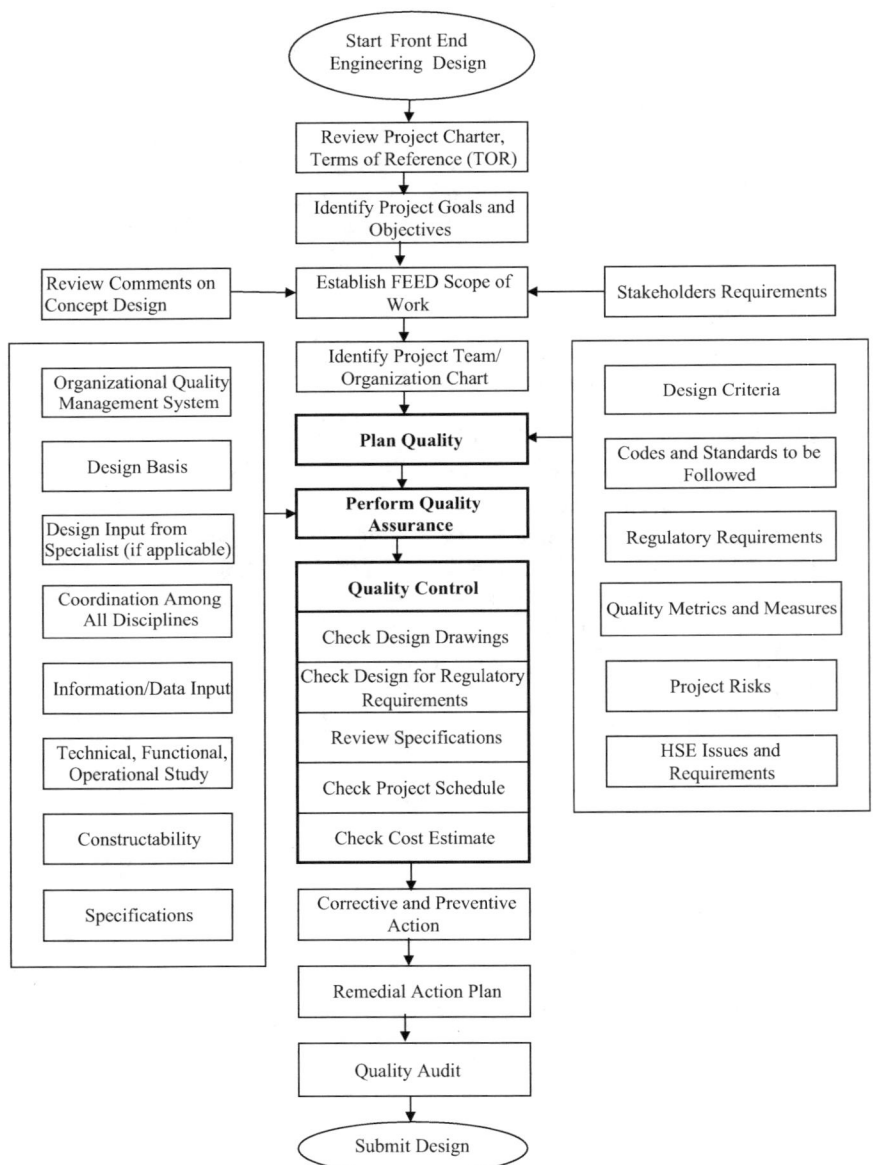

FIGURE 8.12 Quality management procedure for FEED phase.

Quality Management

8.4.9 Estimate Resources

The designer has to estimate the resources required to complete the project. At this stage of the project, more detail about the activities and works to be performed during the construction and testing, commissioning phase is available, the designer has to update the earlier estimated resources and prepare manpower histogram. Also, the designer can estimate total number of design team members to develop design and construction documents.

8.4.10 Develop Communication Plan

The designer has to prepare the communication plan to include mainly the following activities:

- Design progress information to all the stakeholders
- Liaison and coordination among all the stakeholders
- Coordination meetings
- Communication matrix
- Distribution of documents (transmittals, correspondence, etc)

The designer has to specify contractor's communication plan requirements in the contract documents.

8.4.11 Manage Risks

The following are typical risks which normally occur during FEED phase:

- The designer not taking into considerations the concept design deliverables and review comments are not taken into consideration while preparing the FEED
- Regulatory authorities' requirements are not taken into consideration
- Incomplete scope of work for preparing FEED
- The related project information and data collected are incomplete
- The related project data and information collected are likely to be incorrect and wrongly estimated
- Site investigations for existing conditions not carried out
- Fire and safety considerations are not taken into account
- Environmental considerations are not taken into account
- Incomplete design
- Inadequate and ambiguous specifications
- Wrong selection of process, materials, and systems
- Undersize equipment selection
- Estimated total electrical load is much lower than the expected actual consumption
- Errors in calculating traffic study for conveying system
- Errors in estimating the project schedule

- Errors in cost estimation
- Number of drawings not as per TOR requirements

Designer has to take into account the above-mentioned risk factors while developing the FEED.

8.4.12 Manage HSE Requirements

The designer has to develop HSE plan and identify issues to manage the activities during the design stage. The designer has to do Environmental Impact Assessment study and take into consideration while developing FEED. The designer has to include the HSE Management Plan requirements in the contract documents to be considered by the EPC contractor during the execution of the project.

8.4.13 Develop FEED Documents

During this phase, a complete set of working drawings of all the disciplines, site plans, technical specifications, material tale off, schedule, and related graphic and written information to bid the project is prepared. It is necessary that utmost care is taken to develop and assemble all the documents and ensure the accuracy and correctness to meet the owner's objectives.

8.4.13.1 Specifications

Specifications indicating project-specific features of major equipment, systems, and material are prepared during this phase. Normally, specifications are prepared per MasterFormat® contract documents produced jointly by the Construction Specification Institute (CSI) and Construction Specifications Canada (CSC) which are widely accepted as standard practice for the preparation of contract documents.

Table 8.14 lists division numbers and titles of MasterFormat® 2020 published by the Construction Specifications Institute and CSC.

8.4.13.2 Material Take Off List

Table 8.15 is an example of the Bill of Quantities (BOQ) form.

8.4.13.3 Contract Documents

Based on the type of contracting arrangements, the owner would like to handle the project; necessary documents are prepared by establishing a framework for execution of the project. The documents are prepared taking into consideration the following:

- Contract conditions
- Technical conditions

8.4.14 Perform VE

Value Engineering (VE) studies can be conducted at various phases of construction project life cycle; however, the studies conducted in the early stage of project tend to

TABLE 8.14
MasterFormat® 2020

Division Numbers and Titles
PROCUREMENT AND CONTRACTING REQUIREMENTS GROUP
Division 00 Procurement and Contracting Requirements
SPECIFICATIONS GROUP

General Requirements Subgroup		**Site and Infrastructure Subgroup**	
Division 01	General Requirements	Division 30	Reserved
Facility Construction Subgroup		Division 31	Earthwork
Division 02	Existing Conditions	Division 32	Exterior Improvements
Division 03	Concrete	Division 33	Utilities
Division 04	Masonry	Division 34	Transportation
Division 05	Metals	Division 35	Waterway and Marine Construction
Division 06	Wood, Plastics, and Composites	Division 36	Reserved
Division 07	Thermal and Moisture Protection	Division 37	Reserved
		Division 38	Reserved
Division 08	Openings	Division 39	Reserved
Division 09	Finishes		
Division 10	Specialties	**Process Equipment Subgroup**	
Division 11	Equipment	Division 40	Process Interconnections
Division 12	Furnishings	Division 41	Material Processing and Handling Equipment
Division 13	Special Construction	Division 42	Process Heating Cooling, and Drying Equipment
Division 14	Conveying Equipment	Division 43	Process Gas and Liquid Handling, Purification, and Storage Equipment
Division 15	Reserved	Division 44	Pollution and Waste Control Equipment
Division 16	Reserved	Division 45	Industry-Specific Manufacturing Equipment
Division 17	Reserved	Division 46	Water and Waste water Equipment
Division 18	Reserved	Division 47	Reserved
Division 19	Reserved	Division 48	Electric Power Generation
Facility Services Subgroup		Division 49	Reserved
Division 20	Reserved		
Division 21	Fire Suppression		
Division 22	Plumbing		
Division 23	Heating, Ventilation, and Air Conditioning (HVAC)		
Division 24	Reserved		
Division 25	Integrated Automation		

(Continued)

TABLE 8.14 (*Continued*)
MasterFormat® 2020

Division Numbers and Titles
PROCUREMENT AND CONTRACTING REQUIREMENTS GROUP
Division 00 Procurement and Contracting Requirements
SPECIFICATIONS GROUP

General Requirements Subgroup   **Site and Infrastructure Subgroup**

Division 26	Electrical
Division 27	Communications
Division 28	Electronic Safety and Security
Division 29	Reserved

Source: The Construction Specifications Institute and CSC. Reprinted with permission from CSI.

Specifications indicating project specific features of major equipment, systems, material are prepared during this phase. Normally, contract document specifications are prepared using the MasterFormat® taxonomy, widely accepted as the standard taxonomy for preparing contract documents. Table 8.14, listing MasterFormat division numbers and titles (©2020 The Construction Specifications Institute, Inc. (CSI)., is reproduced in this book with CSI's permission.

TABLE 8.15
Bill of Quantities

Owner name
Project name
Project number:

Item	Description	Qty	Unit	Unit Rate	Total Amount	Remarks
	Division 3 – Concrete					
	The Contractor is referred to the Specifications and Drawings for all details related to this section of the Works and he is to include for complying with all the requirements contained therein, whether or not they are specifically mentioned within the item descriptions					
	03300: Cast in-place concrete					
	Substructure					
	Plain concrete 17.5MPa using Sulfate-Resisting Cement (Type V) including formwork and additives					
A	100 mm Blinding		m^3			
	Plain concrete 17.5MPa using Ordinary Portland Cement (Type I) including formwork and additives					

(*Continued*)

TABLE 8.15 (*Continued*)
Bill of Quantities

Owner name
Project name
Project number:

Item	Description	Qty	Unit	Unit Rate	Total Amount	Remarks
B	Concrete filling		m³			
C	To kerb foundation		m³			
	Reinforced concrete 28MPa using sulfate-resisting cement (Type V) including formwork, reinforcement, water stops, expansion and construction joints, joint filler and additives					
D	Raft foundation		m³			
E	300 mm Walls		m³			
F	400 mm Ditto		m³			
	Carried to collection					
	Division 3 – concrete					
	03480: Precast concrete specialties					
	Depressed curbs including reinforcement, finish, anchors, fixings, grouting, fair face finish and painting					
A	250 x 180 mm barrier kerb		m			
B	Ditto, curve on plan		m			
C	350 x 150mm Flush kerb		m			
	Wheel stoppers, fair face finish, and painting					
D	2,000 x 140 x 100 mm Overall size, Parking Area		no			
	03520: Lightweight concrete					
	Lightweight concrete laid in bays of 25 m² minimum, maximum length of bay 7m and expansion joints filled with impregnated compressible foam and resilient packing					
E	50 mm thick, laid to falls, to roof		m²			
F	170 mm thick, laid to falls, to pathways		m²			
	Collection					
	Total of page No. 1/2					
	Total of page No. 2/2					
	Carried to summary					

provide greatest benefit. In most oil and gas projects, VE studies are performed during FEED phase of the project. At this stage, the design professionals have considerable flexibility to implement the recommendations made by the VE Team, without significant impacts on the project schedule or design budget. Normally, VE study is performed by a specialist VE consultant. The number of team members to perform VE Study depends on the client's/owner's requirement. It is advisable that SAVE International registered Certified Value Specialist is assigned to lead this study.

Figure 8.13 illustrates VE Process Activities.

8.4.14.1 Update FEED

After completing the workshop, the VE team prepares a report to support its findings and the implementation program. Its recommendations are reviewed by the owner/PMC and the design team before implementation. The FEED is updated taking into considerations the recommendation by Value Engineering team.

8.4.15 Review FEED

Table 8.16 illustrates major points for the review of FEED.

8.4.16 Finalize FEED

Final FEED is prepared incorporating the comments, if any, found during analysis and review of the drawings and documents for submission to the owner/client.

8.4.17 Develop Tender Documents

Most of the cost of the construction project is expended during the construction phase. In oil and gas projects, the contractor is responsible for engineering, procurement of all the material, providing construction equipment and tools, and supplying the manpower to complete the construction project in compliance with the contract documents.

The following documents are developed by the designer for inviting the EPC contractor to execute the project:

1. Invitation to Bid (ITB)
2. Contract documents
    2.1 Contract conditions
    2.2 Technical conditions
        2.2.1 Drawings
        2.2.2 Specification
        2.2.3 BOQ

In certain cases, a list of equipment/systems/materials is included in the contract documents for submission of technical details for the evaluation of bid.

Table 8.17 lists typical contract documents consisting of tendering procedures, contract conditions, and technical conditions of major oil and gas construction projects.

Quality Management

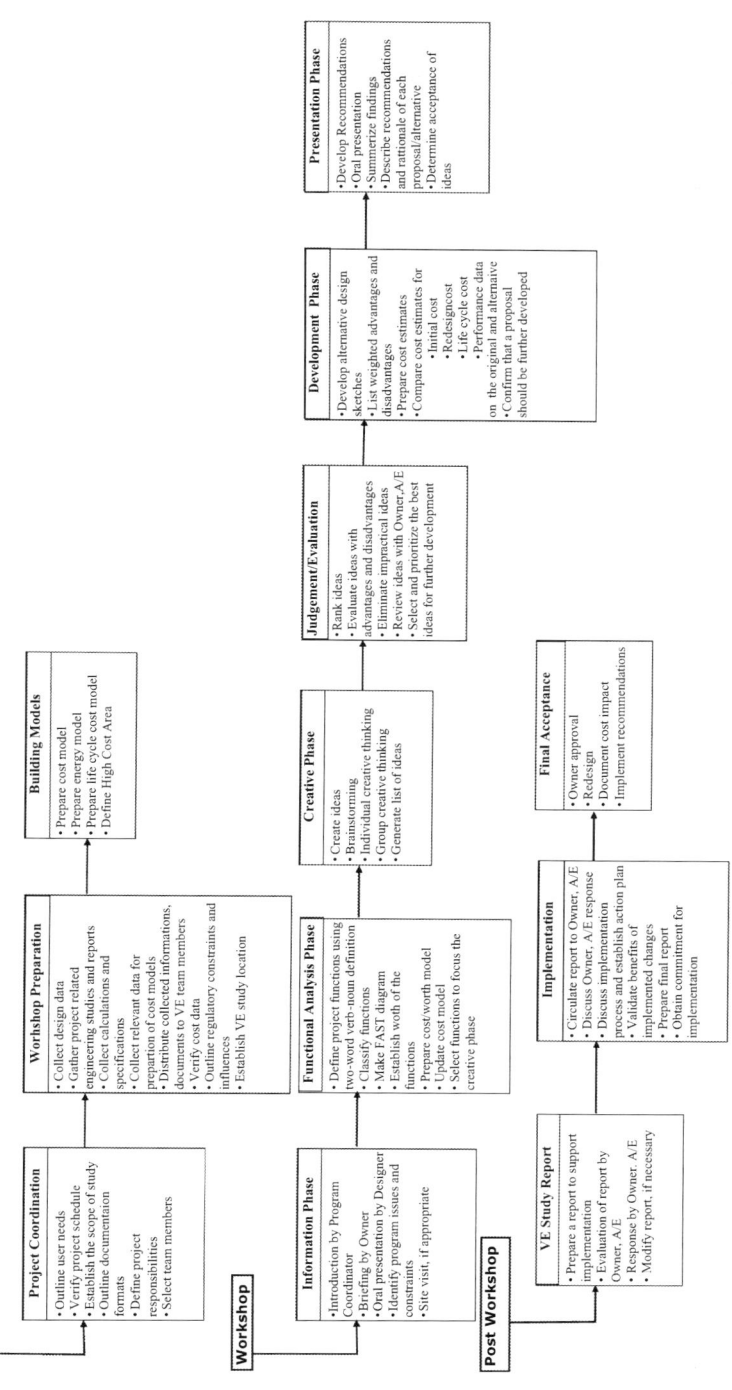

FIGURE 8.13 VE study process activities. (Abdul Razzak Rumane (2010). *Quality Management in Construction Projects*. Reprinted with permission from Taylor & Francis Group.)

TABLE 8.16
Review of FEED

Serial Number	Description	Yes	No	Notes
	A: FEED (General)			
A.1	Does the design support owner's project goals and objectives?			
A.2	Does the design meets all the elements specified in TOR?			
A.3	Whether comments on concept design taken care while preparing FEED?			
A.4	Whether owner's requirements updated and considered?			
A.5	Does the design meet all the performance requirements?			
A.6	Whether constructability has been taken care?			
A.7	Whether technical and functional capability considered?			
A.8	Whether functional capability of process is considered while selecting the process technology?			
A.9	Whether the information/data collected is considered while preparing the design?			
A.10	Whether design confirm with fire and egress requirements?			
A.11	Does the design have provision for all the low-voltage and control system requirements?			
A.12	Whether design risks been identified, analyzed, and responses planned for mitigation?			
A.13	Whether health and safety requirements in the design are considered?			
A.14	Whether environmental constraints considered?			
A.15	Does energy conservation is considered?			
A.16	Does sustainability considered in design?			
A.17	Whether Regulatory approvals obtained?			
A.18	Whether cost-effectiveness over the entire project life cycle is considered?			
A.19	Whether the design meets LEED requirements?			
A.20	Whether accessibility is considered?			
A.21	Does the design meet ease of maintenance?			
A.22	Whether all the drawings are numbered			
A.23	Does the design support proceeding to next design development stage?			
A.24	Whether documents for ITB (ITB) included in the package?			
	B:Buildings			
B.1	Whether architectural plans, elevations, finishes…etc., for building are drawn as per local building codes?			
B.2	Whether structural details for the building are properly shown?			
B.3	Whether traffic analysis is performed to identify number of elevators, escalators?			
B.4	Whether all the required MEP Drawings prepared?			
B.5	Whether emergency diesel generator is considered?			
B.6	Whether electrical substation is as per regulatory requirements?			

(*Continued*)

TABLE 8.16 (*Continued*)
Review of FEED

Serial Number	Description	Yes	No	Notes
	C: Automation and Control			
C.1	Whether all the control systems are considered?			
C.2	Whether ICT drawings are prepared?			
C.3	Whether all the low-voltage systems as per TOR considered?			
	D: Landscape			
D.1	Whether landscape plans prepared?			
D.2	Whether plants selected?			
D.3	Whether irrigation system layout prepared?			
	F: External			
F.1	Whether site layout plans showing roads, walkways, parking areas, prepared?			
F.2	Whether street lighting plans prepared?			
F.3	Whether project site boundaries are properly marked and demarketed?			
	G: Schedule			
G.1	Project schedule is practically achievable?			
	H: Financial			
H.1	Project cost is properly estimated?			
	I: VE			
I.1	Whether VE study performed and recommendations taken care?			
	J: Reports			
J.1	Whether the narrative description complete and include adequate information about the project?			
J.2	Whether Outline specifications include all the works			
J.3	Whether preliminary contract documents have taken care all the TOR requirements			
	K: Drawings, SKETCHES			
K.1	Whether drawings and sketches for all trades prepared as per TOR?			
	L: Models			
L.1	Whether the Models meet the design objectives?			
	M: ITB			
M.1	Whether documents for bidding prepared?			
	N: Submittals			
N.1	Whether numbers of sets prepared as per TOR?			

TABLE 8.17
Typical Contract Documents

Document No (I) Tendering Procedures
Consisting of the following

I.1	Tendering invitation
I.2	General instructions to bidders
I.3	Particular instructions to bidders
I.3	Form of tender and appendix
I.4	List of contractor's staff
I.5	List of equipment and machinery
I.6	Initial bond (form of bank guarantee)
I.7	List of subcontractor(s) or specialist(s)
I.8	

    Performance bond (form of bank guarantee)
    Form of contract agreement
    List of tender documents
    Contractor's certificate of work statement
    Declaration

Document No. (II) Contract Conditions
Consisting of the following

II.1	General Conditions (Legal Clauses and Conditions) of Contract for Engineering and Construction
II.2	Particular Conditions of Contract
II.3	Public Tender Law (As Applicable)

Document No. (III) Technical Conditions and Amendments Consisting of the following

III.1	Technical Condition of Contract
	III.1.1 Contract Specifications
	III.1.1.1 Conditions for Design and Engineering Works
III.2	Technical Specifications
	III.2.1 Scope of Works
	III.2.1.1 Process Design Basis
	III.2.1.2 Scope of Works
III.3	General Specifications
III.4	Particular Specifications
III.5	List of Drawings
III.6	Schedule of Unit Rates
III.7	Price Analysis Schedule
III.8	Addenda (if any)
III.9	Technical requirements (if any), and any other instructions issued by the Project Owner
III.10	HSE Guidelines

8.4.18 Submit FEED Package

Normally, the following items are submitted to the owner for their review and approval in order to proceed with the procedure for the selection of EPC contractor:

1. FEED drawings
2. Project schedule
3. Cost estimate
4. Specifications
5. Reports
6. Tender documents

8.5 BIDDING AND TENDERING PHASE

Bidding and tendering is the fourth phase of the oil and gas construction project life cycle. The following are the primary objectives of this phase:

1. Selection of bidders (prequalification)
2. Invitation to Bid (ITB)
3. Tender preparation and submission
4. Appraisal of tenders, negotiation, and decision

Oil and gas construction projects are capital-investment projects. Most of the cost of the construction project is expended during the construction phase. In most cases, the contractor is responsible for the procurement of all the materials, providing construction equipment and tools and supplying the manpower to complete the project in compliance with the contract documents.

In many countries, it is a legal requirement that government-funded projects employ the competitive bidding method. This requirement gives an opportunity to all qualified contractors to participate in the tender, and normally, the contract is awarded to the lowest bidder. Private-funded projects have more flexibility in evaluating the tender proposal. Private owners may adopt the competitive bidding system, or the owner may select a specific contractor and negotiate the contract terms. Negotiated contract systems have the flexibility of pricing arrangement as well as the selection of the contractor based on his expertise or the owner's past experience with the contractor successfully completing one of his or her projects.

In the case of oil and gas projects, EPC contracting method is followed. The contractor is selected based on the contract documents developed at FEED phase. The contract documents include basic engineering design drawings, contract conditions, and technical conditions for bidding purposes.

For most construction projects, selection of a bidder is based on the lowest tender price. Tenders received are opened and evaluated by the owner/owner's representative. Normally, tender results are declared in the official gazette or by some sort of notifications. The successful bidder is informed of the acceptance of the proposal and is invited to sign the contract. The bidder has to submit the performance bond before the formal contract agreement is signed. If a successful bidder fails to submit

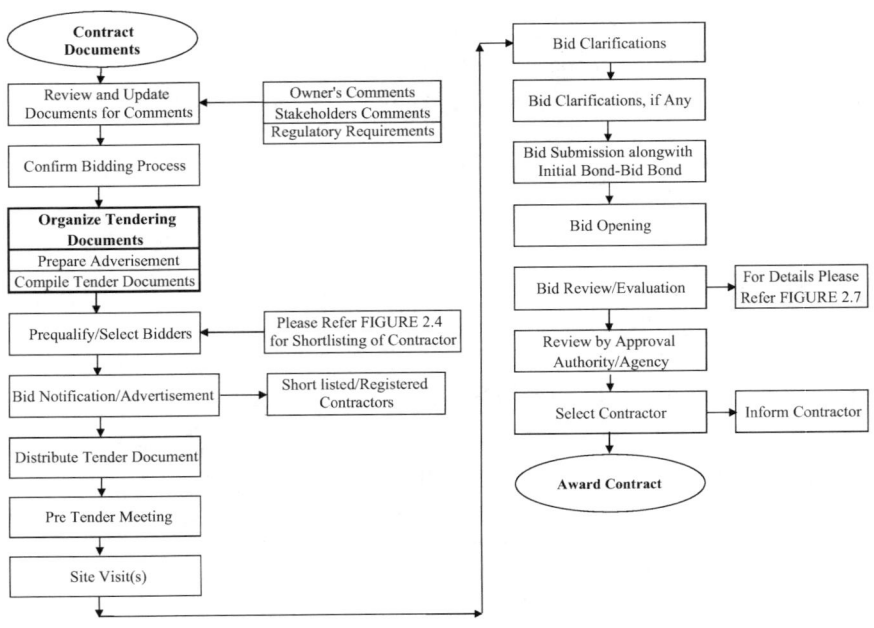

FIGURE 8.14 Logic flowchart for bidding and tendering phase.

the performance bond within the specified period or withdraws his tender, then the contractor loses the initial bond and may be subjected to other regulatory applicable conditions.

The signing of contract agreement between the owner/owner's representative and the contractor binds both parties to fulfill their contractual obligations.

Figure 8.14 illustrate the logic flowchart for bidding and tendering phase.

8.5.1 Organize Tender Documents

The owner hand overs the approved construction documents/tender documents to the tender committee for further action. The bid documents are prepared as per the procurement method and contract strategy adopted during the early stage of the project. Tendering procedure documents submitted by the designer are updated, and the necessary owner related information is inserted in the tender documents. The bid advertisement material is prepared, and upon approval from the owner, the bid notification is announced through different media as per the Organization's/Agency's policy.

8.5.2 Identify Stakeholders/Project Team

The following project team members have direct involvement in the bidding and tendering phase:

Quality Management

TABLE 8.18
Responsibilities of Various Participants during Bidding and Tendering Phase

	Responsibilities		
Phase	Owner/Tender Committee	Consultant (Designer)	Contractor
Bidding and tendering	Advertise bid	Review/evaluate bid	Collection of bid documents
	Distribute bid	Bid conference/pre tender meeting	Attend meeting
	Collect bid (proposal)	Bid clarification	Site visit(s)
	Negotiation	Arrange site visit (s)	Preparation of proposal
	Approve EPC contractor	Issue addendum, if any	Submission of proposal
	Award contract	Recommend successful Bidder	
		Prepare contract documents	

- Owner
- Tender committee
- Designer (consultant)
- PMC
- Bidders

Table 8.18 illustrates responsibilities of various participants during the bidding and tendering phase.

8.5.3 Identify Tendering Procedure

The owner has to identify the tendering procedure. Different types of tendering methods are discussed under Section 7.3.1.

8.5.3.1 Define Bidder Selection Procedure

Owner has to select the bidder selection procedure such as low bid and quality-based. Refer Section 7.3.1 for different types of contacting methods and contractor selection procedure.

8.5.3.2 Establish Bid Review Procedure

The designer (consultant) establishes bid evaluation procedure in consultation with the owner.

8.5.4 Identify Bidders

Short-listing of bidders is done with prequalification questionnaires and their responses. This has been already discussed under Section 7.3.1.

8.5.4.1 Prequalify Bidders

Prequalification of contractor is already discussed under Section 7.3.1.

8.5.5 Manage Tendering Process

The tendering process involves the following activities:

- Bid notification
- Distribution of tender documents
- Pretender meeting(s)
- Issuing addendum, if any
- Bid submission

8.5.5.1 Advertise Tender

The tender is announced in different types of media such as newspaper, magazines, electronic media, as per the Organization's/Agency's policy.

8.5.5.2 Distribute Tender Documents

Normally, tender documents are distributed to eligible bidders against the payment of fee announced in the bid notification, which is nonrefundable.

8.5.5.3 Conduct Prebid Meeting

The owner conducts pretender meeting to provide an opportunity for the contractors bidding the project to review and discuss the construction documents and to discuss the following:

- General scope of the project
- Any particular requirements of bidders that may have been difficult to specify
- Explain details of complex matters
- Engagement of subcontractor and specialist subcontractors
- Particular risks
- Any other matters that will contribute to the efficient delivery of project

The meeting is attended by the designer (consultant), bidders (contractors), PMC, and tender committee member. Queries from the contractors pertaining to contract documents are noted, and the designer (consultant) provides written response to these queries by clarifying all the points. The bidders have to consider the clarification points and incorporate the requirements while calculating the bid price. The responses recorded in the meeting become the part of contract documents (part of addendum) which is signed by the owner and successful bidder. Figure 8.15 illustrates the Bid Clarification Form which becomes part of contract documents.

8.5.5.4 Submit/Receive Bids

Bids are received in accordance with the Instructions to Bidders section of the tender documents. The bid should be accompanied with an initial bond in favor of owner/tender committee to be valid for a period mentioned in tendering procedures. All the bids received are documented and notified. The tender, which is submitted as a sealed document, is opened as mentioned in tendering procedures.

Quality Management

Serial Number	Name of Contractor	Item No. and Clause Reference	Queries	Owner/Consultant's Clarification	Remarks

Project Name
Project Number
Bid Clarification Form

FIGURE 8.15 Bid clarification.

8.5.6 Manage Bidding and Tendering Phase Quality

In order to minimize errors during bidding and tendering phase and minimize omissions, the designer has to plan quality (planning of bidding work), perform quality assurance, and control quality for preparing bidding and tendering activities. This will mainly consist of the following:

1. Plan Quality:
    - Organize tender documents
    - Identify regulatory requirements
    - Identify bidders
    - Bid notification
    - Tender documents distribution system
    - Establish responsibility matrix
    - Bid clarification method
    - Bid collection method
    - Bid evaluation procedure
    - Contract award procedure
2. Quality Assurance:
    - Collect regulatory information
    - Owner approval for bidding
    - Distribution of bid documents
    - Pretender meetings
    - Site visits
    - Bid clarifications
    - Bid collection

3. Control quality:
   - Bid documents to prequalified/shortlisted/registered bidders
   - Contents of bid documents as required
   - Bid document distribution as per announced date
   - Check for specified number of documents
   - Check bid bond amount
   - Bid evaluation
   - Award contract to selected/qualified contractor

8.5.7 Manage Risks

Table 8.19 lists the risks the contractor has to manage during bidding and tendering phase.

It is essential that the contractor verifies material take off (BOQ) list for correctness with respect to contract documents while estimating the bid value.

The following are typical risks likely to occur during this phase:

- Not all the qualified bidders taking part in bidding for the project
- Bidders noticing errors and omissions in the construction documents resulting delay in submission of bids
- Amendment to construction documents

TABLE 8.19
Major Risk Factors Affecting Contractor

Serial Number		Risk Factor
1		Bidding and tendering
	1.1	Low bid
	1.2	Poor definition of scope of work
	1.3	Overall understanding of project
	1.4	Review of contract specs with MTO quantities
	1.5	Errors in proposing specified process
	1.6	Errors in selecting specified equipment
	1.7	Errors in not submitting technical details of certain equipment, system requested in the tender documents
	1.8	Errors in resource estimation
	1.9	Errors in resource productivity
	1.10	Errors in resource availability
	1.11	Errors in material price
	1.12	Improper schedule
	1.13	Quality standards
	1.14	Exchange rate
	1.15	Review of contract document requirements with regulatory requirements
	1.16	Unenforceable conditions or contract clauses

- Addendum
- Delay in submission of bids than the notified one
- Bid value exceeding the estimated definitive cost
- Successful bidder fails to submit performance bond

The owner/designer has to consider these risks and plan the phase duration accordingly.

8.5.8 Review Bid Documents

The designer (consultant) reviews the bids for compliance to tender requirements.

8.5.8.1 Evaluate Bids

The designer (consultant) evaluates the bid documents of each of the submitted bids. Table 8.20 illustrates checklist for bid evaluation.

8.5.8.2 Select Contractor

The contractor is selected based on the procurement strategy adopted by the owner.

8.5.9 Award Contract

Figure 8.16 illustrates contract award process.

8.6 CONSTRUCTION PHASE

Construction is the fifth phase of the oil and gas construction project life cycle. Construction is translating owner's goals and objectives, by the contractor, to build the facility as stipulated in the contract documents, plans, and specifications within the budget and on schedule. Construction is an important phase in construction projects. A majority of total project schedules and the budget are expended during construction. Similar to costs, the time required to construct the project is much higher than the time required for preceding phases. Construction usually requires large number of work force and variety of activities. Most oil and gas projects follow EPC type of contracting system. The EPC contractor is responsible for

1. detailed engineering
2. procurement
3. construction

Construction activities involve erection, installation, or construction of any part of the project and start-up, testing, and commissioning activities. Construction activities are actually carried out by the contractor's own work force or by subcontractors. Construction therefore requires more detailed attention of its planning, organizations, monitoring and control of project schedule, budget, quality, safety, and environmental concerns.

Figure 8.17 illustrates organizational structure of EPC contracting.

TABLE 8.20
Checklist for Bid Evaluation

Serial Number	Description	Yes	No	Notes
	A: Documents			
A.1	Bid submitted before closing time on the date specified in the bid documents			
A.2	Bidders identification is verified			
A.3	Bid is properly signed by the authorized person			
A.4	Bid bond is included			
A.5	Required certificates are included			
A.6	Bidders confirmation to the validity period of bid			
A.7	Confirmation to abide by the specified project schedule			
A.8	Bid documents have no reservation or conditions (limitation or liability)			
A.9	Preliminary method statement			
A.10	List of equipment and machinery			
A.11	Technical details for specific equipment and machinery			
A.12	List of proposed core staff as listed in the tender documents			
A.13	Complete responsiveness to the commercial terms and conditions			
A.14	All the required information is provided (completeness of information)			
A.15	All the supporting documents required to determine technical responsiveness is submitted			
	B: Financial			
B.1	All the items are priced			
B.2	Bid Amount clearly mentioned			
B.3	Prices of provision items			

The oil and gas construction phase consists of various activities such as mobilization, development of detailed engineering, procurement, execution of work, planning and scheduling, control and monitoring, management of cost/resources/procurement, quality, and inspection. Figure 8.18 illustrates major activities to be performed during the construction phase.

These activities are performed by various parties having contractual responsibilities to complete the specified work. Coordination among these parties is essential to ensure that the constructed facility meets the owner's objectives.

8.6.1 Development of Project Site Facilities

Once the contract is awarded to a successful bidder (contractor), then it is the responsibility of the contractor to respond to the needs of the client (owner) by constructing the project as specified in the contract documents, drawings, and specifications within the specified time and budget. The contractor is given few weeks to start

Quality Management

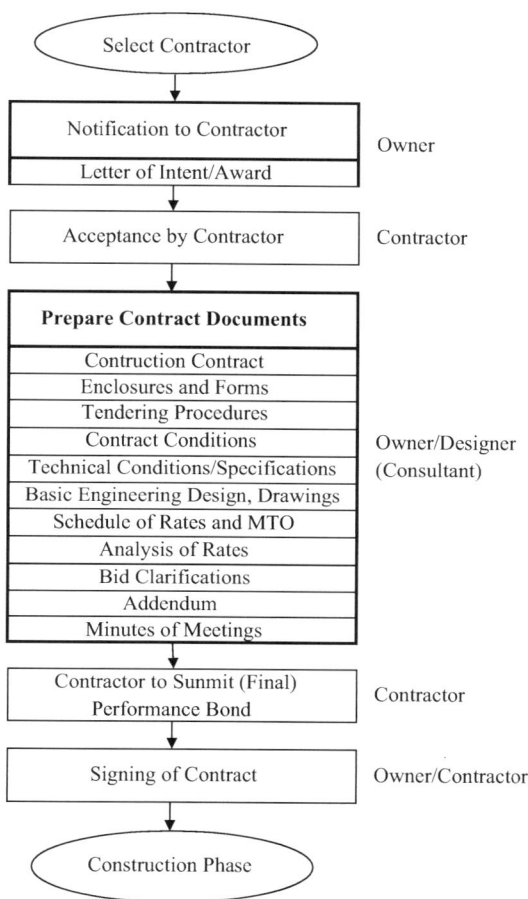

FIGURE 8.16 Contract award procedure. (Abdul Razzak Rumane. (2016). *Handbook of Construction Management*. Reprinted with permission from Taylor & Francis Group.)

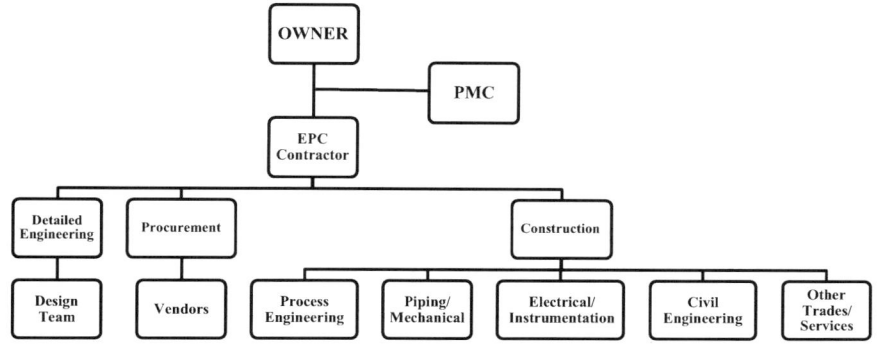

FIGURE 8.17 Organizational structure of EPC contracting.

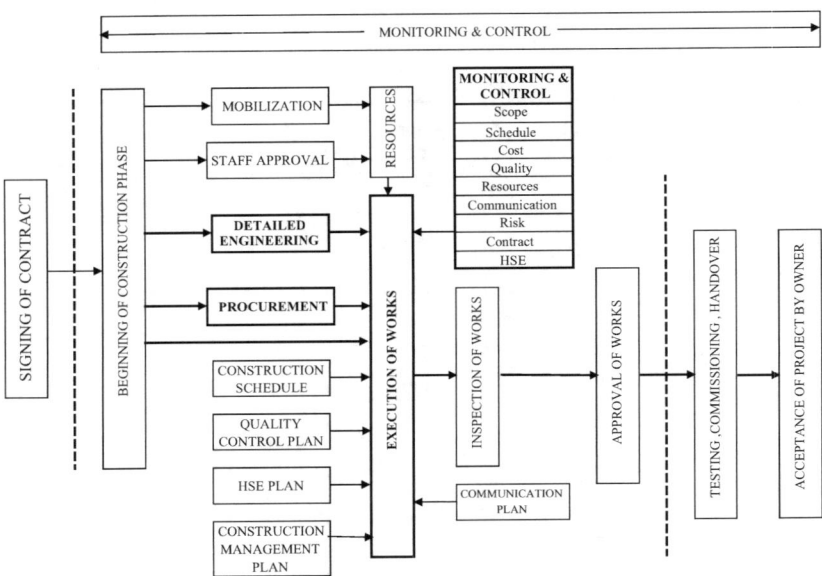

FIGURE 8.18 Major activities during construction phase (EPC contracting system).

the construction work after signing of the contract. A letter from the client/owner is issued to the contractor to begin the project work subject to the conditions of the contract. This letter is known as "Notice to Proceed" letter.

8.6.1.1 Notice to Proceed

The Notice to Proceed authorizes contractor to proceed with work to construct the project/facility as per the agreement. Prior to issuance of Notice to Proceed, the owner has to ensure that

- necessary permits have been obtained from the relevant authorities/agencies to hand over the construction site to the contractor
- Supervision staff/Project Manager/Construction Manager is selected to supervise the work
- relevant departments have been informed about signing of contract for availability of funds
- supervision undertaking guarantee is signed by the supervision consultant
- owner's representative is selected
- notice-to-proceed date is mutually discussed and agreed as per the conditions of contract
- all the copies of construction documents are distributed to the concerned stakeholders
- authorization letter to owner's representative and engineer's representative is already issued

Figure 8.19 is a sample notice to proceed.

Quality Management

LETTER HEAD

Ref: --------------------
Date: ----------------------

<u>**NOTICE TO PROCEED**</u>

To,

Contractor Name: SAMPLE LETTER

Address:

    Subject: Contract Number------------

    Attention: ------------------

Sir/Madam

You are hereby authorized to proceed with Project No. ------------ in accordance with construction contract dated -----------.This contract calls all the contracted works to be completed within --------- calendar days. The date of enterprise shall be -------------.

Sincerely,

Enclosures:

CC:

FIGURE 8.19 Notice to proceed.

8.6.1.2 Kick-off Meeting

The "Kick-off Meeting" is the first meeting with the owner/client and project team members. It is also called preconstruction meeting. The Kick-off Meeting provides opportunity to all project team members to interact and knowing each other.

Figure 8.20 is an example of Kick-off Meeting agenda.

8.6.1.3 Mobilization

The activities to be performed during the mobilization period are defined in the contract documents. During this period, the contractor is required to perform many of the activities before the beginning of actual construction work at the site. Necessary permits are obtained from the relevant authorities to start the construction work at the site. After being granted access to the construction site by the owner, the contractor starts mobilization work, which consists of preparation of site offices/field offices for the owner, supervision team (consultant), PMC, and for the contractor himself.

PROJECT NAME

Contract Number:			
Type of Meeting:		Date of Meeting:	
Place of Meeting:		Time of Meeting:	
Owner:			
PMC:			
Contractor:			
Others (As Applicable)			

AGENDA   *SAMPLE AGENDA*

1.0 Points to be Discussed

1.1 Intoduction

1.2 Project goals and objectives

1.3 Scope of work

1.4 Permit, Bonds, Insurance

1.5 Site handover procedure

1.6 Mobilization

1.7 Contractor's organization chart (Design, Construction)

1.8 Construction Schedule

1.9 Communication and Correspondance

1.10 Transmittals and Submittal Procedure

1.11 Meetings (Progrees, Coordination)

1.12 Construction Management Plan

1.13 Quality Management Plan

1.14 Risk Management Plan

1.15 HSE Management Plan

1.16 Payment

1.7 Nominated sub contractors

2.0 Any other business

Signed by: ………………………..   Position: …………………………..

Deate: …………………………..

FIGURE 8.20 Kick-off meeting agenda.

This includes all the necessary on-site facilities and services necessary to carry out specific tasks. Mobilization activities usually occur at the beginning of a project but can occur anytime during a project when specific on-site facilities are required. During this time, the project site is handed over to the contractor. The contractor performs site survey and testing of soil, etc., to facilitate the start of construction work.

In anticipation of the award of contract, the contractor begins the following activities much in advance, but these are part of contract documents, and the contractor's action is required immediately after signing off the contract in order to start construction:

- Mobilization of construction equipment and tools
- Workforce to execute the project

8.6.1.3.1 Bonds, Permits, and Insurance

As per contract documents, the contractor has to

1. submit advance payment guarantee
2. permit from local authority (municipality)
3. insurance policies covering following areas
    a. contractor's all risks and third party insurance policy
    b. contractor's plant and equipment insurance policy
    c. workmen's compensation insurance policy
    d. site-storage insurance policy

Normally, the submitted originals are retained by the owner and the copies are kept with the Resident Engineer or Project/Construction Manager (Consultant).

8.6.1.3.2 Temporary Facilities

The requirements to set up temporary facilities are specified in the contract. The contractor has to submit layout plans, dimensions, and other pertinent details for temporary facilities to be constructed. These include the following:

1. Site offices for owner, PMC, supervisor (consultant)
2. Storage facilities
3. Toilets and wash rooms
4. Sanitary and drainage system
5. Drinking water facility
6. Safety and healthcare facilities
7. Site electrification
8. Temporary fire-fighting system
9. Site fence
10. Site access road
11. Necessary utilities for construction
12. Communication system
13. Signage

14. Fuel storage area
15. Guard room
16. Testing laboratory
17. Waste dumping area
18. Area for storage of hazardous material

Upon approval of plans, the contractor proceeds with the construction of temporary facilities and necessary utilities.

The contractor has to designate authority-approved dumping area for waste material.

Figure 8.21 illustrates the logic flowchart for EPC contracting.

8.6.2 IDENTIFY STAKEHOLDERS

Most stakeholders listed in Figure 7.2 have involvement in the project during construction phase. These are as follows:

1. Owner
    - Owner's representative/Project Manager
2. PMC
3. Construction supervisor
    - Consultant (designer)
4. Specialist consultant
5. Contractor
    - Main contractor
    - Subcontractor(s)
    - Vendors
6. Regulatory authorities
7. End user

8.6.2.1 Identify Owner's Representative/Project Manager

The owner from his/her office deputes or hire from outside, Owner's Representative (OR), to administer the overall project. The OR should have relevant experience and knowledge about the construction processes of similar nature of project. He/she should be able to manage the project with the help of supervision team members.

8.6.2.2 Identify Contractor's Project Team Members

In EPC type contracting system, the team members during the construction phase consist of

1. Design Team
2. Construction Team

Figure 8.22 is the staff approval request form used by the contractor to propose the staff to work for these positions and is submitted along with the qualification and experience certificates of the proposed staff.

Quality Management

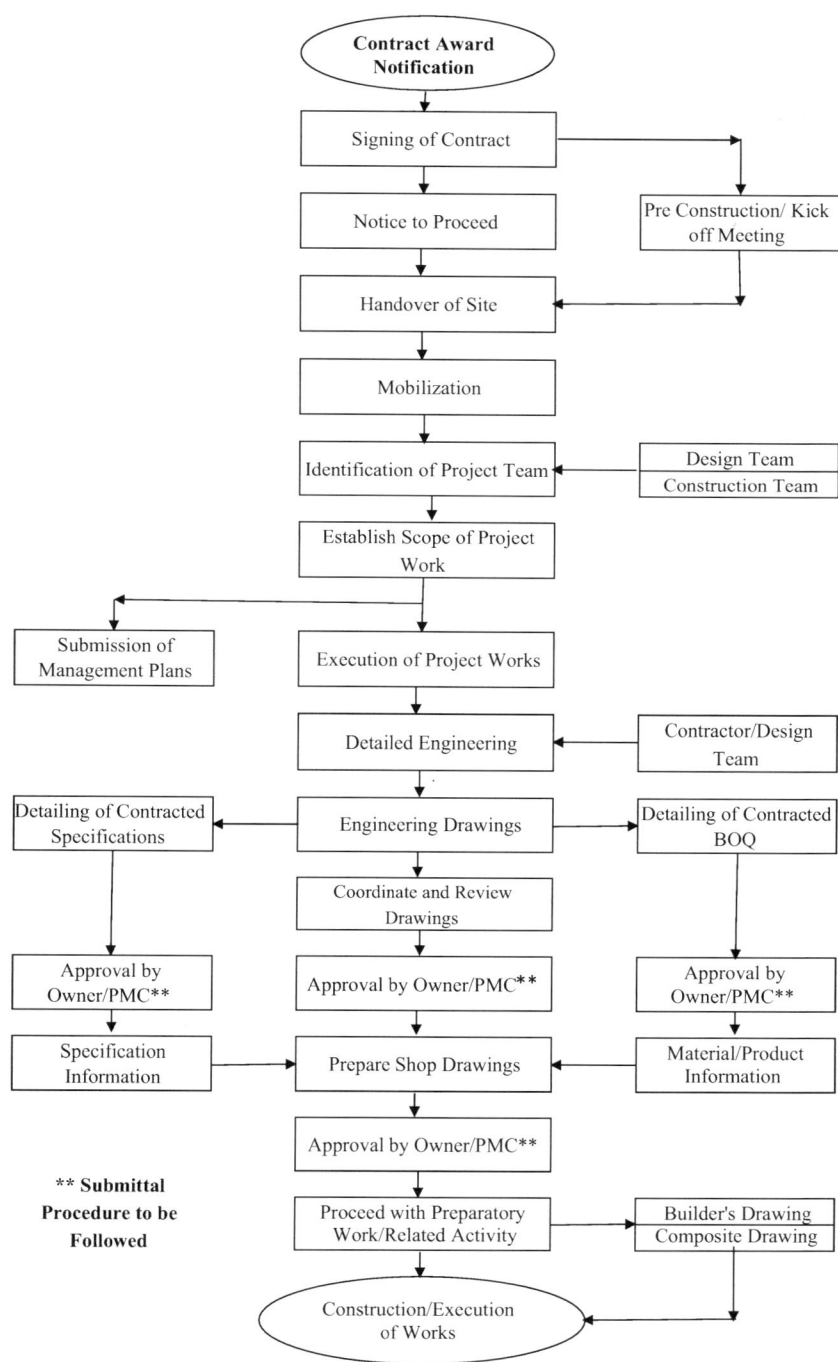

FIGURE 8.21 Logic flowchart for EPC construction process.

	Project Name
	Consultant Name
	REQUEST FOR SITE STAFF APPROVAL

CONTRACT NO. :	NO. :
CONTRACTOR :	DATE:

SAMPLE FORM

To : Owner

1.	Name	:
2.	Profession	:
3.	Position No. in Document-I	:
4.	No. of years of Experience	:
5.	Membership of Professional Body	: Valid ☐ Not Valid ☐
6.	Requested Date of Commencement	:
7.	Remarks	:

Signature Contractor's Project Manager

OWNER COMMENTS   APPROVED ☐   NOT APPROVED ☐

Owner Rep. Signature   Date

Distribution   OWNER   A/E   CONTRACTOR

FIGURE 8.22 Request for staff approval.

Contractor has to submit the organization chart based on the typical minimum core staff list and approved staff.

8.6.2.2.1 Contractor's Design Team Members

Figure 8.23 illustrates design management team and their major responsibilities.

Each of the managers has many other team members. These members are selected based on the organizational structure and suitable skills required to perform the job. These include

Quality Management

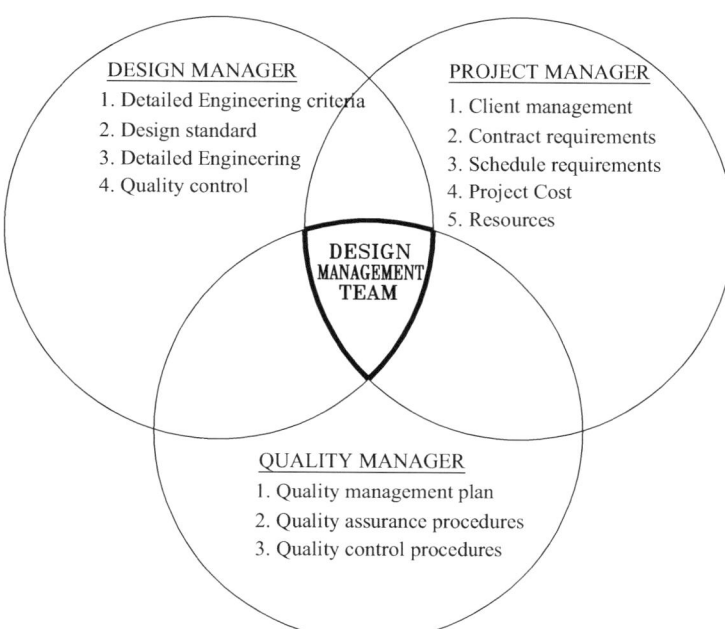

FIGURE 8.23 Design management team. (Abdul Razzak Rumane (2013) *Quality Tools for Managing Construction Projects*. CRC Press, Boca Raton, Fl. Reprinted with permission from Taylor & Francis Group.)

1. Team Leader (Principal Engineer) – each trade
2. Team Members (Engineers and CAD technicians for each discipline)
3. Quantity surveyor (cost engineer)
4. Owner's representative
5. End user

8.6.2.2.2 Contractor's Construction Team Members

Contract documents normally specify a list of minimum number of core staff to be available on site during the construction period. Absence of contractor's core staff requirements and their qualifications are listed under tendering procedures. The absence of these staff from the project site without prior permission attracts penalty to the contractor. Normally, the penalty amount is specified in the contract documents. Upon signing of the contract, the contractor has to submit the names of the staff for the positions described in the contact documents for approval from the owner/consultant to work on the project. Contractor has to select an appropriate candidate to propose for the specified position. Figure 8.24 shows Site Staff Selection Procedure and Figure 8.25 shows Project Staffing Process.

The following is a typical list of contractor's minimum core staff needed during the construction period for execution of work of a major building construction project.

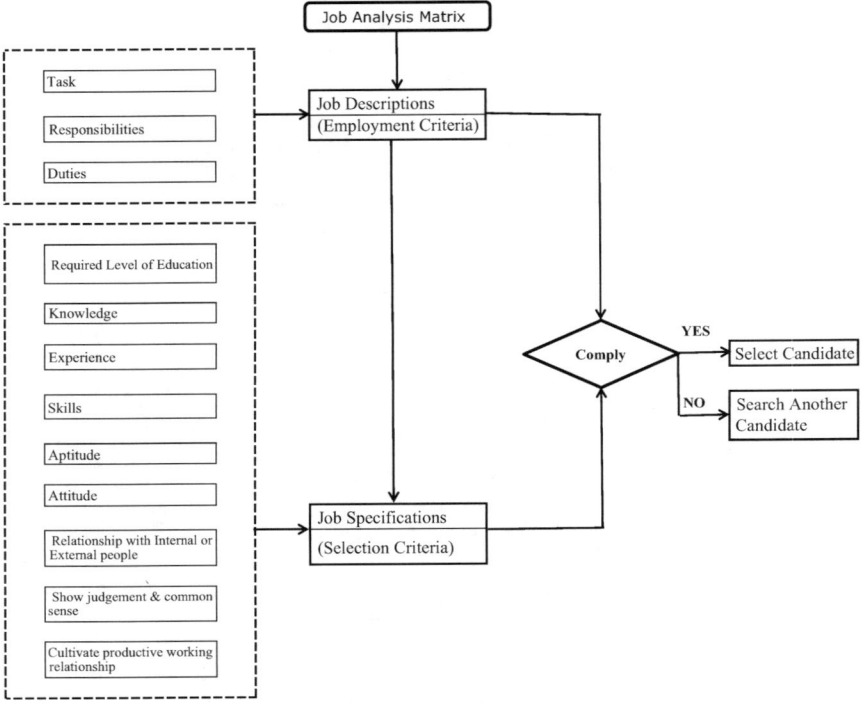

FIGURE 8.24 Site staff selection procedure. (Abdul Razzak Rumane. (2016). *Handbook of Construction Management: Scope, Schedule, and Cost Control*. Reprinted with permission from Taylor & Francis Group.)

1. Project Manager
2. Senior Engineer for process engineering works
3. Senior Engineer for piping works
4. Senior Engineer for mechanical works
5. Senior Engineer for utility works
6. Senior Engineer for civil works
7. Senior Architect
8. Senior Engineer for electrical works
9. Senior Engineer for infrastructure works
10. Senior Engineer for low-voltage systems
11. Planning and Control Engineer
12. QA/QC Manager/Engineer
13. Senior Safety Engineer
14. Senior Quantity Surveyor/Cost Engineer
15. Contract Administrator
16. Laboratory Technician
17. Foremen (different trades)

Quality Management

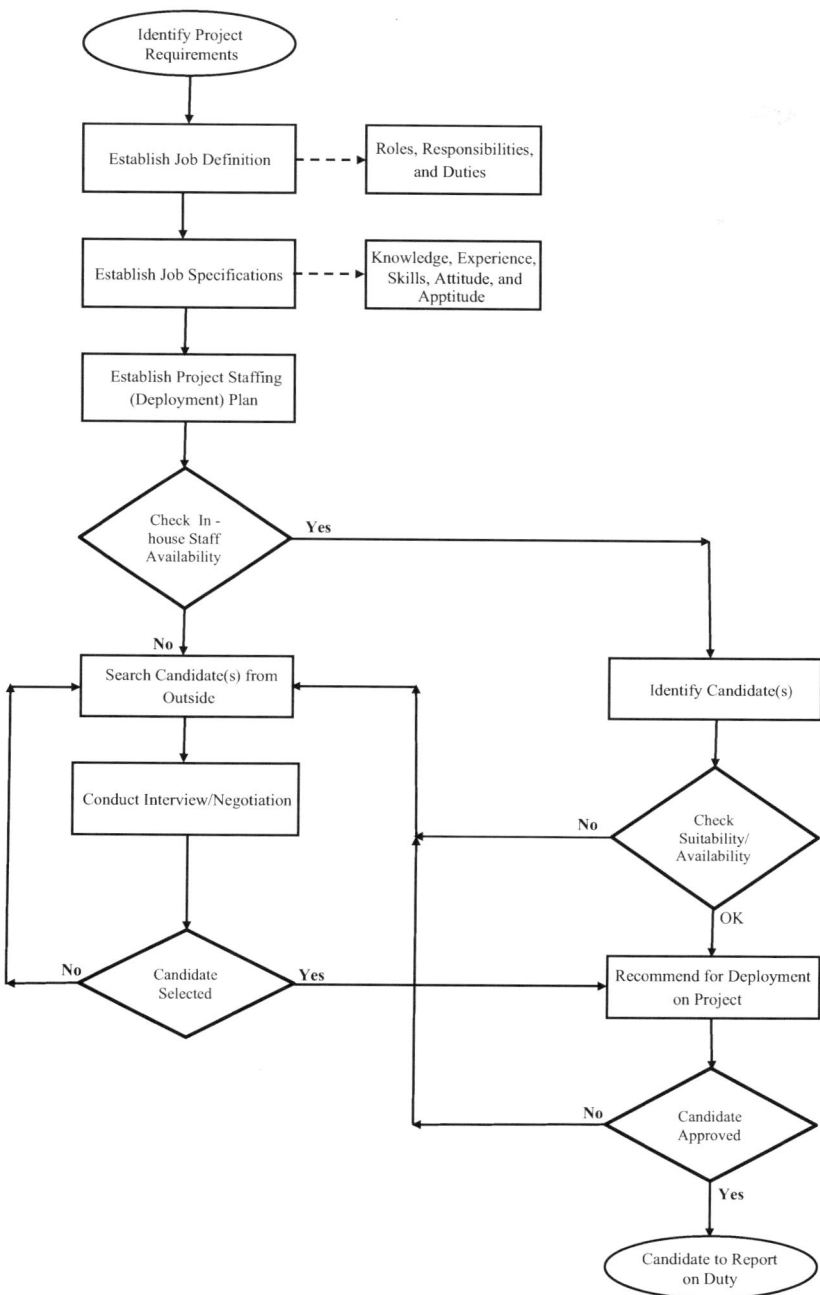

FIGURE 8.25 Project staffing process. (Abdul Razzak Rumane. (2016). *Handbook of Construction Management: Scope, Schedule, and Cost Control.* Reprinted with permission from Taylor & Francis Group.)

8.6.2.3 Identify PMC Team

Normally, the owner/client engages PMC firm from the inception of the project through to issuance of substantial completion certificate to manage the entire construction project on behalf of the owner. Project management services are generally offered by registered consultant firms/professionals having the ability and expertise to manage construction projects. Table 8.21 lists typical responsibilities of the PMC.

8.6.2.4 Identify Supervision Team

Normally, the client selects the same firm which has designed the project. The firm known as "Consultant" is responsible to supervise the construction process and achieve project quality goals. The firm appoints a representative, who is acceptable and approved by the owner/client, to be on site and is often called as Resident Engineer (RE). The Resident Engineer along with supervision team members is responsible to supervise, monitor, and control, implement the procedure specified in the contract documents, and ensure completion of project within specified time, budget, and per defined scope of work. Table 2.12 discussed earlier under Chapter 2 illustrates typical responsibilities to be performed by the supervisor during the construction phase.

8.6.2.5 Identify Regulatory Authorities

In certain countries, there is a regulation to submit electrical, mechanical, HVAC drawing for review and approval by the authorities. Contractor has to identify which drawings/documents are to be submitted to authorities during the construction phase. Necessary letter to the regulatory authorities/agencies is issued by the owner upon request for such letter from the contractor.

8.6.2.6 Identify Subcontractors

In most construction projects, the contractor engages special subcontractors to execute certain portion of contracted project works. Areas of subcontracting are generally listed in the particular conditions of the contract document. The contractor has to submit their names for approval to the owner prior to their engagement to perform any work at the site.

Figure 8.26 is a request for subcontractor approval to be submitted by the contractor for getting approval of any subcontractor proposed to work on the construction project.

The request includes all the related information to prove the subcontractor's capability of providing services that meet the required project quality, the resources available to meet the specified schedule, and past performance; also, if any quality system was implemented by the subcontractor, that fact is noted. Table 8.22 is an example subcontractor selection questionnaire.

8.6.2.6.1 Nominated Subcontractors

Sometimes, the owner nominates subcontractors to execute a portion of a contract; such a subcontractor is known as a nominated subcontractor.

TABLE 8.21
Typical Responsibilities of PMC

1-During Design Phases

1.1	Oversee and coordinate design
1.2	Review design documents
1.3	Constructability review
1.4	VE suggestions
1.5	Review construction schedule/execution plan
1.6	Review construction cost estimates
1.8	Schedule and budget management
1.9	Project risk management
1.10	Project HSE management
1.11	Review of FEED and specifications
1.12	Review of ITB documents

2-During Construction Phases

2.1	Ensure that contractor submitted performance bond
2.2	Ensure that contractor submitted worker's insurance policy
2.3	Selecting and recommending contractor's core staff
2.4	Selecting and recommending subcontractor
2.5	Review and approval of detailed engineering
2.6	Review and approval of detailed specifications
2.7	Review and approval of detailed BOQ
2.8	Construction schedule approval
2.9	Scope control
2.10	Construction scheduling and monitoring
2.11	Cost tracking and management
2.12	Evaluation of payment request and recommending progress payments
2.13	Manage project progress record
2.14	Monitor daily progress
2.15	Monitoring contractor's performance and ensure that the work is performed as specified, as per approved shop drawings, and as per applicable codes
2.16	Monitor project quality management system
2.17	Coordination of on-site, off-site inspection
2.18	Vendor/Material Review and approvals
1.19	Conduction periodic progress meetings
2.20	Preparation of minutes of meeting and distribution as per agreed upon matrix
2.21	Technical correspondence between contractor and owner
2.22	Monitor project risk
2.23	Change order management
2.24	Managing contractor's RFI
2.25	Monitoring contractor's HSE plans (Health, Safety, and Environment)
2.26	Review and approval of HAZID, HAZOP, EIA reports
2.27	Review, evaluation, and documentation of claims
2.28	Review and approve inspection and testing plan
2.29	Oversea testing of systems
2.30	Review of punch list

Project Name	
Contract No.:	
Contractor :	

**CONTRACTOR REQUEST
FOR SUB-CONTRACTOR APPROVAL**

Serial No.: _____

Kindly approve the following as a SUB-CONTRACTOR in the above mentioned Project

Sub-Contract Works _____
Sub-Contractor _____
Address (Head Office) _____

Reference Letter

Attachments:

Commercial Register ☐   Foundation Contract ☐
Experience ☐   Resources ☐
Current Work ☐   Financial Status ☐
Others (List, if any) ☐

_____
_____
_____

Performance Bond (Yes/No) _____ No. _____

Signed by
Contractor's Representative _____ Date: _____

OWNER'S NAME

Received by: _____ Signature: _____ Date: _____

PMC/SITE SUPERVISION CONSULTANT

Received by: _____ Signature: _____ Date: _____

PMC/Consultant's recommendation : _____
_____
_____
_____

Signed by
PM/Resident Engineer _____ Date: _____

Distribution :   OWNER(Original)   PMC/Supervision Consultant (Copy)   Contractor (Copy)

Note :   Contractor shall submit Original to OWNER with copy to PMC/Site Supervision Consultant.

FIGURE 8.26 Request for subcontractor approval.

TABLE 8.22
Subcontractor Prequalification Questionnaire

Instructions
Please type or write all your replies legibly. Attach additional sheets, if required.

PART I
I.1 Company Information

I.1.1 Name of Organization:
I.1.2 Commercial Registration no.:
I.1.3 Year of Establishment:
I.1.4 Type of Company:
I.1.5 Company Address:
I.1.6 Affiliate company name(s) and Address:

I.2 Subcontract Works (Please Tick Mark All Interested)

Sr.No.	Work Description	Detail Design	Preparation of Shop Drawing	Construction	Inspection/Auditing
1	Process Engineering	☐			
2	Piping				
3	Mechanical				
4	Civil				
5	Instrumentation				
6	Electrical				
7	Buildings				
8	Low-voltage systems				
9	Landscape work				
10	External works				
11					
12					

PART II
II.1 Financial Information

II.1.1 Provide copy of audited Balance Sheet:
II.1.2 Provide Bonding Capacity:
II.1.3 Provide Insurance Capacity:
II.1.4 Provide Bank Reference:

PART III
III.1 Organization Details

III.1.1 Core Business Area:
III.1.2 Organization Chart:
III.1.3 ISO Certification:
III.1.4 Year of Experience:

(Continued)

TABLE 8.22 (*Continued*)
Subcontractor Prequalification Questionnaire

III.2 Project Details

III.2.1 Project History for last 10 years

Sr.No.	Name of Project	Type of Work	Value	Peak Workforce	Start Date	Finish Date
1						
2						
3						
4						

III.2.2 Current Projects

Sr.No.	Name of Project	Type of Work	Value	Peak Workforce	Start Date	Expected Finish Date
1						
2						
3						

PART IV

IV.1 Management Staff

IV.1.1 Provide list of Project Managers, Project Engineers, Engineers

IV.2 Workforce

Sr.No.	Work Description	Technicians	Foreman	Skilled	Unskilled
1	Process Engineering				
2	Piping				
3	Civil				
4	Mechanical				
5	Instrumentation				
6	Electrical				
7	Buildings				
8	Low-voltage systems				
9	Landscape work				
10	External works				
11					
12					

PART V

V.1 Quality Management System

V.1.1 Provide copy of ISO Certificate:
V.I.2 Person in-charge of QA/QC Activities:
V.I.3 Number of Quality Auditors:

PART VI

VI.1 HSE System

VI.1 Does company has HSE Policy?
VI.1.2 Provide Site Accident records for last 2 years

(*Continued*)

TABLE 8.22 (*Continued*)
Subcontractor Prequalification Questionnaire

Declaration

We hereby declare that the information provided herein is true to our knowledge.
Note:
All relevant documents attached.

Signature of Authorised Person

8.6.2.6.2 Identify Vendors

The contract documents list the names of approved vendor for procurement of material/equipment/systems. The contractor has to follow the submittal procedure and obtain approval prior to ordering for the material.

8.6.2.7 Develop Responsibility Matrix

Table 8.23 illustrates the contribution of various participants during the construction phase.

8.6.3 ESTABLISH CONSTRUCTION PHASE REQUIREMENTS

Once the contract is awarded to a successful bidder (contractor), then it is the responsibility of the contractor to construct the project as specified in the contract documents, drawings, and specifications within the specified time and budget to meet the owner requirements to produce specified products.

TABLE 8.23
Responsibilities of Various Participants during the Construction Phase

		Responsibilities	
Phase	Owner	PMC/Supervisor	Contractor
Construction	Approve subcontractor(s)	Review and approval of detailed engineering	Development of detailed engineering
	Approve contractor's core staff	Supervision of construction works	Development of detailed specifications
	Legal/regulatory clearance	Approve plan	Development of detailed BOQ
	SWI	Monitor work progress	Development of procurement plan
	Variation Order (VO)	Approve material	Execution of work
	Payments	Approve shop drawings	Contract management
		Monitor schedule	Selection of subcontractor(s)
		Control budget	Planning
		Recommend payment	Resources
			Procurement
			Quality
			Safety

8.6.3.1 Identify Contract Requirements

In order to develop contract requirements (scope of project work), the contractor has to review all the documents which are part of the contract that contractor has signed with the owner of the project. This includes the following:

- Front End Engineering Design Drawings
- Contract conditions
  - General conditions
  - Particular conditions
- Technical specification
  - General specifications
  - Particular specification
  - Bill of Quantities (BOQ)
- Other related documents

8.6.3.1.1 Review Contract Drawings

The contractor has to review the contract drawings to understand the project requirements and development of detailed engineering. Each trade engineer has to review the drawing and understand the detailed engineering development requirements for smooth construction process to be followed in order to avoid omission and rework. The trade engineers have to

- prepare a list of issues to be resolved and information needed from the owner regarding FEED (contract drawings) to develop detailed engineering
- prepare a list of conflicting items in the drawings
- identify critical quality attributes
- identify high-risk items
- check constructability
- check contracted FEED deliverables and design criteria
- identify discrepancies between drawings and specifications to be incorporated in the detailed engineering
- identify any missing drawing required to execute the project
- identify complete scope of work
- prepare construction material take-off (MTO) for procurement

8.6.3.1.2 Review Specifications

The contractor has to review specifications and check for the following major points:

- Matching of construction drawings and specification requirements
- Matching of construction drawings and material requirements
- If there are discrepancies between specifications and drawings
- Any missing specifications
- Issues that need to be resolved
- Codes and standards to be followed
- Whether contracted schedule is achievable
- Quality requirements

Quality Management

- Submittal requirements
- Safety requirements
- Environmental requirements

8.6.3.1.3 Review Contract Documents

Contractor has to carefully study all the clauses of contract and identify

- high-risk clauses
- items having price difference between bid price and the actual price for the items to be procured and installed for specified performance of work/system
- items need to be resolved by raising Request for Information (RFI)
- Payment procedure
- Schedule constraints
- Ambiguous clauses
- Clauses that favor owner
- Any hidden clause that will entitle owner for compensation claim from the contractor
- Coordination among all the parties involved in the project
- Priorities among various construction documents
- Clauses having conflict with regulatory requirements
- Discrepancies and conflicting clauses
- Change order process
- Claim for extra work
- Damage and penalty for delay clauses
- Force majeure clause

8.6.3.2 Develop Construction Phase Scope

The contract requirements are established taking into consideration items identified under Section 8.6.3.1. The following are the major requirements:

- Development of detailed engineering
- Material procurement as per specified codes and standards from the approved vendors
  - Identification of log lead items
  - List of prioritization items
  - Material/equipment to be at site as per construction schedule
- Construction of the facility/project as per the specifications to produce required products
  - Complete the facility within the specified schedule
  - Complete the facility within the contracted amount
  - Complete the facility as per the quality requirements

8.6.3.2.1 Identify Detailed Engineering Deliverables

Detailed engineering follows the contracted requirements taking into consideration the configuration and the allocated baseline from the development of FEED. Detail engineering involves the process of successively breaking down and analyzing and designing the process, piping, instrumentation, civil works, infrastructure, and other specified works/systems and their components so that it complies with the recognized codes and standards of safety and performance while rendering the design in the form of drawings and specifications that will tell the contractors exactly how to construct the facility to meet the owner's need.

While developing detailed engineering, the contractor (designer) has to ensure that all suggested changes/detailing are re-evaluated to ensure that the changes will not detract from the contracted/budgeted deliverables, contracted cost, and project design goals/objectives.

Detailed engineering activities are more in-depth than the design activities in the FEED stage. The size, shape, levels, performance characteristics, technical details, and requirements of all the individual components are established and integrated into the detailed engineering. The detailed engineering drawings involve development of detailed plans, sections, and elevations that are drawn to scales, principle dimensions are noted, and design calculations are checked to conform the accuracy of design and its compliance to the codes and standards.

For the development of detailed engineering, the contractor (designer) has to consider the following points:

1. Review of contract drawings
2. Review of contract specifications
3. Review of site investigations and survey reports
4. Preparation of detailed engineering for all the trades
5. Interdisciplinary coordination to resolve the conflict
6. Availability of material/equipment during construction
7. High-risk items
8. Environmental considerations
9. Obtain regulatory approval, if any
10. Constraints such as contracted project cost

Figure 8.27 illustrates the logic flowchart for detailed engineering

8.6.3.2.2 Identify Procurement Deliverables

Procurement is acquisition of material, equipment, and services to successfully fulfill the need for a specific purpose/use. In EPC-type contracting system, contractor is responsible for procurement of material, equipment, and systems to be installed on the project. The contractors have their own procurement strategies.

The following are the major procurement deliverables:

1. Quantity of material, equipment, and system as per the approved detailed engineering drawings and the approved BOQ and specification
2. All the material, equipment, and system to be as per the specified codes and standards
3. All the material, equipment, and system should conform to the latest technology and suitable for networking through internet of things (IOT).

Quality Management

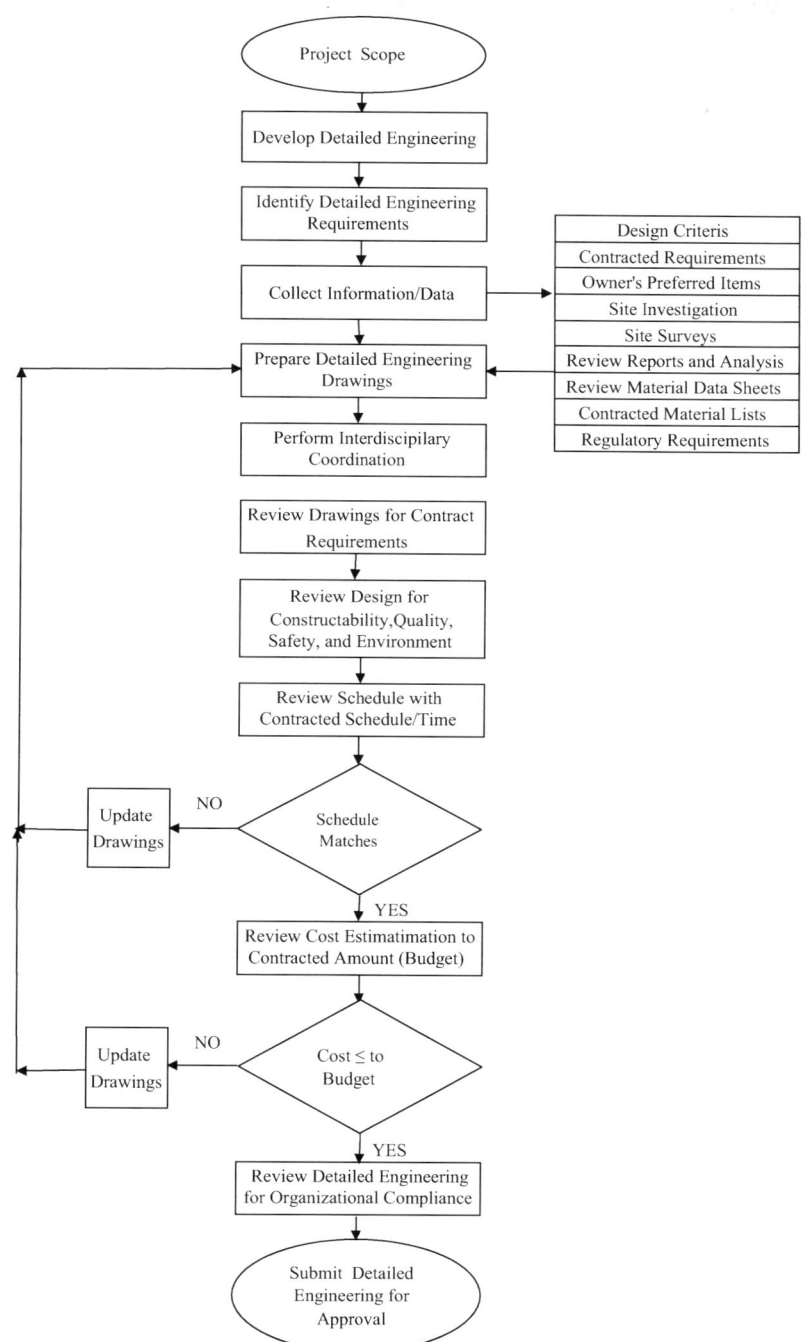

FIGURE 8.27 Logic flowchart for detailed engineering.

4. All the material, equipment, and system to be purchased from approved vendors
5. Contractor should get approval from the owner prior to procurement of all the materials, equipment, and systems specified in the contract documents
6. All the material, equipment, and system to be available for installation/execution as per approved schedule
7. Inspection of all material, equipment, and system to be as per approved quality control plan
8. Contractor is responsible for follow-up program with the vendor to ensure timely delivery
9. Contractor to identify long lead items and take action accordingly
10. Contractor to arrange for safe storage of all the materials, equipment, after receipt at site

The contractor has to consider the following as a minimum while finalizing the material procurement:

- Contractual commitment
- Specification compliance
- Statutory obligations
- Delivery time
- Cost
- Quality
- Performance

8.6.3.2.3 Identify Construction Works Deliverables

The construction phase deliverables of oil and gas project are established based on construction phase requirements/scope discussed under Section 8.6.3. These are as follows:

A. Construction–installation/execution and inspection of following works:
  1. Site work such as cleaning and excavation of project site
  2. Grading works
  3. Earth work
  4. Process engineering works
  5. Piping works including pipe support and structural works
  6. Storage tanks
  7. Instrumentation and control works
  8. Electrical works
  9. Construction of foundation for equipment and machinery
  10. Installation of equipment and machinery
  11. Utility services
  12. Supervisory control system
  13. Process automation system
  14. Fire- and gas-detection system
  15. Fire-fighting system

16. Leak-detection system
17. Corrosion monitoring system
18. Cathodic protection system
19. Construction of buildings with specified finishes and furnishings
20. Structural works for buildings
21. MEP services for buildings
22. Elevators/conveying system
23. Emergency power/generator
24. Installation of fire-fighting systems
25. Lightning protection system
26. Fire alarm system
27. Communication system
28. Electronic security and access control system
29. Data security system
30. Landscape works
31. External works
32. Parking

B. Planning and scheduling
   1. Construction schedule
   2. S-Curve
   3. Contractor's quality control plans
   4. Management plans

C. Monitoring and control during construction
   1. Stakeholder's requirements
   2. Project schedule
   3. Project cost
   4. Project quality
   5. Resources
   6. Contract
   7. Safety during construction

D. Inspection of executed works
   1. Inspection of executed works
   2. Validate executed works

E. Drawings
   1. Record drawings/update working drawings (shop drawings)

F. Specifications
   1. Update specifications

8.6.4 Project Planning and Scheduling

Project planning is a logical process to ensure that the work of the project is carried out:

- In an organized and structured manner
- Reducing uncertainties to a minimum
- Reducing risk to a minimum

- Establishing quality standards
- Achieving results within budget and scheduled time

Prior to the start of execution of a project or immediately after the actual project starts, the contractor prepares the project construction plans based on the contracted time schedule of the project. Detailed planning is needed at the start of construction to decide how to use resources such as laborers, plant, materials, finance, and subcontractors economically and safely to achieve the specified objectives. The plan shows the periods for all sections of the works and activities, indicating that everything can be completed by the date specified in the contract and ready for use or for installation of equipment by other contractors.

Effective project management requires planning, measuring, evaluating, forecasting, and controlling all aspects of project quality and quality of work, cost, and schedules. The purpose of the project plan is to successfully control the project to ensure completion within the budget and schedule constraints. Project planning is the evolution of the time and efforts to complete the project.

Upon signing of the contract, the contractor has to submit mainly the following to PMC/supervisor (consultant) for their review and approval:

1. Contractor's construction schedule
2. S-Curve
3. Contractor's quality control plan (CQCP)

Among these plans, construction schedule is most important. This is the first and foremost program contractor has to submit for approval. The contractor cannot proceed with construction unless preliminary construction schedule is approved. In certain cases, the progress payment is having relation with the approval of contractor's construction schedule. The contractor is not paid unless contractor's construction schedule is approved.

Contractor has to prepare the following plans and submit along with construction schedule. These are as follows:

- Resource management plan
- Project S-Curve

Apart from the above plans, the contractor has also to prepare and submit the following plans:

- Stakeholder Management Plan
- Resource Management Plan
- Communication Management Plan
- Risk Management Plan
- Contract Management Plan
- HSE Management Plan

8.6.4.1 Develop Contractor's Construction Schedule

Prior to the start of execution of project or immediately after the actual project starts, the contractor prepares the project construction plans based on the contracted time schedule of the project. Detailed planning is needed at the start of construction in order to decide how to use the resources such as labors, plant, materials, finance, and subcontractors economically and safely to achieve the specified objectives. The plan shows the periods for all sections of the works and activities indicating that everything can be completed by the date specified in the contract and ready for use or for installation of equipment by other contractors.

Project planning is a logical process to determine what work must be done to achieve project objectives and ensure that the work of project is carried out

- in an organized and structured manner
- reducing uncertainties to minimum
- reducing risk to minimum
- establishing quality standards
- achieving results within budget and scheduled time

Depending on the size of the project, the project is divided into multiple zones and relevant activities are considered for each zone to prepare the construction program. While preparing the program, the relationships between project activities and their dependency and precedence are considered by the planner. These activities are connected to their predecessor and successor activity based on the way the task is planned to be executed. There are four possible relationships that exist between various activities. These are finish-to-start relationship, the start-to-start relationship, the finish-to-finish relationship, and start-to-finish relationship.

Once all the activities are established by the planner and estimated duration of each activity has been assigned, the planner prepares a detailed program fully coordinating all the construction activities.

The first step in the preparation of construction program is to establish project activities, and the next step is to sequence project activities and establish estimated time duration of each activity. The deadline for each activity is fixed, but it is often possible to reschedule by changing the sequence in which the tasks are performed, while retaining the original estimated.

Figure 8.28 illustrates schedule development process.

The activities to be performed during the execution of project are grouped in a number of categories. Each of these categories has a number of activities. The following are the major categories of construction projects schedule:

A. General activities
   1. Mobilization
   2. Staff approval
   3. Subcontractor approval

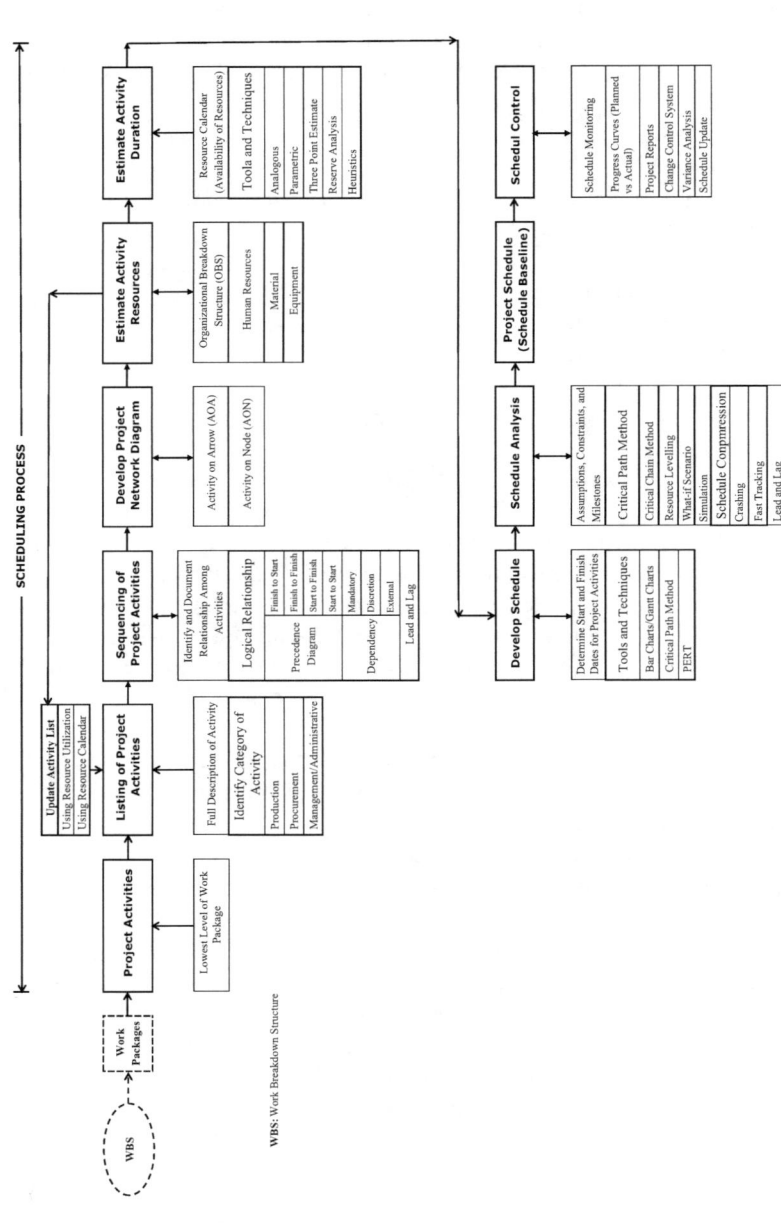

FIGURE 8.28 Schedule development process. (Abdul Razzak Rumane. (2016). *Handbook of Construction Management: Scope, Schedule, and Cost Control*. Reprinted with permission from Taylor & Francis Group.)

Quality Management

B. Engineering
  1. Detailed engineering
  2. Procurement of materials submittal and approval
  3. Shop drawing submittal and approval
C. Procurement
  1. Approval of vendor/supplier
  2. Submittal and approval of material, equipment, and systems
  3. Procurement of material, equipment, and systems
  4. Inspection of material, equipment, and systems
D. Construction (few activities are listed below)
  a. Site earth works
  b. Dewatering and shoring
  c. Excavation and backfilling
  d. Process engineering works
  e. Piping works
  f. Instrumentation works
  g. Mechanical works
  h. Electrical works
  i. Buildings
  j. Civil works
  k. Drainage system
  l. Utility
  m. Equipment and machinery installation
  n. Elevator/conveying systems works
  o. Fire suppression works
  p. Plumbing and public health works
  q. HVAC works
  r. Fire alarm system works
  s. Information and communication system works
  t. Low-voltage systems works
  u. Landscape works
  v. External works
E. Close-out
  1. Testing and commissioning
  2. Completion and handover

The contractor also submits the following along with the construction schedule:

1. Resources (equipment and manpower) schedule
2. Cost-loading (schedule of item's pricing based on BOQ)

Figure 8.29 illustrates an example of construction schedule summary.

8.6.4.2 Develop Project S-Curve

S-Curve is a graphical display of cumulative cost, resources, or other quantities plotted against time. S-Curve is used for forecasting cash flow which is based on

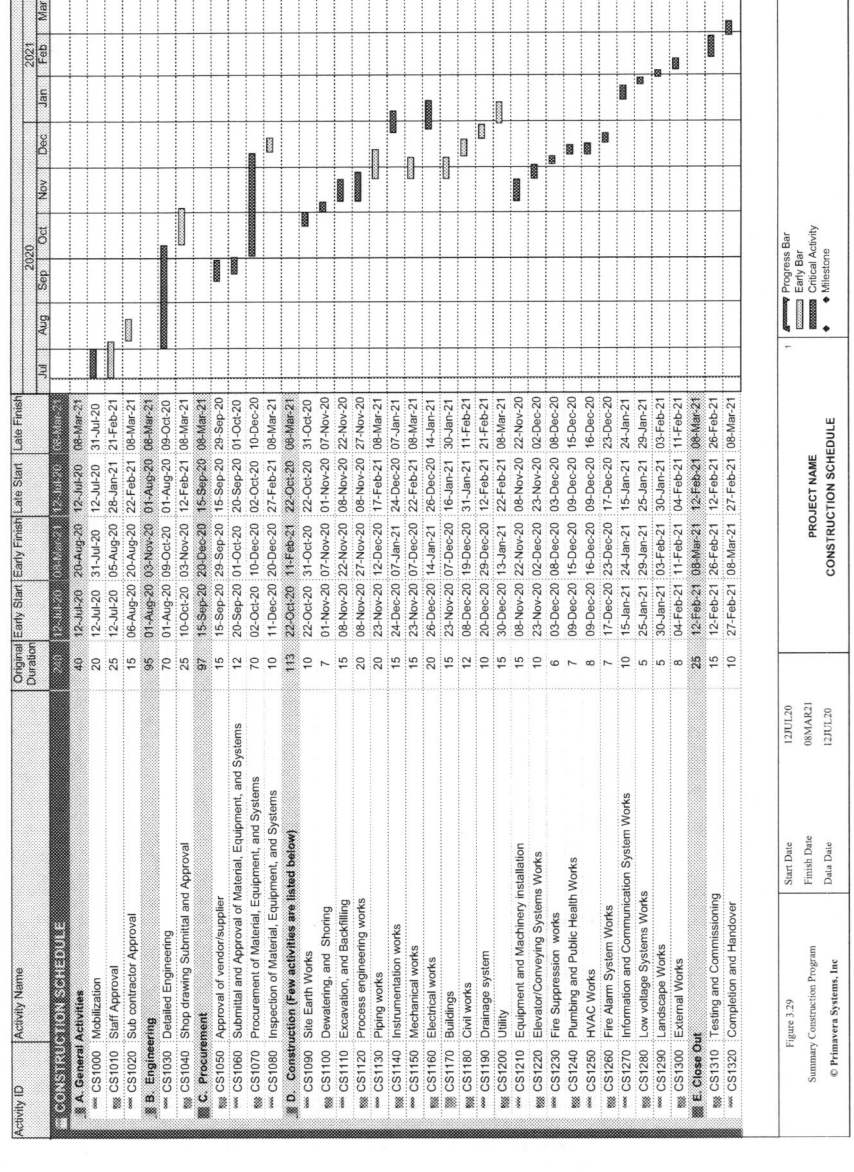

FIGURE 8.29 Summary construction program.

Quality Management

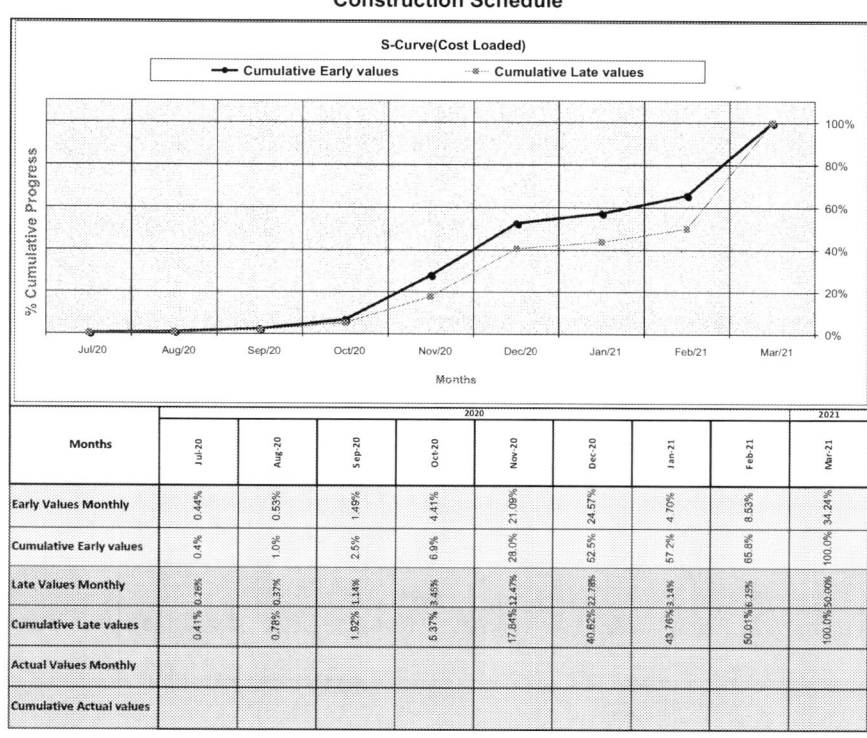

FIGURE 8.30 S-Curve (cost-loaded).

the work (activities) the contractor is expected to complete and how much amount (payment) the contractor will receive and predicts how much amount the contractor will be spending over the established project schedule (time). The S-Curve is used to measure project performance and predict the expenses over project duration. The S-Curve helps owner/client to know the project funding requirements. It is also an indication of the progress of work to be completed in a project. Funding requirements, total, and periodic are derived from the S-Curve. This also represents the planned progress of a project. Figure 8.30 illustrates the Planned Project S-Curve prepared by the contractor which is based on the construction schedule.

8.6.4.2.1 Cost-Loaded Curve

Conceptually, cash flow is a simple comparison of when revenue will be received and when the financial obligations must be paid. This is obtained by loading each activity in the approved schedule with the budgeted cost in the BOQ. The process of inputting schedule of values is known as cost-loading. The graphical representation of the above is obtained as a curve and is known as the cost-loaded curve. Figure 8.31 illustrates cost-loaded S-curve prepared by the contractor.

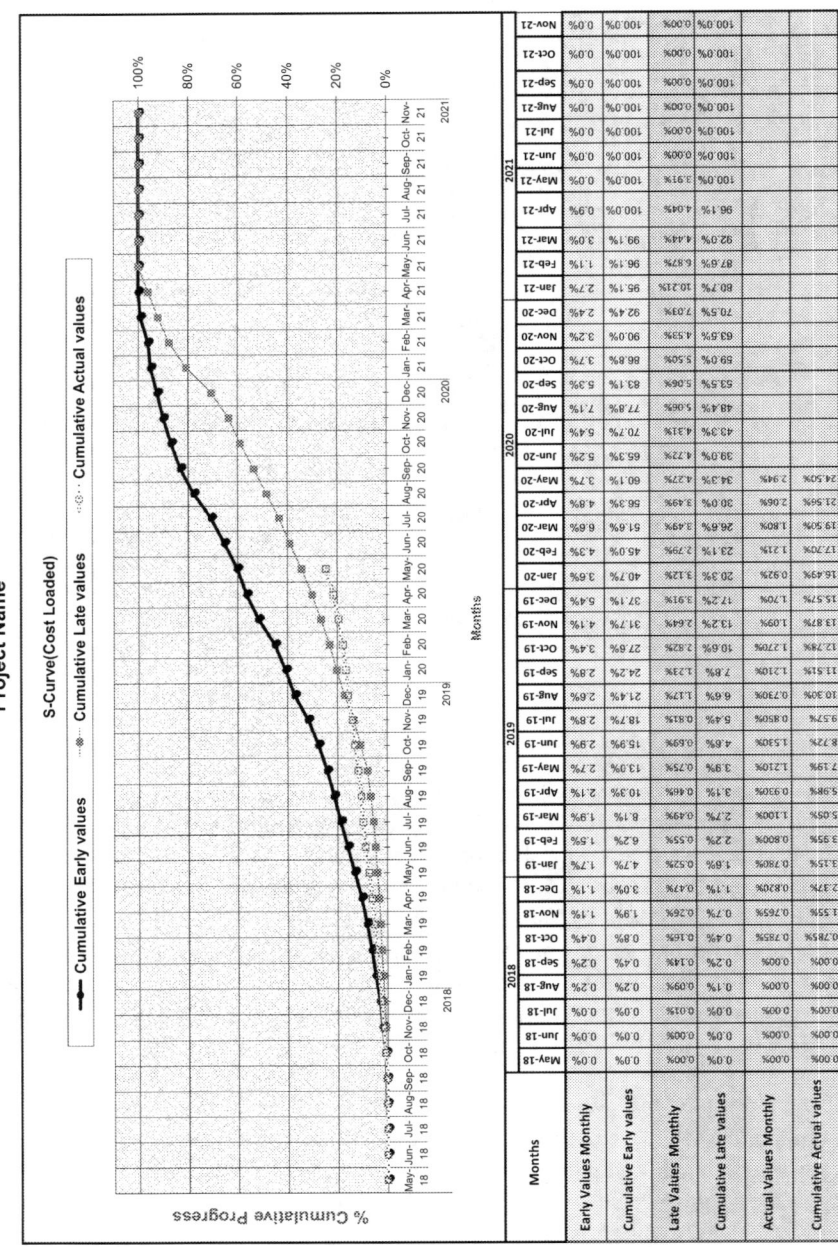

FIGURE 8.31 S-Curve (cost-loaded).

8.6.4.3 Develop Contractor's Quality Control Plan

The CQCP is the contractor's everyday tool to ensure meeting the performance standards specified in the contract documents. The adequacy and efficient management of CQCP by contractor's personnel have great impact on both the performance of contract and owner's quality assurance surveillance of the contractor's performance.

CQCP is the documentation of contractor's process for delivering the level of construction quality required by the contract. It is a framework for the contractor's process for achieving quality construction. CQCP does not endeavor to repeat or summarize contract requirements. It describes the process which contractor will use to assure compliance with the contract requirements. The quality plan is virtually manual tailor-made for the project and is based on contract requirements.

The CQCP is prepared based on the project specific requirement as specified in the contract documents. The plan outlines the procedures to be followed during the construction period to attain the specified quality objectives of the project fully complying with the contractual and regulatory requirements.

In the quality plan, the generic documented procedures are integrated with any necessary additional procedures peculiar to the project in order to attain specified quality objectives. Application of various quality tools, methods, and principles at different stages of construction projects is necessary to make the project qualitative, economical, and meet the owner needs/specification requirements.

Based on contract requirements, the contractor prepares his quality control plan and submits the same to consultant for their approval. Figure 8.32 illustrates the logic flow for CQCP. This plan is followed by the contractor to maintain the project quality.

The quality control plan outlines the procedures to be followed during the construction period to attain the specified quality objectives of the project fully complying with the contractual and regulatory requirements. The plan provides the mechanism to achieve the specified quality by identifying the procedures, control, instructions, and tests required during the construction process to meet the owner's objectives.

Table 8.24 illustrates contents of CQCP. However, the contractor has to take into consideration the requirements listed under contract documents depending on the nature and complexity of the project.

8.6.5 Develop Management Plans

Apart from the construction schedule, S-Curve, and CQCP, the contractor has to submit mainly the following plans to PMC/Supervisor (consultant) for their review and approval:

1. Stakeholder Management Plan
2. Resource Management Plan
3. Communication Management Plan
4. Risk Management Plan
5. Contract Management Plan
6. HSE Management Plan

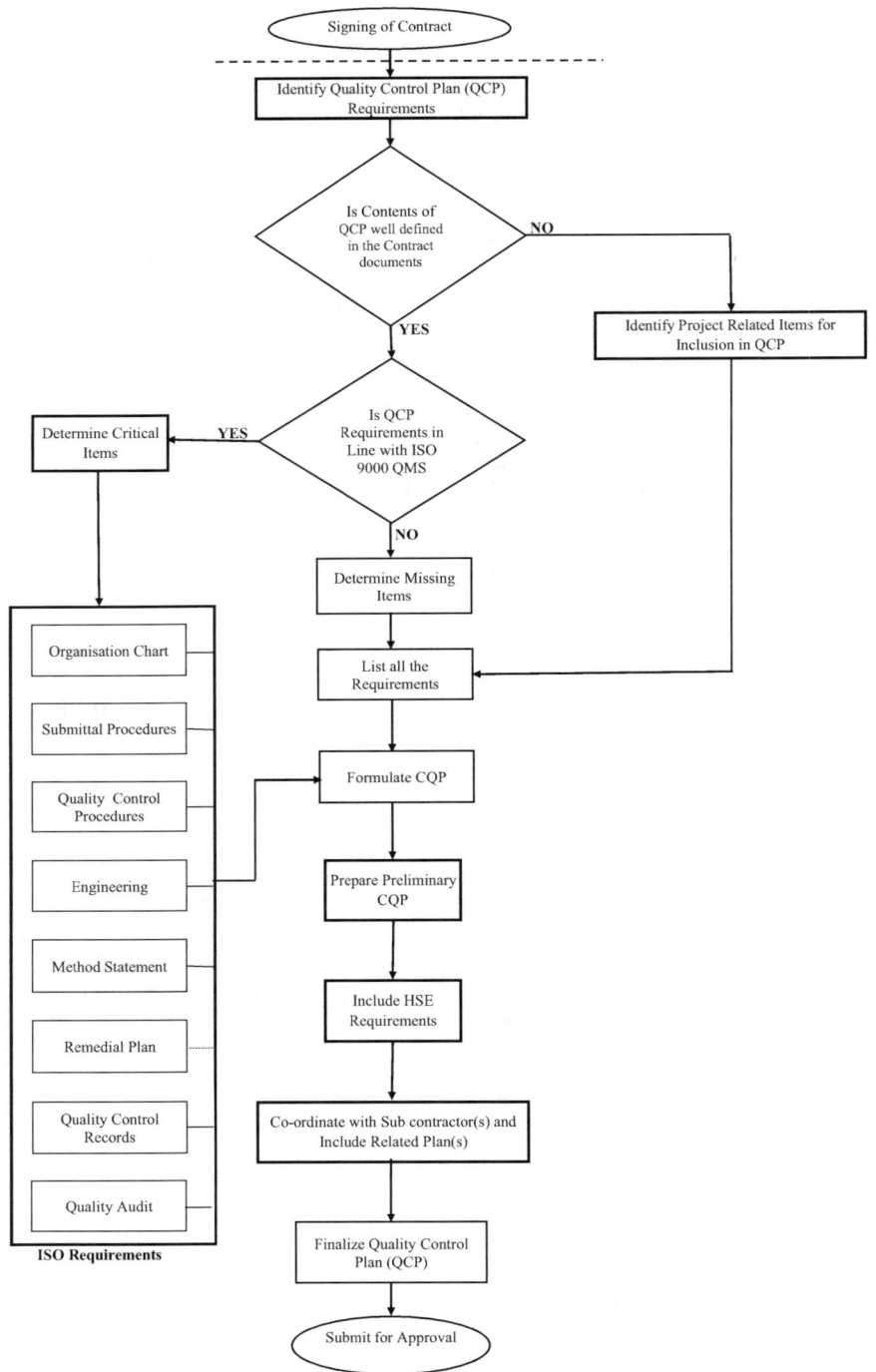

FIGURE 8.32 Logic flowchart for development of CQCP.

TABLE 8.24
Contents of Contractor's Quality Control Plan

Section	Topic
1	Introduction
2	Description of project
3	Quality control organization
4	Qualification of QC staff
5	Responsibilities of QC personnel
6	Procedure for submittals
7	Quality control procedure
	7.1 Detailed engineering
	7.1.1 Accuracy and correctness of detailed engineering
	7.1.2 Quality of drawings
	7.1.3 Check specifications
	7.1.4 Check contract documents
	7.1.5 Number of drawings
	7.2 Procurement
	7.2.1 Vendor selection
	7.2.2 Cost of material, equipment, system
	7.2.3 Off-site manufacturing, inspection, and testing
	7.2.4 Inspection of material received at site
	7.2.5 Material storage and handling
	7.3 Construction
	7.3.1 Inspection of site activities (checklists)
	7.3.2 Inspection of installed equipment (checklists)
	7.3.3 Inspection and testing procedure for systems
	7.3.4 Procedure for laboratory testing of material
	7.3.5 Corrective and Preventive action
	7.3.6 Protection of works
8	Method statement for various installation activities
9	Project-specific quality procedures
10	Owner-specific processes
11	Company's quality manual and procedure.
12	Periodical testing of construction equipment
13	Subcontractor's quality plan
	13.1 Subcontractor's QA/QC personnel and resources
14	Vendor's quality plan
15	Quality control records
16	Resource management
17	Communication and meetings
18	Risk management
19	Health, safety, and environment
20	Testing, Commissioning, and handover
	20.1 Inspection and testing plan
	20.2 Inspection and testing of major equipment
	20.3 Performance test
21	Quality updating program
22	Quality auditing program

8.6.5.1 Develop Stakeholder Management Plan

In order to run a successful project, it is important to address the needs of project stakeholders effectively predicting how the project will be affected and how the stakeholders will be affected. Stakeholder management planning is a process to develop stakeholder engagement plan depending on the roles and responsibilities of the stakeholders and their needs, expectations, and influence on the project. Table 8.25 shows an example matrix for site administration of a construction project. A distribution matrix based on the interest and involvement of each stakeholder is prepared, and the appropriate documents are sent for their action/information as per the agreed upon requirements.

8.6.5.2 Develop Resource Management Plan

Resource management in construction mainly relates to the management of following processes:

1. Human resources (Project Teams)
    i. Project owner team (Project Manager)
    ii. PMC
    iii. Designer (consultant)
2. Construction resources (contractor)
    i. Contractor's staff
        – Design team
        – Construction core team
    ii. Manpower for construction works
    iii. Material for project
        – Equipment
        – Materials
        – Systems

In most construction projects, contractor is responsible to engage subcontractors, specialist installers, suppliers, arrange for materials, equipment, construction tools, and all types of human resources to complete the project as per the contract documents and to the satisfaction of owner/owner's appointed supervision team. Workmanship is one of the most important factors to achieve the quality in construction; therefore, it is required that the construction workforce is fully trained and have full knowledge of all the related activities to be performed during the construction process.

Once the contract is awarded, the contractor prepares a detailed plan for all the resources needed to complete the project.

Contract documents normally specify a list of minimum number of core staff to be available on site during the construction period. Absence of any of these staff may result in penalty to be imposed on contractor by the owner.

Contractor's human resources mainly consist of two categories:

1. Contractor's own staff and workers
2. Subcontractor's staff and workers

TABLE 8.25
Matrix for Site Administration and Communication

Serial Number	Description of Activities	Contractor	Consultant/ PMC	Owner
1	General	-	-	P
	Notice to proceed	P	R	A
	Bonds and guarantees	-	P	A
	Consultant staff approval	P	R/B	A
	Contractor's staff approval	P	R	A
	Payment guarantee	P	R	A
	Master schedule	-	P	A
	Stoppage of work	-	P	A
	Extension of Time	P	R	A
	Deviation from contract documents			
	Material			
	Cost			
	Time			
2	Communication			
	2.1 General correspondence	P	P	P
	2.2 Job site instruction	D	P	C
	2.3 Site works instruction	D	P/B	A
	2.4 RFI	P	A	C
	2.5 Request for modification	P	B	A
3	Submittals			
	3.1 Subcontractor	P	B/R	A
	3.2 Materials	P	A	C
	3.3 Shop drawings	P	A	C
	3.4 Staff approval	P	B	A
	3.5 Premeeting submittals	P	D	C
4	Plans and programs			
	4.1 Construction schedule	P	R	C
	4.2 Submittal logs	P	R	C
	4.3 Procurement logs	P	R	C
	4.4 Schedule update	P	R	C
5	Quality			
	6.1 Quality control plan	P	R	C
	6.2 Checklists	P	D	C
	6.3 Method statements	P	A	C
	6.4 Mock-up	P	A	B
	6.5 Samples	P	A	B
	6.6 Remedial notes	D	P	C
	6.7 Nonconformance report	D	P	C
	6.8 Inspections	P	D	C
	6.9 Testing	P	A	B
6	Site safety			

(*Continued*)

TABLE 8.25
Matrix for Site Administration and Communication

Serial Number		Description of Activities	Contractor	Consultant/ PMC	Owner
	7.1	Safety Program	P	A	C
	7.2	Accident Report	P	R	C
7		Monitor and control			
	5.1	Progress	D	P	C
	5.2	Time	D	P	C
	5.3	Payments	P	R/B	A
	5.4	Variations	P	R/B	A
	5.5	Claims	P	R/B	A
8		Meetings			
	8.1	Progress	E	P	E
	8.2	Coordination	E	P	C
	8.3	Technical	E	P	C
	8.4	Quality	P	C	C
	8.5	Safety	P	C	C
	8.6	Close-out	-	P	
9		Reports			
	9.1	Daily report	P	R	C
	9.2	Monthly report	P	R	C
	9.3	Progress report	-	P	A
	9.4	Progress photographs	-	P	A
10		Close-out			
	10.1	Snag list	P	P	C
	10.2	Authorities approvals	P	C	C
	10.3	As-built drawings	P	D/A	C
	10.4	Spare parts	P	A	C
	10.5	Manuals and documents	P	R/B	A
	10.6	Warranties	P	R/B	A
	10.7	Training	P	C	A
	10.8	Hand over	P	B	A
	10.9	Substantial completion certificate	P	B/P	A

P- Prepare/initiate
B- Advise/assist
R- Review/comment
A- Approve
D- Action
E- Attend
C- Information

Source: Abdul Razzak Rumane. (2010). *Quality Management in Construction Projects.* Reprinted with permission from Taylor & Francis Group

Quality Management

The main contractor has to manage all these personnel by

1. assigning the daily activities
2. observing their performance and work output
3. daily attendance
4. safety during construction process

Figure 8.33 illustrates the contractor's planned manpower chart for the construction project.

Likewise, the contract documents specify that a minimum equipment set is to be available on site during the construction process to ensure smooth operation of all the construction activities. They are normally listed in the contract documents. These are as follows:

- Tower crane
- Mobile crane
- Normal mixture
- Concrete mixing plant
- Dump trucks
- Compressor
- Vibrators
- Water pumps
- Compactors
- Concrete pumps
- Trucks
- Concrete trucks
- Diesel generator sets

Figure 8.34 illustrates the equipment schedule, which lists the equipment the contractor has to make available for major construction projects.

In EPC-type of construction projects, the contractor is responsible for procurement of material, equipment, and systems to be installed on the project. The specifications are based on approved detailed engineering. The contractor also prepares a procurement log based on the project completion schedule. Figure 8.35 illustrates the material management process for construction projects.

8.6.5.3 Develop Communication Management Plan

Construction project has involvement of many stakeholders. The project team must provide timely and accurate information to identified stakeholders that will receive communications. Effective communication is one of the most important factors contributing to the success of the project. For smooth flow of construction process activities during construction phase, proper communication and submittal procedure need to be established between all the concerned parties at the beginning of the construction activities.

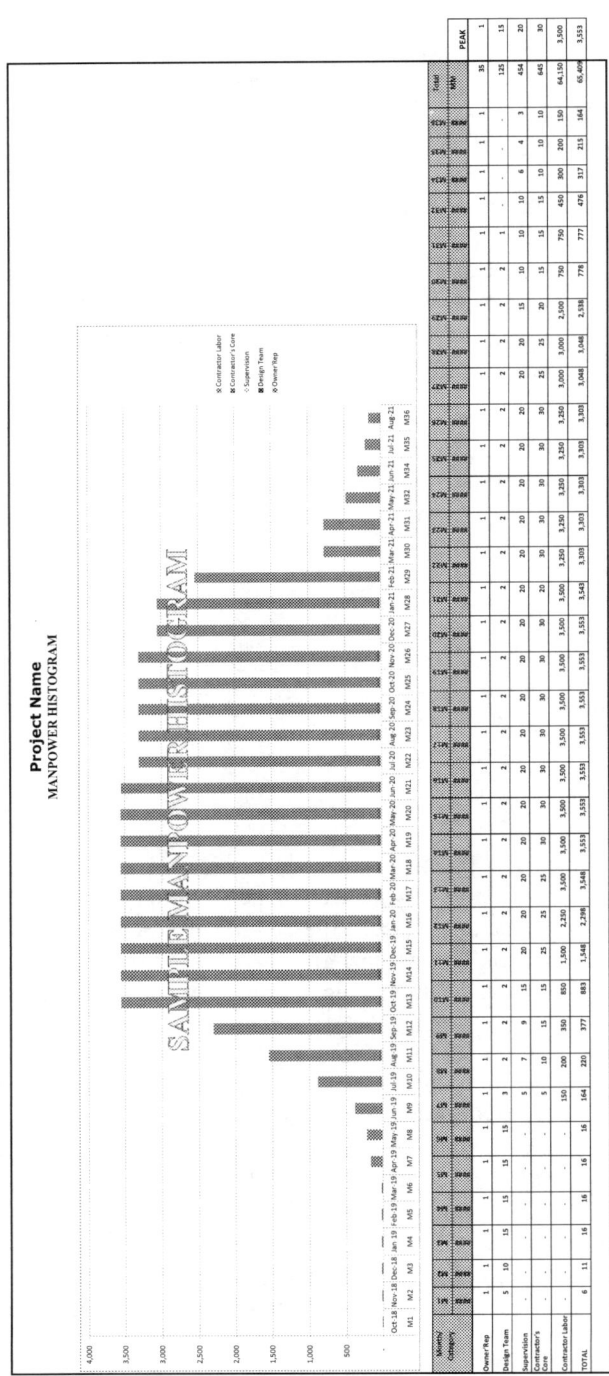

FIGURE 8.33 Project manpower histogram.

EQUIPMENT LIST AND UTILIZATION SCHEDULE FOR MAJOR CONSTRUCTION PROJECT

NUMBERS PER MONTH

Sr. No.	Item Description	Unit	2018 AUG	2018 SEP	2018 OCT	2018 NOV	2018 DEC	2019 JAN	2019 FEB	2019 MAR	2019 APR	2019 MAY	2019 JUN	2019 JUL	2019 AUG	2019 SEP	2019 OCT	2019 NOV	2019 DEC
1	Tower Crane	Nos.	_	_	4	9	9	9	9	9	9	9	9	9	9	9	9	9	9
2	Construction Hoist	Nos.						_	_	_	_	_	_	_	_	_	_	_	_
3	Mobile Crane	Nos.	_	_	1	1	2	2	2	4	4	4	4	4	4	4	4	4	4
4	Bulldozer	Nos.	_	3	4	4	2	2	2	_	_	_	_	_	_	_	_	_	_
5	Dump Truck	Nos.	_	4	4	4	4	4	4	_	_	_	_	_	_	_	_	_	_
6	Excavator	Nos.	_	2	2	2	2	2	2	2	2	_	_	_	_	2	_	_	_
7	JCB	Nos.	_	1	4	4	4	2	2	_	_	_	_	_	_	_	_	_	_
8	Concrete Pump	Nos.	_	_	2	6	6	6	6	4	4	4	4	4	4	4	4	4	4
9	Transit Mixer	Nos.	_	_	10	30	30	30	30	30	15	15	15	15	15	15	15	15	15
10	Fork Lift	Nos.	_	_	2	2	2	2	2	2	2	2	2	2	2	2	2	2	2
11	Trailer	Nos.	_	_	2	4	4	4	4	4	2	2	2	2	2	2	2	2	2
12	Total Station	Nos.	2	2	2	2	2	2	2	2	2	2	2	2	2	2	2	2	2
13	Theodoliite	Nos.	2	2	2	2	2	2	2	2	2	2	2	2	2	2	2	2	2
14	Level Instrument	Nos.	4	4	4	4	4	4	4	4	4	4	4	4	4	4	4	4	4
15	Precision Level	Nos.	2	2	2	2	2	2	2	2	2	2	2	2	2	2	2	2	2
16	Bar Bending Machine	Nos.	_	_	4	6	6	6	6	6	6	6	6	6	6	6	6	6	6
17	Bar Cutting Machine	Nos.	_	_	4	6	6	6	6	6	6	6	6	6	6	6	6	6	6
18	Bar Screw - Tread	Nos.	_	_	2	2	2	2	2	2	2	2	2	2	2	2	2	2	2
19	Diesel Generator	Nos.	_	_	4	4	4	4	4	4	4	4	4	4	4	4	4	4	4
20	Compactor	Nos.	_	_	2	2	2	2	2	2	2	2	2	2	2	2	2	2	2
21	Crawler Compactor	Nos.	_	_	2	2	2	2	2	2	2	2	2	2	2	2	2	2	2
22	High Pressure Pump	Nos.	_	_	2	2	2	2	2	2	2	2	2	2	2	2	2	2	2
23	Manual Hoist	Nos.	_	_	5	5	5	5	5	5	5	5	5	5	5	5	5	5	5
24	ConcreteVibrator	Nos.	_	_	25	25	25	25	25	25	25	25	25	25	25	25	25	25	25
25	Mortar Mixer	Nos.	_	_	_	_	_	_	_	_	_	_	_	_	6	6	6	6	6
26	Welding Machine	Nos.	_	_	_	4	4	4	4	4	4	50	50	50	50	50	50	50	50
27	Circular Saw	Nos.	_	_	6	6	6	6	6	6	6	6	6	6	6	6	6	6	6
28	Pipe Painting Machine	Nos.	_	_	_	_	_	_	_	_	_	_	2	2	2	2	2	2	2
29	Main Ditribution Panel	Nos.	_	_	2	2	2	2	2	2	2	2	2	2	2	2	2	2	2
30	Duct Fabrication Machine	Nos.	_	_	_	_	_	_	_	_	_	_	_	_	_	2	2	2	2
31	Electric Hammer	Nos.	_	_	4	4	4	4	4	4	4	4	4	4	4	4	12	12	12
32	Power Float	Nos.	_	_	16	16	16	16	16	16	16	16	16	16	16	16	16	16	16
33	Angle Grinder	Nos.	_	_	_	_	4	4	4	4	4	4	4	4	4	4	4	4	4
34	Electric Hand Drill	Nos.	_	_	12	12	12	12	12	12	12	12	12	12	12	12	12	12	12
35	Lighting Equipment	Nos.	_	_	5	5	10	10	10	10	10	10	10	10	10	10	10	10	10
36	Air Jack	Nos.	_	_	_	_	_	_	2	2	2	2	2	2	2	2	6	6	6
37	Angle Saw	Nos.	_	_	_	6	6	6	6	6	6	6	6	6	6	6	6	6	6
	Total		10	20	140	185	193	191	193	185	168	212	214	214	220	224	234	234	234

Sr. No.	Item Description	2020 JAN	2020 FEB	2020 MAR	2020 APR	2020 MAY	2020 JUN	2020 JUL	2020 AUG	2020 SEP	2020 OCT	2020 NOV	2020 DEC	2021 JAN	2021 FEB	2021 MAR	2021 APR	2021 MAY	2021 JUN	2021 JUL	2021 AUG	2021 SEP	2021 OCT	2021 NOV	2021 DEC
1	Tower Crane	9	9	9	9	8	8	8	8	8	8	8	4	4	_	_	_	_	_	_	_	_	_	_	_
2	Construction Hoist	4	9	9	9	9	9	9	9	9	9	9	9	9	9	9	9	9	9	9	9	9	9	9	9
3	Mobile Crane	4	4	4	4	4	4	4	4	4	4	4	4	4	4	4	4	4	4	4	4	4	4	4	4
4	Bulldozer	_	_	_	_	_	_	_	_	_	_	_	_	_	_	_	_	_	_	_	_	_	_	_	_
5	Dump Truck	4	4	4	_	_	_	_	_	_	_	_	_	_	_	_	_	_	_	_	_	_	_	_	_
6	Excavator	_	_	_	_	_	_	_	_	_	_	_	_	_	_	_	_	_	_	_	_	_	_	_	_
7	JCB	2	2	2	_	_	_	_	_	_	_	_	_	_	_	_	_	_	_	_	_	_	_	_	_
8	Concrete Pump	4	4	4	4	4	4	4	4	4	4	4	4	4	_	_	_	_	_	_	_	_	_	_	_
9	Transit Mixer	15	15	15	15	15	15	15	15	15	15	15	15	15	_	_	_	_	_	_	_	_	_	_	_
10	Fork Lift	2	2	2	2	2	2	2	2	2	2	2	2	2	2	2	2	2	2	2	2	2	2	2	2
11	Trailer	2	2	2	2	2	2	2	2	2	2	2	2	2	2	2	2	2	2	2	2	2	2	2	2
12	Total Station	2	2	2	2	2	2	2	2	2	2	2	2	2	2	2	2	2	2	2	2	2	2	2	2
13	Theodoliite	2	2	2	2	2	2	2	2	2	2	2	2	2	2	2	2	2	2	2	2	2	2	2	2
14	Level Instrument	4	4	4	4	4	4	4	4	4	4	4	4	4	4	4	4	4	4	4	4	4	4	4	4
15	Precision Level	2	2	2	2	2	2	2	2	2	2	2	2	2	2	2	2	2	2	2	2	2	2	2	2
16	Bar Bending Machine	6	6	6	6	6	6	6	6	6	6	6	6	6	_	_	_	_	_	_	_	_	_	_	_
17	Bar Cutting Machine	6	6	6	6	6	6	6	6	6	6	6	6	6	_	_	_	_	_	_	_	_	_	_	_
18	Bar Screw - Tread	2	2	2	2	2	2	2	2	2	2	2	2	2	_	_	_	_	_	_	_	_	_	_	_
19	Diesel Generator	4	4	4	4	4	4	4	4	4	4	4	4	4	4	4	4	4	4	4	4	4	4	4	4
20	Compactor	4	4	4	4	2	2	2	2	2	2	2	2	2	_	_	_	_	_	_	_	_	_	_	_
21	Crawler Compactor	2	2	2	2	2	2	2	2	2	2	2	2	2	_	_	_	_	_	_	_	_	_	_	_
22	High Pressure Pump	2	2	2	2	2	2	2	2	2	2	2	2	2	2	2	2	2	2	2	2	2	2	2	2
23	Manual Hoist	5	5	5	5	5	5	5	5	5	5	5	5	5	5	5	5	5	5	5	5	5	5	5	5
24	ConcreteVibrator	25	25	25	25	25	25	25	25	25	25	25	25	25	_	_	_	_	_	_	_	_	_	_	_
25	Mortar Mixer	6	6	6	6	6	6	6	6	6	6	6	6	6	6	6	6	6	6	6	6	6	6	6	6
26	Welding Machine	50	20	20	20	20	50	50	50	50	50	50	50	20	20	20	20	20	20	20	20	20	20	20	20
27	Circular Saw	6	6	6	6	6	6	6	6	6	6	6	6	6	6	6	6	6	6	6	6	6	6	6	6
28	Pipe Painting Machine	2	2	2	2	2	2	2	2	2	2	2	2	2	2	2	2	2	2	2	_	_	_	_	_
29	Main Ditribution Panel	2	2	2	2	2	2	2	2	2	2	2	2	2	2	2	2	2	2	2	2	2	2	2	2
30	Duct Fabrication Machine	2	2	2	2	2	2	2	2	2	2	2	2	2	2	2	2	2	2	2	_	_	_	_	_
31	Electric Hammer	12	12	12	12	12	12	12	12	12	12	12	12	12	12	12	12	12	12	12	12	12	12	12	12
32	Power Float	16	16	16	16	16	16	16	16	16	16	16	16	16	16	16	16	16	16	16	_	_	_	_	_
33	Angle Grinder	4	4	4	4	4	4	4	4	4	4	4	4	4	4	4	4	4	4	4	4	4	4	4	4
34	Electric Hand Drill	12	12	12	12	12	12	12	12	12	12	12	12	12	12	12	12	12	12	12	12	12	12	12	12
35	Lighting Equipment	10	10	10	10	10	10	10	10	20	20	20	20	20	20	20	20	20	20	20	20	20	20	20	20
36	Air Jack	6	6	6	6	6	6	6	6	6	6	6	6	6	6	6	6	6	6	6	6	6	6	6	6
37	Angle Saw	6	6	6	6	6	6	6	6	6	6	6	6	6	6	6	6	6	6	6	6	6	6	6	6
	Total	246	221	221	215	212	242	242	242	252	252	252	248	218	152	152	152	152	152	152	132	132	132	132	132

FIGURE 8.34 Equipment schedule.

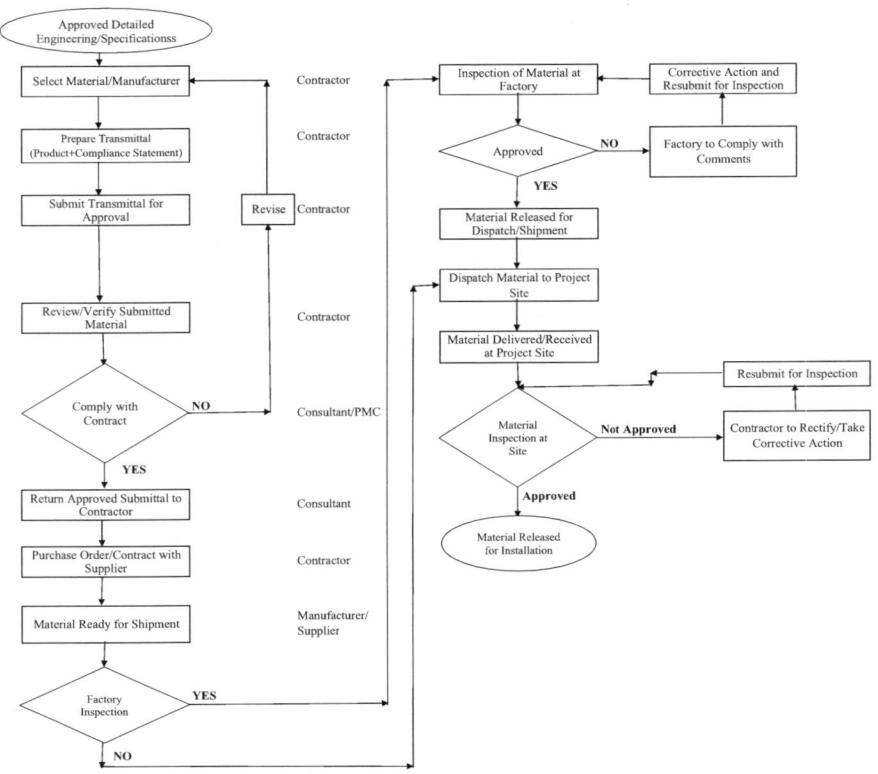

FIGURE 8.35 Material management process for construction project.

Table 8.26 illustrates an example guideline to prepare communication matrix for site administration during the construction phase, and Table 8.27 illustrates contents of communication management plan.

The communication between contractor and supervisor takes place through transmittal form. Table 8.28 lists various types of project control documents used during the construction project. The number of copies, distribution, and communication method are specified in the contract documents. These documents are normally transmitted by the contractor using the transmittal form. Figure 8.36 illustrates the sample transmittal form.

The correspondence between consultant and contractor is normally though letters or job-site instructions. Figure 8.37 is a job-site instruction form used by the consultant to communicate with the contractor.

8.6.5.4 Develop Risk Management Plan

The probability of occurrence of risk during the construction phase is very high compared to design phases. During the construction phase, uncertainty comes from various sources as this phase has involved various participants. Since the duration of the construction phase is longer than earlier phases, the contractor has to also consider

TABLE 8.26
Guidelines to Prepare Communication Matrix

Project Name

Name of Construction/Project Manager:	Name of Consultant:
Contractor Name:	Project Number:

Serial Number	Type of Document	Originator	Receiver(s)	Purpose	Frequency	Method	Responsible Person for Action	Comments

occurrence of financial, economical, commercial, political, and natural risks. The contractor has to develop risk management plan. Risk management plan identifies how risk associated with the project will be identified, analyzed, managed, and controlled. Risk management is an integral part of project management as the risk is likely to occur at any stage of the project. Therefore, the risk has to be continually monitored and response actions to be taken immediately. The risk management plan outlines how risk activities will be performed, recorded, and monitored throughout the life cycle of the project. It is intended to maximize the positive impact for the benefit of the project and decrease/minimize or eliminate the impact of events adverse to the project. The risk management must commence early in the project development stage (study stage) and proceed as the project evolves and more and more information about the project is available. The project plan should

- define risk management strategy/approach
- define project objectives, goals related to risk management
- identify risk owner and team members
- define risk decisions
- detail about risk resources

TABLE 8.27
Contents of Contractor's Communication Management Plan

Section	Topic
1	Introduction
	1.1 Project description
2	Stakeholders
	2.1 Project team directory
	2.2 Project organization chart
	2.3 Roles and responsibilities
	2.4 Stakeholders requirement
3	Communication methods
4	Communication management constraints
5	Communication matrix
6	Distribution of communication documents
	5.1 General correspondence
	5.2 Submittals
	5.3 Status reports
	5.4 Meetings
	5.5 Management plans
	5.6 Change orders
	5.7 Payments
6	Regulatory requirements
7	Communication plan update

TABLE 8.28
List of Project Control Documents

Serial Number	Document Name	
I	Administrative	
	I-1	Material entry permit
	I-2	Material removal permit
	I-3	Vehicular entry permit
	I-4	Site entry permit
	I-5	Visitor entry permit
	I-6	Municipality permit
	I-7	Request for overtime
	I-8	Theft and damage report
	I-9	Performance bonds
	I-10	Advance payment guarantee
	I-11	Insurance
	I-12	Accident report
	I-13	Sample tag

(Continued)

TABLE 8.28 (*Continued*)
List of Project Control Documents

Serial Number		Document Name
II		Contracts related
	II-1	Notice to proceed
	II-2	Job site instruction
	II-3	SWI
	II-4	Attachment to site works
	II-5	Request for staff approval
	II-6	Request for subcontractor approval
	II-7	Request for vendor approval
	II-8	Material delivered at site
	II-9	Variation order
	II-10	Attachment to variation order
	II-11	Baseline change request form
	II-12	Extension of time
	II-13	Suspension of work
	II-14	Attendees
	II-15	Minutes of meeting
	II-16	Transmittal for minutes of meeting
	II-17	Submittal form
III		Engineering submittal
	III-1	Master schedule
	III-2	Cost loaded schedule
	III-3	Engineering drawings
	III-4	Material approval
	III-5	Specification comparison statement
	III-6	Product data
	III-7	Product sample
	III-8	Work shop drawings
	III-8	Builders drawings
	III-10	Composite drawings
	III-11	Method statement
	III-12	RFI
	III-13	Request for modification
	III-14	Variation order (proposal)
	III-15	Request for alternative or substitution
IV		PCS reporting forms
	IV-1	Contractor's submittal status Log E-1
	IV-2	Contractor's procurement Log E-2
	IV-3	Contractor's shop drawing status log
	IV-4	Daily progress report
	IV-5	Weekly progress report
	IV-6	Look ahead schedule
	IV-7	Monthly progress report

(*Continued*)

TABLE 8.28 (*Continued*)
List of Project Control Documents

Serial Number		Document Name
	IV-8	Progress photographs
	IV-9	Daily checklist status
	IV-10	Progress payment request
	IV-11	Payment certificate
	IV-12	Submittal Schedule
	IV-13	Schedule update report
V		Management plans
	V-1	Quality control plan
	V-2	SMP
	V-3	Environmental protection plan
VI		Quality control forms
	VI-1	Checklist (request for inspection)
	VI-2	Checklist for form work
	VI-3	Notice for daily concrete casting
	VI-4	Checklist for concrete casting
	VI-5	Quality control of concreting
	VI-6	Report on concrete casting
	VI-7	Notice for testing at lab
	VI-8	Concrete quality control form
	VI-9	Checklist for process work
	VI-10	Checklist for piping work
	VI-11	Checklist for instrumentation work
	VI-12	Checklist for utility work
	VI-13	Checklist for steel fabrication work
	VI-14	Checklist for detection and protection works
	VI-15	Checklist for fire-fighting work
	VI-16	Checklist for architectural work
	VI-17	Checklist for civil work
	VI-18	Checklist for mechanical work
	VI-19	Checklist for HVAC work
	VI-20	Checklist for electrical work
	VI-21	Checklist for elevator work
	VI-22	Checklist for low-voltage system work
	VI-23	Checklist for external work
	VI-24	Checklist for landscape
	VI-25	Remedial note
	VI-26	Nonconformance/compliance report
	VI-27	Material inspection report
	VI-28	Safety violation notice
	VI-29	Notice of commencement of new activity
	VI-30	Removal of rejected material
	VI-31	Testing and commissioning

(*Continued*)

TABLE 8.28 (*Continued*)
List of Project Control Documents

Serial Number	Document Name
VII	Closeout forms
VII-1	As-built drawings
VII-2	Substantial completion certificate
VII-3	Handing-over certificate
VII-4	Taking-over certificate
VII-5	Manuals
VII-6	Handing over of spare parts
VII-7	Defect liability certificate

Source: Modified from: Abdul Razzak Rumane. (2013). *Quality Tools for Managing Construction Projects.* Reprinted with permission from Taylor & Francis Group.

- include risk management process
    i. methods of risk identification
    ii. methods of risk assessment
    iii. level of risk
    iv. response to risk
    v. management of risk
    vi. control of risk
- process of integrating risk management activities into project scope, schedule, cost, and quality
- document and record risks
- communicate procedure for risk reporting
- update risk management plan

8.6.5.5 Develop Contract Management Plan

Contract management in construction projects is an organizational method, process, and procedure to obtain the required construction project/products. It includes the process to acquire construction project completely with all the related material, equipment, system, and services from outside contractors/companies to the satisfaction of the owner/client/end user.

Contract management in construction project involves the following:

- Identification of
    i. What are the services in-house available
    ii. What services to be procured from outside agencies/organizations
    iii. How to procure (direct contract and competitive bidding)
    iv. How much to procure
    v. How to select a supplier/contractor
    vi. How to arrive at appropriate price, terms, and conditions

	Project Name		
	Consultant Name		
	SUBMITTAL TRANSMITTAL FORM		
Contractor Name :			
Contract No. :			
To.	Resident Engineer		
Transmittal No.:		Date :	

Submittal Type:		Action Requested:	
ED	Engineering Drawings	1	For Approval
DG	Shop Drawings	2	For Review and Comment
SK	Sketches	3	For Information
PR	Material/Product/System	4	For Construction
MD	Manufacturer's Data	5	For Incorporation Within the Design
SM	Sample	6	For Costing
MM	Minutes of Meeting	7	For Tendering
RP	Reports		
LG	Logs		
OT	Others (please specify)		

We are sending herewith the following:

ENCLOSURES

Item	Qty	Ref. No.	Description	Type	Action

Comments:
_____
_____
_____

Issued by: Received by:

Signature: Signature:

Date: Date:

FIGURE 8.36 Transmittal form.

- Signing of contract
- Timely delivery
- Receiving the right type of material/system
- Timely execution of work
- Inspection of work to maintain quality of the project
- Completion of project within agreed upon schedule
- Completion of project within agreed upon budget
- Documenting reports and plans

Quality Management

	Project Name	
	Consultant Name	

JOB SITE INSTRUCTION (JSI)

CONTRACTOR: _____   JSI No. : _____

CONTRACT No.: _____   DATE : _____

The work shall be carried out in accordance with the Contract Documents without change in Contract Sum or Contract Time. Proceeding with the work in accordance with these instructions indicates your acknowledgement that there will be no change in the Contract Sum or Contract Time.

Subject:

SAMPLE FORM

ATTACHMENTS: (List attached documents that support description.)

Signed: _____   Received by Contractor : _____
Resident Engineer   Date:

Distribution: ☐ Owner   ☐ Consultant (Supervision)/PMC   ☐ Contractor

FIGURE 8.37 Job-site instruction.

The administration of contract is the process of formal governance of contract and changes to the contract document. It is concerned with managing contractual relationship between various participants to successfully complete the facility to meet owner's objectives. It includes tasks such as

- Administration of project requirement
- Administration of project team members
- Communication and Management reporting

- Execution of contract
- Monitoring contract performance (scope, cost, schedule, quality, risk)
- Inspection and quality
- Variation order process
- Making changes to the contract documents by taking corrective action as needed
- Payment procedures

It is required that the contract administration procedure is clearly defined for the success of the contract and that the parties to the contract understand who does what, when, and how. The following are some typical procedures that should be in place for management of the contract management activities:

1. Contract document maintenance and variation
2. Performance review system
3. Resource management and planning
4. Management reporting
5. Change control procedure
6. Variation order procedure
7. Payment procedure

Table 8.29 lists the contents of contract management plan.

8.6.5.6 Develop HSE Management Plan

The construction industry has been considered to be dangerous for a long time. The nature of work at site always presents some dangers and hazards. There are a relatively high number of injuries and accidents at construction sites. Safety represents an important aspect of construction projects. In construction projects, the requirements to prepare Safety Management Plan (SMP) by the contractor are specified under contract documents. The contractor has to submit the plan for review and approval by supervisor/consultant during mobilization stage of the construction phase. The following are the guidelines normally specified in the contract documents which are to be considered by the contractor to establish HSE management plan:

1. Project scope detailing description of project and safety requirements
2. Safety policy statement documenting the contractor's/subcontractor's commitment and emphasis on safety
3. Regulatory requirements about safety
4. Roles and responsibilities of all individuals involved
5. Safety management of different activities
6. Site communication plan detailing how safety information will be shared
7. Hazard identification, risk assessment, and control
8. Emergency evacuation plan
9. Accident reporting system
10. Accident investigation to document root causes and determine corrective and preventive actions

TABLE 8.29
Contents of Contract Management Plan

Serial Number	Topics
1	Contract summary, deliverables, and scope of work
2	Type of contract
3	Contract schedule
4	Contract cost
5	Project team members with roles and responsibilities
6	Core staff approval procedure
7	Contract communication matrix/Management reporting
8	Coordination process
9	Liaison with regulatory authorities
10	Engineering drawings submission/approval process
11	Vendor selection process
12	Material/Product/System review/approval process
13	Shop drawing review/approval process
14	Project monitoring and control process

15	Contract change control process
	   a. Scope
	   b. Material
	   c. Method
	   d. Schedule
	   e. Cost

16	Review of variation/change requests
17	Project hold-up areas
18	Quality of performance
19	Inspection and acceptance criteria
20	Risk identification and management
21	Progress payment process
22	Safety management
23	Claims, disputes, conflict and litigation resolution
24	Contract documents and records
25	Post contract liabilities
26	Contract closeout and independent audit.

Source: Abdul Razzak Rumane. (2013). *Quality Tools for Managing Construction Projects*. Reprinted with permission from Taylor & Francis Group.

11. Measures for emergency situations
12. Plant, equipment maintenance, and licensing
13. Routine inspections
14. Continuous monitoring and regular assessment
15. Health surveillance

16. System feedback and continuous improvements
17. Safety assurance measures
18. Evaluation of subcontractor's safety capabilities
19. Site neighborhood characteristics and constraints
20. System education and training
21. Safety audit
22. Documentation
23. Records
24. Procedure for Project HSE Review
25. System update

The following are the main responsibilities of a Safety Engineer/Officer:

1. Conducting safety meetings
2. Monitoring on-the-job safety
3. Inspecting the work and identifying hazardous areas
4. Initiating a safety awareness program
5. Ensuring availability of first aid and emergency medical services per local codes and regulations
6. Ensuring that personnel are using protective equipment such as hard hat, safety shoes, protective clothing, life belt, and protective eye coverings
7. Ensuring that the temporary firefighting system is working
8. Ensuring that work areas are free from trash and hazardous material
9. Housekeeping

8.6.6 Execute Project Works

The following are the contract requirements as per the scope of project work to be executed by the contractor:

1. Development of detailed engineering
2. Material procurement as per specified codes and standards from the approved vendors
3. Construction the facility/project as per the specifications to produce required products

8.6.6.1 Develop Detailed Engineering

Detailed engineering helps ensure project quality through several activities. These include

- meeting the owner's requirements and the project goals and objectives
- developing project activity plan for the project
- estimating accurately the hours of effort and costs involved in achieving a quality project
- developing a realistic schedule with appropriate milestones to confirm progress

- building flexibility into the project baseline to allow for changes and update the schedule and budget
- monitoring and controlling of project progress

Detailed engineering activities are similar, although more in-depth than the design activities in the FEED. The size, shape, levels, performance characteristics, technical details, and requirements of all the individual components are established and integrated into the detailed engineering.

The contractor (designer) has to include the following elements while developing detailed engineering:

- Contract/specification requirements
- Plans
- Elevations
- Levels
- Axis
- Grids
- Sizing
- Zoning
- Constructability
- Technical capability
- Functional capability
- Operational capability
- Engineering economy
- Energy conservation requirements
- Sustainability
- Product quality
- Reliability
- Operational safety
- Safety and environmental requirements
- Calculation for accuracy of the engineering
- Schematics, wherever required
- Regulatory requirements
- Supportability during maintenance (maintainability)

These elements have to be considered as a base for the development of detailed engineering to meet customer requirements and will help achieve the qualitative project.

8.6.6.1.1 Drawings

The contractor (designer) has to develop detailed engineering based on the criteria discussed in Section 8.6.5.1. Following is the list of major areas for the development of detailed engineering for oil and gas project:

1. Plot Plan
    1.1 General Plot plan
    1.2 Unit Plot Plan

2. Process
    2.1   Process Design Basis and Criteria
    2.2   Process Description
    2.3   Shutdown Philosophy
    2.4   Operation Philosophy
    2.5   Flare and Venting Philosophy
    2.6   Heat and Material Balance
    2.7   Process and Instrumentation Diagrams
    2.8   Process Flow Diagram
    2.9   Process Automation Diagram
    2.10 Equipment Selection Criteria
    2.11 Process Equipment Datasheet
    2.12 Heat Exchanger's Thermal Design
    2.13 Cause and Effect Diagram
3. Process SMP
4. Piping
    4.1   Piping Design Basis
    4.2   Piping General Arrangement Drawing
    4.3   Piping Isometric
    4.4   Piping and Instrumentation Diagram
    4.5   Piping Routing Design
    4.6   Piping System Requirements
    4.7   Pipe Racks/Supports
5. Pumping Station
6. Mechanical
    6.1   Mechanical Design Basis
    6.2   Storage Tank Design
    6.3   Equipment Mechanical Design
7. Utility Flow Diagram
    7.1   Utility Balance Study
    7.2   Underground Piping Layout
8. Civil
    8.1   Civil and Structural Design Basis
    8.2   Grading Plan
    8.3   Equipment Foundation
    8.4   Septic Tanks
    8.5   Steel Structure Design
    8.6   Loading Platform
    8.7   Shelter
9. Electrical
    9.1   Electrical Design Basis
    9.2   LV Switchgear
    9.3   HV Switchgear
    9.4   Cable Schedule
    9.5   Earthing/Grounding System
    9.6   Variable Frequency Drives

Quality Management

10. Electrical Substation
11. Instrumentation and Control
    11.1 Instrumentation and Control Design Basis
    11.2 Overall Instrumentation System
    11.3 Control Rooms
12. Inline Valve Datasheet
    12.1 Control Valves
13. Equipment Handling Layout
14. Tie-in Drawing
15. Fire and Gas Detection System
    15.1 Fire and Gas System Layout Drawing
16. Fire-fighting system
    14.1 Fire Water Distribution Layout
17. Supervisory Control System
18. Leak Detection System
19. Cathodic Protection System
20. Corrosion monitoring system
21. Security and Access Control System
22. ICT
23. Building
    23.1 Building Architectural Drawings
    23.2 Structural Drawings
    23.3 MEP Drawings
    23.4 Drainage System
    23.5 Fire Suppression System
    23.6 Elevator/Conveying System
    23.7 Low-Voltage System
    23.8 Schematic for Water Distribution System
    23.9 Schematic for Building Management System
    23.10 Schematic for Electrical Works
    23.11 Switchgear and Panels
24. Landscape
    24.1 Plants
    24.2 Irrigation System
25. External Works
    25.1 Street/Road
    25.2 Pedestrian Walkways
    25.3 Street Lighting
26. Parking
    26.1 Parking
    26.2 Parking Control System
27. Coating and Painting
28. Specifications
29. Calculations
    29.1 Pump Sizing
    29.2 Storage Tanks

- 29.3 Pressure Vessels
- 29.4 Exchangers
- 29.5 Effluent Water Transfer Pumps
- 29.6 Effluent Water Injection Pumps
- 29.7 Relief Valves
30. Datasheet
    - 30.1 Mechanical Equipment Data Sheet
    - 30.2 Control Valves
    - 30.3 Piping Class Data Sheet
    - 30.4 Instrument Datasheet
31. Lists
    - 31.1 Mechanical Material List
    - 31.2 Instrument List
    - 31.3 Tie-in Material
    - 31.4 Line List
32. Reports and Analysis
    - 32.1 Flare Analysis
    - 32.2 Corrosion and Risk Assessment
    - 32.3 Piping Stress Analysis
    - 32.4 Control System Report
    - 32.5 Environmental Impact Assessment
    - 32.6 HAZID
    - 32.7 HAZOP
    - 32.8 Geographical Survey
    - 32.9 Topographical Survey
    - 32.10 Pulsation Study for Pumps
    - 32.11 Dynamic Analysis

8.6.6.1.1.1 Perform Interdisciplinary Coordination The contractor (designer) has to perform interdisciplinary coordination of drawings among all the trades to ensure that all the requirements have been incorporated in the detailed engineering.

8.6.6.1.2 Detailed BOQ

The contracted BOQ is revised taking into consideration the detailed engineering.

8.6.6.1.3 Detail Specifications

The contracted specifications are revised taking into consideration the detailed engineering.

8.6.6.1.4 Estimate Project Schedule

The project schedule is estimated taking into consideration the detailed engineering. The contractor (designer) has to ensure that the revised schedule is as per the contracted schedule to complete the project.

8.6.6.1.5 Estimate Cost

The project cost is estimated taking into consideration the detailed engineering. The contractor (designer) has to ensure that the estimated cost does not exceed the contracted amount.

8.6.6.1.6 Manage Detailed Engineering Quality

In order to reduce errors and omission in detailed engineering, it is necessary to review and check the engineering development for quality assurance by the quality control personnel from the project team through itemized review checklists to ensure that detailed engineering drawings fully meet owner's objectives/goals and are technically and functionally compatible.

The designer has to plan quality (planning of detailed engineering work), perform quality assurance, and control quality for preparing detailed engineering. This will mainly consist of the following.

8.6.6.1.6.1 Plan Quality

- Identify requirements listed under contract documents
- Establish scope of work for preparation of detailed engineering drawings
- Determine number of drawings to be produced
- Identify quality standards and codes to be complied
- Establish design criteria
- Identify regulatory requirements
- Identify safety and environmental requirements
- Establish quality organization with responsibility matrix
- Develop engineering (drawings and documents) review procedure

8.6.6.1.6.2 Quality Assurance

- Collect data
- Investigate site conditions
- Prepare engineering drawings
- Ensure functional and technical compatibility
- Ensure the design is constructible
- Ensure operational objectives are met
- Ensure drawings are fully coordinated with all disciplines
- Ensure the design is cost-effective
- Ensure selected/recommended materials meet owner objectives
- Ensure that design fully meets the owner's objectives/goals

8.6.6.1.6.3 Control Quality

- Check the quality of engineering drawings
- Check accuracy and correctness of design
- Check for regulatory compliance

TABLE 8.30
Mistake Proofing Chart to Eliminate Detailed Engineering Errors

Serial Number	Items	Points to be Considered to Avoid Mistakes
1	Information	1. Contract documents 2. Contract SPECIFICATIONS 3. FEED bais 4. Client's preferred requirements 5. Data/information collection 6. Regulatory requirements 7. Codes and standards 8. Historical data 9. HAZID report 10. Organizational requirements
2	Mismanagement	1. Compare detailed engineering with actual requirements 2. Inter disciplinary coordination 3. Application of different codes and standards 4. Drawing size of different trades/specialist consultants
3	Omission	1. Review and check design with contract requirements 2. Review and check detailed engineering with client requirements 3. Review and check detailed engineering with Regulatory requirements 4. Review and check detailed engineering with codes and standards 5. Check for all required documents
4	Selection	1. Qualified team members 2. Available material 3. Nonhazardous material 4. Installation methods

- Check project schedule
- Check project cost
- Check interdisciplinary requirements
- Check the required number of drawings prepared drawing Table 8.30 illustrates mistake proofing chart to eliminate detailed engineering errors.

8.6.6.1.7 Estimate Project Resources

The contractor (designer) has to estimate the resources based on detailed engineering required to complete the project. While developing detailed engineering, more information in details is available to estimate manpower resources during the construction phase.

8.6.6.1.8 Manage Risks

The following are typical risks which normally occur during the development of detailed engineering:

Quality Management

- All the contract requirements are not taken into consideration while preparing the detailed engineering
- Scope of work-detailed engineering is not properly established and is incomplete
- The related project data and information collected are incomplete
- The related project data and information collected are likely to be incorrect and wrongly estimated
- Site investigations for existing conditions not verified
- Regulatory authorities' requirements are not taken into consideration
- Fire and safety considerations recommended by the authorities not incorporated in the detailed engineering
- Environmental consideration
- Incomplete detailed engineering drawings and related information
- Inappropriate construction method
- Conflict with different trades
- Interdisciplinary coordination not done
- Wrong selection of materials, equipment, and systems
- Undersize equipment selection
- Project schedule not updated as per detailed data and project assumptions
- Errors in detail cost estimation
- Number of drawings not as per contract requirements

The contractor (designer) has to take into account the above-mentioned risk factors while developing the detail design.

8.6.6.1.9 Manage HSE Requirements

The contractor (designer) has to take into account environmental impact analysis (EIA) and HAZID reports while developing the detail design.

8.6.6.1.10 Review Detailed Engineering

The success of a project is highly correlated with the quality and depth of the detailed engineering prepared during this phase. Coordination and conflict resolution are important factors during the development of design to avoid omissions and errors. The detailed engineering has to be reviewed for the accuracy of drawings, interdisciplinary coordination, and documents before these are submitted to the owner/consultant/project manager (PMC).

Table 8.31 illustrates major points for review of detailed engineering.

8.6.6.1.10.1 Review Drawings (Quality Check) The contractor (designer) has to check the drawings for formatting, annotation, and interpretation. Table 8.32 lists the items to be checked for quality check (correctness) of detailed engineering drawings.

8.6.6.1.10.2 Review Drawings (Coordination) Review all the drawings to ensure interdisciplinary coordination.

TABLE 8.31
Review of Detailed Engineering

Serial Number	Description	Yes	No	Notes
	A: Detailed Engineering (General)			
A.1	Does the detailed engineering support owner's project goals and objectives?			
A.2	Does the detailed engineering meets all the elements specified in contract documents?			
A.3	Whether the information/data collected is considered while preparing the detailed engineering?			
A.4	Whether owner's requirements updated and considered?			
A.5	Does the design meet all the performance requirements?			
A.6	Whether constructability has been taken care?			
A.7	Whether technical and functional capability considered?			
A.8	Whether functional capability of process is checked while preparing the detailed engineering?			
A.9	Whether design confirm with fire and egress requirements?			
A.10	Whether engineering/installation risks been identified, analyzed, and responses planned for mitigation?			
A.11	Whether health and safety requirements are considered while preparing detailed engineering?			
A.12	Whether environmental constraints considered?			
A.13	Does energy conservation is considered?			
A.14	Does sustainability considered while preparing detailed engineering?			
A.15	Whether Regulatory requirements are considered?			
A.16	Whether accessibility is considered?			
A.17	Does the detailed engineering meet ease of maintenance?			
A.18	Whether all the drawings are numbered?			
	B: Process Engineering			
B.1	Whether all the drawings related to process engineering prepared?			
	C: Mechanical			
C.1	Whether storage tank drawings prepared as per contract requirements?			
C.2	Whether steel fabrication drawings prepared as per contract requirements?			
	D: Electrical and Instrumentation			
D.1	Whether electrical and instrumentation drawings prepared as per contract requirements?			
	E: Utility			
E.1	Whether all the utility drawings prepared as per contract requirements?			
	F: Civil			
F.1	Whether civil drawings prepared as per contract requirements?			

(*Continued*)

TABLE 8.31 (*Continued*)
Review of Detailed Engineering

Serial Number	Description	Yes	No	Notes
	G: Buildings			
G.1	Whether architectural plans, elevations, finishes…etc., for building are drawn as per local building codes?			
G.2	Whether structural details for the building are properly shown?			
	H: Automation and Control			
H.1	Whether all the detailed engineering for instrumentation and control systems are prepared?			
H.2	Whether ICT drawings are prepared?			
H.3	Whether detailed engineering for all the low-voltage systems as per contract considered?			
	I: Landscape			
I.1	Whether landscape plans prepared			
I.2	Whether plants selected?			
I.3	Whether irrigation system layout prepared?			
	J: External			
J.1	Whether site layout plans showing roads, walkways, parking areas, prepared?			
J.2	Whether street lighting plans prepared?			
J.3	Whether project site boundaries are properly marked and demarcated?			
	K: Schedule			
K.1	Whether the schedule matches as per contract execution plan and is achievable?			
	L: Financial			
L.1	Whether estimated cost is equal or less than the contracted amount?			
	M: Drawings			
M.1	Whether drawings for all trades prepared as per contract requirements?			
	N: Reports			
N.1	Whether reports are prepared as per contract requirements objectives?			
	O: Submittals			
O.1	Whether numbers of sets prepared as per contract requirements?			

8.6.6.1.10.3   Review BOQ   The contractor has to prepare BOQ based on detailed engineering and compare with the contracted BOQ to evaluate if there are major changes in the BOQ.

8.6.6.1.10.4   Review Specifications   The contractor (designer) has to review the contracted specifications to find out any new activity is considered while developing detailed engineering.

TABLE 8.32
Quality Check for Detailed Engineering Drawings

Serial Number	Points to be Checked	Yes/No
1	Check for use of approved version of AutoCAD	
2	Check drawing for • Title frame • Attribute • North orientation • Key plan • Issues and Revision Number	
3	Client name and logo	
4	Designer (consultant name)	
5	Drawing title	
6	Drawing number	
7	Contract reference number	
8	Date of drawing	
9	Drawing scale	
10	Annotation; • Text size • Dimension style • Fonts • Section and elevation marks	
11	Layer standards including line weights	
12	Line weights, line type (continuous, dash, dot, etc.)	
13	Drawing continuation reference and Match line	
14	Plot Styles (CTB-color-dependent plot style tables)	
15	Electronic CAD File Name and Project location	
16	XREF (X reference) attachments (if any)	
17	Image reference (if any)	
18	Section references	
19	Symbols	
20	Legends	
21	Abbreviations	
22	General notes	
23	Drawing size as per contract requirements	
24	List of drawings	

Source: Abdul Razzak Rumane. (2013). *Quality Tools for Managing Construction Projects.* Reprinted with permission from Taylor & Francis Group.

8.6.6.1.11 Finalize Detailed Engineering

A final detailed engineering design is prepared incorporating the comments, if any, found during analysis and review of the drawings and documents for submission to the owner/client.

8.6.6.2 Develop Procurement Documents

In EPC type of contracting system, procurement of material, equipment, and system plays an important role due to high cost (economy) of material and its usage/performance and operation during the life cycle of the project. This includes the following:

- Preparing list of items to be procured
- Selection of vendor/supplier
- Approval of supplier by the Owner/Consultant/PMC
- Approval of material, equipment, and system
- Finalize purchase order/contract
- Progress follow-up production/manufacturing and delivery of material
- Inspection and testing method prior to dispatch
- Material receipt at site
- Inspection of material on site
- Storage of material
- Release for installation/construction

8.6.6.2.1 Prepare Material Procurement List

The following is the list of major items for procurement derived from detailed engineering and BOQ:

1. Pipes
2. Valves
3. Compressors
4. Pressure vessels
5. Heat exchangers
6. Pumps
7. Booster pumps
8. Water treatment pumps
9. Chemical pumps
10. Water pumps
11. Drainage pumps
12. Nitrogen generators
13. Steam generators
14. Filters
15. Scrubbers
16. Coolers
17. Separators
18. Accumulators
19. Flare and incinerator
20. Refrigeration
21. Generators
22. Instrumentation
23. Storage tanks

24. Fire-fighting system
25. Fire- and gas-detection system
26. Corrosion monitoring system
27. Building materials
28. Shelters
29. Painting and coating
30. Civil works-related material
31. Steel structure items
32. Ready mix concrete
33. Reinforcement material
34. HVAC
35. Plumbing
36. Electrical
37. Cables
38. LV switchgear
39. Panels
40. Distribution boards
41. VFD
42. Emergency generator
43. Transformer
44. HV Switchgear
45. UPS
46. Lightning protection
47. Elevator/conveying system
48. EOT cranes
49. Material handling equipment
50. Fire alarm system
51. Information and communication system
52. Security and safety system
53. Data security
54. Cathodic protection system
55. Process control and metering
56. Supervisory control system
57. Leak-detection system
58. Public address system
59. Audio visual system
60. Plants
61. Parking control

8.6.6.2.2 Identify Vendors

In most construction projects, the contractor is responsible for procurement of material, equipment, and systems to be installed on the project. The contractors have their own procurement strategies. While submitting, the bid contractor obtains the quotations from various vendors/subcontractors who are listed/registered with the company. However, it is likely that these vendors may not be included under the approved list of manufacturers/vendors in the contract specifications. Since the

Quality Management

product specifications for the oil and gas projects emphasize that certain items should be procured from specific manufacturers, it is required that the contractor follows proper vendor selection procedure to meet the contract requirements and also the organization's procurement strategy. Figure 8.38 illustrates the logic flowchart for the vendor selection procedure.

8.6.6.2.3 Manage Procurement Quality

The contract document identifies specific inspection and tests to be carried out on the material, equipment, and systems prior to installation/execution of works. An implicit assumption in the traditional quality control practices is the notion of an acceptable quality level, which is an allowable fraction of defective items. The level of inspection for different types of products is specified in the contract documents. An implicit assumption in the traditional quality control practices is the notion of an acceptable quality level, which is an allowable fraction of defective items. Material suppliers are also required to ensure zero defects in delivered goods. Suppliers with good records, certified, and approved should be selected for supply of materials, equipment, and systems for the project.

In order to minimize receipt of defective material, the contractor has to plan quality (planning of design work), perform quality assurance, and control quality for managing procurement activities. This will mainly consist of the following.

8.6.6.2.3.1 Plan Quality

- Identify (list) the items to be procured
- Identify specification, drawings, sketches, photos, performance criteria, etc., of the items to be procured
- Identify and prequalify vendors
- Select qualified vendor
- Establish procurement criteria
- Establish pricing strategy (quotation) of the items to be procured
- Establish schedule of material requirements at site
- Supply chain management plan
- Procurement log in line with planned requirement at site
- Establish inspection and test plan (during production, transportation, and receipt at site)
- Level of inspection
- Inspection schedule
- Frequency of inspection
- Inspection and testing methods
- Acceptance criteria
- Inspection meetings/schedule

8.6.6.2.3.2 Perform Quality Assurance

- Approve the vendor
- Approve material
- Quotation from the vendor

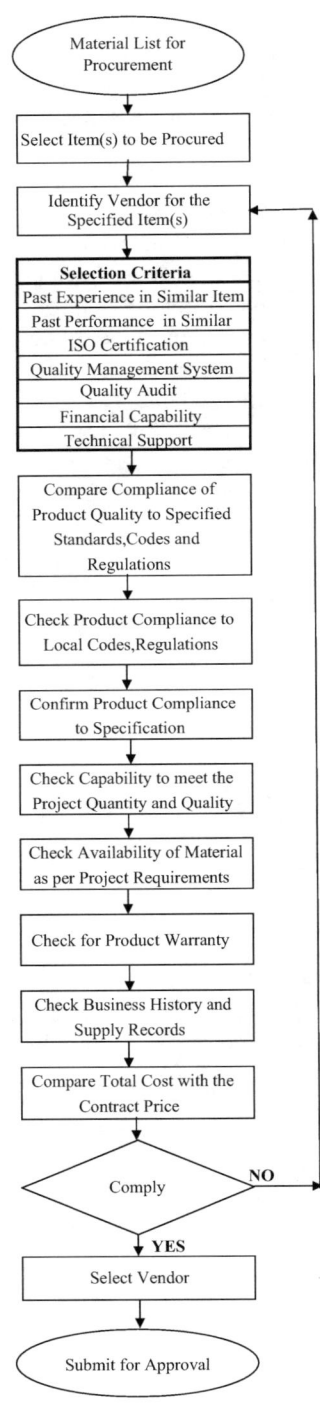

FIGURE 8.38 Logic flowchart for vendor-selection procedure.

Quality Management

- Place purchase order/contract
- Follow up production progress
- Follow up delivery of products
- Ensure logistics requirements
- Approval of third party inspection agency
- Approval of special inspection agency
- Records of inspection and test

8.6.6.2.3.3   Quality Control

- Material inspection at manufacturer's place (inspection/tests during production)
- Material receiving inspection at site
- Stage manufacturing inspection at manufacturer's place as per ITP
- Material quantity
- Manufacturing records, if any
- Delivery date/schedule
- Authentication of inspection report/certificate
- Validate inspection release
- Validate shipping release

8.6.6.2.4   Finalize Procurement

A detailed procedure for submitting materials, samples, and shop drawings is specified under section "SUBMITTAL" of contract specifications. The contractor has to submit the same to owner/consultant/PMC for reviewing their review and approval. The consultant/PMC reviews the submittal and returns the transmittal to the contractor with an appropriate action.

Figure 8.39 illustrates the logic flowchart for vendor, material approval, and procurement procedure.

Figure 8.40 shows a site transmittal form for material approval.

The contractor has to submit the following, as minimum, to the owner/consultant to get their review and approval of materials, products, equipment, and systems. Contractor cannot use these items unless they are approved for use in the project.

The consultant/PMC reviews the submittal to verify that the proposed product/sample complies with the contract specifications and returns the transmittal to the contractor mentioning one of the following actions on the transmittal:

- **A** – Approved
- **B** – Approved As Noted
- **C** – Revise and Resubmit OR Not Approved
- **D** – For Information OR More Information Required

8.6.6.2.4.1   Product Data   The contractor has to submit the following details:

- Manufacturers' technical specifications related to the proposed product
- Installation methods recommended by the manufacturer
- Relevant sheets of manufacturer's catalog(s)

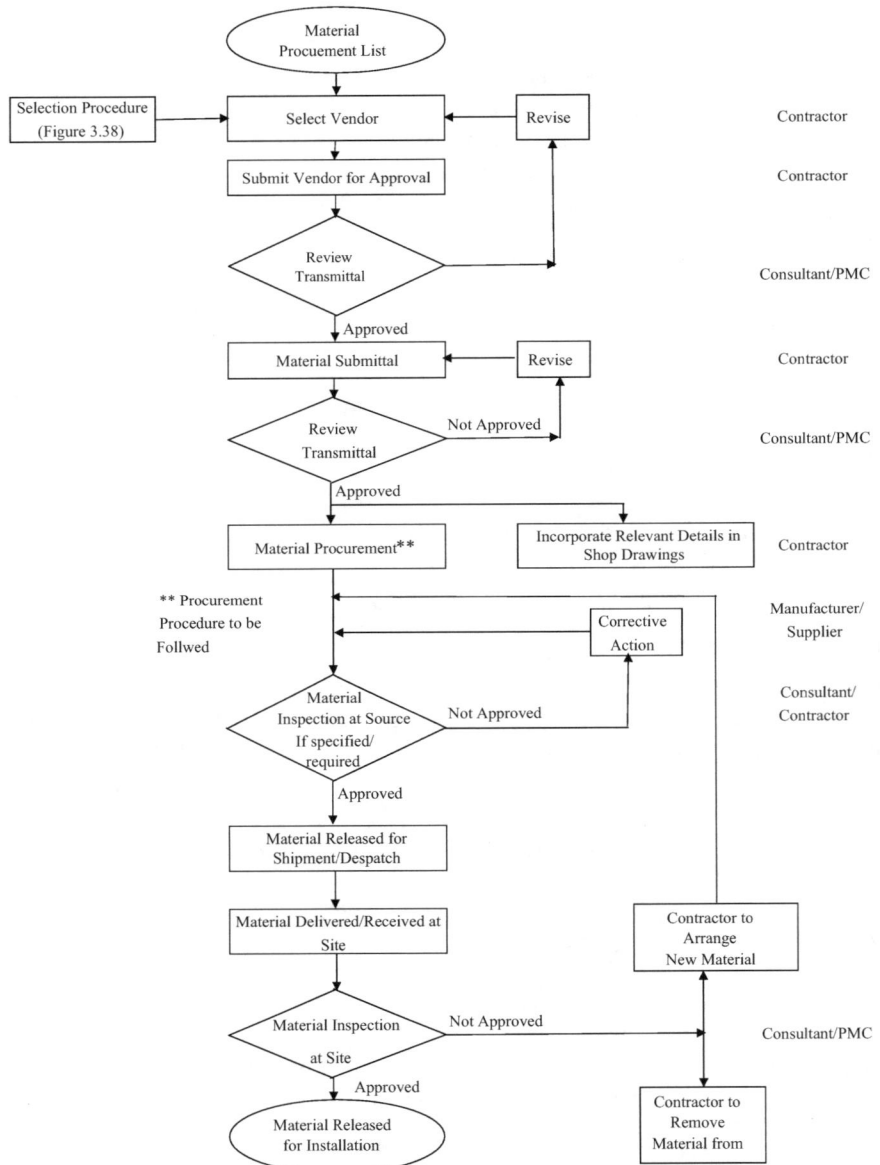

FIGURE 8.39 Logic flowchart for vendor, material approval, and procurement procedure.

- Confirmation of compliance with recognized international quality standards
- Performance characteristic and curves (if applicable)
- Manufacturer's standard schematic drawings and diagrams to supplement standard information related to project requirements and configuration of the same to indicate product application for the specified works (if applicable)

Quality Management

Project Name
Consultant Name
SITE TRANSMITTAL
Request for Material Approval

CONTRACT No. : _____   TRANSMITTAL NO. _____   REV. _____

CONTRACTOR : _____

TO   : _____

WE REQUEST APPROVAL OF THE FOLLOWING MATERIALS/GOODS/PRODUCTS/EQUIPMENT

ITEM NO.	DWG., SPEC. OR BOQ. REF	DESCRIPTION	SUBMITTAL CODE *	ACTION CODE **
		SAMPLE FORM		

DETAILS OF INFORMATION, LITERATURE, CATALOG CUTS, AND THE LIKE ATTACHED ARE:
_____
_____

SAMPLES:
Enclosed [ ]   Submitted under separate cover [ ]   Not applicable [ ]

N.B. We certify that above items have been reviewed in detail & and are correct & in strict performance with the Contract Drawings & Specification except as otherwise stated.

CONTRACTOR'S REP. : _____   DATE : _____

RECEIVED BY CONSULTANT : _____   DATE : _____

cc: Owner Rep.

Resident Engineer to enter ACTION CODE and REMARKS

R.E.'s REMARKS :
_____
_____
_____

                                              Initials _____   Date _____

Corrections or comments made relative to submittals during this review do not relieve Contractor from compliance with the requirements of the Drawings and Specifications. This check is only for review of general conformance with the design concept of the project and general compliance with the information given in the Contract Documents. Contractor is responsible for confirming and correlating all quantities and dimensions, selecting fabrication process and techniques of construction; coordinating his work with that of other trades, and performing his work in a safe and satisfactory manner.

Resident Engineer : _____   DATE : _____

Received by Contractor : _____   DATE : _____

cc: Owner Rep.

* SUBMITTAL CODE:	** ACTION CODE:		
1: Submitted for Approval	A: Approved	C:	Not Approved

FIGURE 8.40 Site transmittal for material.

- Compatibility certificate (if applicable)
- Single-source liability (this is normally required for systems approval where different manufacturer's items are used)

Project Name	
Consultant Name	

SPECIFICATION COMPARISON STATEMENT (SCS)

Contractor: _____   Date: _____

Contract No. _____   A/S No.: _____

Submittal No.: _____ Revision: _____ Transmittal Ref.: _____

Submittal Title: _____   Specification Ref: _____

S No.	SPECIFICATION REQUIREMENTS	CONTRACTOR'S PROPOSAL	REMARKS
		SAMPLE FORM	

FIGURE 8.41 Specification comparison statement.

8.6.6.2.4.2 Compliance Statement The contractor has to submit a specification comparison statement along with the material transmittal. The compliance statement form is normally included as part of contract documents. The information provided in the compliance statement helps consultant/PMC to review and verify product compliance to the contracted specifications.

In case of any deviation to that of specified item, the contractor has to submit a schedule of such deviation(s) listing all the points not conforming to the specification.

Figure 8.41 shows the specification comparison statement to be submitted along with the material transmittal.

Upon approval of vendor and material by the owner/consultant, and obtaining the final price for the material supply, the contractor finalizes the procurement process and the purchase order/contract is made with the vendor.

8.6.6.3 Execute Construction/Installation Works

The contractor is responsible to execute the contracted works in accordance with the approved detailed engineering drawings and specifications as specified in the contract documents. The contractor has to arrange necessary resources to complete the project within the schedule and contracted amount. The contractor has to maintain the executed works until handing over the project to the owner/end user and maintain for additional period if contracted to do so during operation. During construction period, the contractor has to protect executed/installed works to ensure that the works are not damaged. The contractor has to use new and approved material to construct the project/facility.

Construction activities in oil and gas project mainly consist of the following:

1. Site work such as cleaning and excavation of project site
2. Site earth works
3. Dewatering and Shoring
4. Excavation and backfilling
5. Construction of foundation and platform for equipment and machinery
6. Process engineering works
7. Piping works
8. Instrumentation works
9. Process Automation
10. Supervisory controls
11. Fire detection systems
12. Corrosion monitoring system
13. Mechanical works
14. Electrical works
15. Buildings
16. Civil works
17. Drainage system
18. Utility
19. Equipment and machinery installation
20. Elevator and conveying systems works
21. Fire suppression works
22. Plumbing and public health works
23. HVAC works
24. Fire alarm system works
25. Information and communication system works
26. Data security system
27. Low-voltage systems works
28. Landscape works
29. External works

8.6.6.3.1 Manage Execution/Installation of Works

The contractor is responsible to execute the contracted works in accordance with contract drawings and specifications as specified in the contract documents. The contractor has to arrange all the resources and manage the execution as per the approved schedule by constructing the project as per approved shop drawings by installing the approved material, equipment, and systems. To start with the execution of contracted works, contractor has to submit the construction schedule to the consultant/PMC and get the approval for the same. Similarly, contractor has to prepare different management plans as per the contract requirement. The contractor has to prepare shop drawings on the approved detailed engineering drawings and submit the same for approval. Similarly, contractor has to get approval of the vendor and the materials, equipment, and systems that will be used on the project.

8.6.6.3.1.1 Submittals In construction projects, there are various types of documents which are to be sent to different stakeholders. Proper correspondence and reporting method are important to distribute this information.

Construction projects have involvement with many stakeholders. A large number of documents move forward and backward between these stakeholders for information or action. During the design stage, the communication is mainly between the owner (Project/Construction Manager) and designer (consultant). However, once the contractor is selected depending on the type of project delivery system, the contractor is actively involved in the project communication system. Apart from forming these three parties, some other stakeholders, who have interest, expectations, and influence in the project, have also sent the copy of information/documents for their information or action.

During the construction phase, there are many documents sent back and forth between various stakeholders. The following are the types of documents exchanged among owner, supervisor, and contractor:

- Administrative
- Contract-related
- Engineering submittals
- Project monitoring and control
- Quality

Table 8.28 discussed earlier illustrates the list of forms normally used during the construction phase to communicate different types of project-related documents.

These documents are normally transmitted by the transmittal form. Please refer Figure 8.36 sample transmittal form discussed earlier.

A detailed procedure for submitting materials/products/systems, samples, and shop drawings is specified under section "SUBMITTAL" of contract specifications. The contractor has to submit the same to owner/consultant for reviewing their review and approval. The contract documents specify administrative and procedural requirements for submission of submittals and other documents. Contractor has to comply with the contractual requirements for submittal requirements.

Submittal is a quality assurance activity during the construction phase. This activity(ies) performed by the contractor to provide evidence and confidence that the works, materials, and systems executed and installed shall meet owner's objectives as specified in the contract documents and to ensure and guarantee that the performance of the project/facility is fully functional to the satisfaction of owner/end user.

The contractor performs quality assurance process by

1. selecting the materials and systems fully complying with contract specifications and installing the approved materials and systems only;
2. preparing the shop drawings detailing all the requirements included in the working drawings and installing/executing the works as per approved shop drawings;
3. installing the works, materials, and systems as per the specified method statement and as per recommendations from the manufacturer of products.

Submittal process, in construction projects, is essential to ensure that contractor's understanding of product specifications, contract drawings, and installation method matches with the designer's intent of product usage and installation method. The submittal process provides the owner the assurance that the contractor is complying with the design concept, and the installed material will function as required by the contract documents. Submittals are documents that are presented by contractor for approval, review, decision, or consideration.

Generally, these submittals fall into three categories. These are

1. Approval Submittals
2. Review Submittals
3. Information Submittals

8.6.6.3.1.1.1 MATERIAL Prior to the start of execution/installation of work, the contractor has to submit specified material/product/system to the Owner/Consultant/PM, as per the project specification requirements, for approval, review, or information. Please refer Figure 8.40 for Site Transmittal Form for material approval and Figure 8.41 for Specification Comparison Form discussed earlier under Section 8.6.6.2.

8.6.6.3.1.1.2 SHOP DRAWINGS The contract (detailed engineering) drawings and specifications prepared by the designer (contractor) are indicative and are generally meant for determining the planning and pricing of the construction project. In many cases, they are not sufficient for installation or execution of works at various stages. More details are required during the construction phase to ensure specified quality. These details are provided by the contractor on the shop drawings. Shop drawings are used by the contractor as reference documents to execute/install the works. Detailed shop drawings help contractor to achieve zero defect in installation at the first stage itself, thus avoiding any rejection/rework.

The number of shop drawings to be produced is mutually agreed between the contractor and consultant depending on the complexity of the work to be installed and to ensure that adequate details are available to execute the work.

The shop drawings are to be drawn accurately to the scale and shall have project-specific information in it. The shop drawings shall not be a reproduction of contract drawings.

The contractor is required to prepare shop drawings by taking into consideration the following as a minimum but not limited to

1. Reference to contract (detailed engineering) drawings. This helps Consultant/PM to compare and review the shop drawing to that of contract drawing
2. Detailed plans and information based on the contract (detailed engineering) drawings
3. Notes of changes or alterations from the contract documents
4. Detailed information about fabrication or installation of works
5. All dimensions needed to verify at the job site
6. Identification of product (material and equipment)
7. Installation information about the materials and equipment to be used

8. Type of finishes, color, and textures
9. Installation details relating to the axis or grid of the project
10. Roughing in and setting diagram
11. Coordination certification from all other related trades (subcontractors)

The contractor has to consider the following point as the minimum, while developing shop drawings of different trades, to meet the design intents.

1. Review contract specification
2. Review contract drawings/detailed engineering drawings
3. Determine and verify field/site measurements
4. Installation information about the material to be used
5. Installation details relating to the axis or grid of the project
6. Dimensions of the product and equipment to be installed
7. Roughing in requirements
8. Coordination with other trade (disciplines) requirements
9. Clearly marking the changes and deviations to the contract drawings

Figure 8.42 illustrates the Site Transmittal Form for shop drawing approval.

The consultant/PMC reviews the submittal to verify that the proposed product/sample/shop drawings comply with the contract specifications and return the transmittal to the contractor mentioning one of the following actions on the transmittal:

- **A** – Approved
- **B** – Approved As Noted
- **C** – Revise and Resubmit OR Not Approved
- **D** – For Information OR More Information Required

All the works are executed as per the approved shop drawings. The approved materials, systems, and shop drawings are used by the supervision team to inspect the executed works and control the quality of works.

Figure 8.43 illustrates the shop drawing preparation and approval procedure.

Immediately after the approval of individual trade shop drawings, the contractor has to submit builder's workshop drawings and composite/coordinated shop drawings taking into consideration the following as a minimum.

8.6.6.3.1.1.2.1 BUILDERS WORKSHOP DRAWINGS Builders workshop drawings indicate the openings required in the civil or architectural work for services and other trades. These drawings indicate the size of openings, sleeves, level references with the help of detailed elevation and plans. Figure 8.44 illustrates the builders' workshop drawing preparation and approval procedure.

8.6.6.3.1.1.2.2 COMPOSITE/COORDINATION SHOP DRAWINGS The composite drawings indicate relationship of components shown on the related shop drawings and indicate required installation sequence. Composite drawings shall show the interrelationship of all services with each other and with the surrounding civil and architectural work.

Quality Management

Project Name
Consultant Name
SITE TRANSMITTAL
Request for Shop Drawings Approval

CONTRACT No.: _____ TRANSMITTAL NO. _____ REV. _____

CONTRACTOR : _____

TO : _____

WE REQUEST APPROVAL OF THE FOLLOWING ENCLOSED DRAWINGS

ITEM NO.	DWG., SPEC. OR BOQ. REF	DRAWING TITLE	DWG. NOS.	Rev.	SUBMITTAL CODE *	ACTION CODE **

N.B. We certify that above items have been reviewed in detail & and are correct & in strict performance with the Contract Drawings & Specification except as otherwise stated.

CONTRACTOR'S REP. : _____ DATE : _____

RECEIVED BY CONSULTANT : _____ DATE : _____

cc: Owner Rep.

Resident Engineer to enter ACTION CODE and REMARKS

R.E's REMARKS :

_____
_____
_____
_____
_____
_____ Initials _____ Date _____

Corrections or comments made relative to submittals during this review do not relieve Contractor from compliance with the requirements of the Drawings and Specifications. This check is only for review of general conformance with the design concept of the project and general compliance with the information given in the Contract Documents. Contractor is responsible for confirming and correlating all quantities and dimensions, selecting fabrication process and techniques of construction; coordinating his work with that of other trades, and performing his work in a safe and. satisfactory manner.

Resident Engineer : _____ DATE : _____

Received by Contractor : _____ DATE : _____

cc: Owner Rep.

cc: Project Manager	** ACTION CODE:			
1: Submitted for Approval	A:	Approved	C:	Not Approved
2: Submitted for your Information	B:	Approved as Noted	D:	For Information
3:				

FIGURE 8.42 Site transmittal for workshop drawings (Figure 8.43).

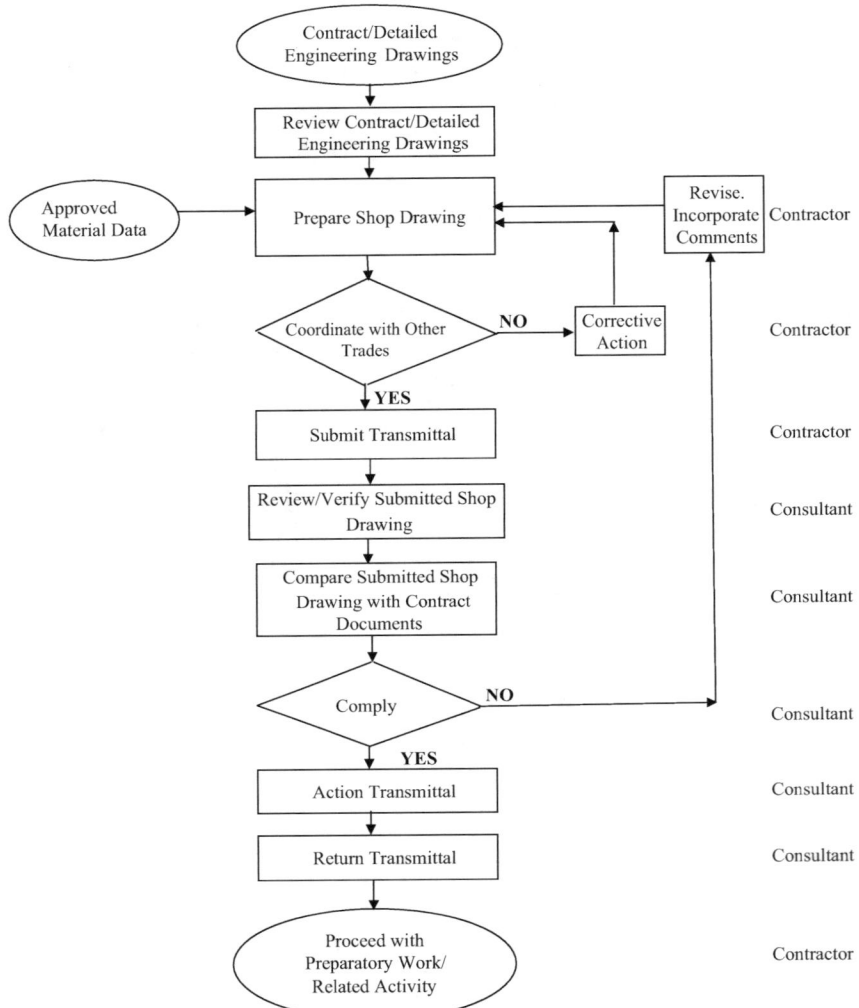

FIGURE 8.43 Shop-drawing preparation and approval procedure.

Composite drawings shall also show the detailed coordinated cross sections, elevations, reflected plans, etc., resolving all conflicts in levels, alignment, access, space, etc.... These drawings are to be prepared taking into consideration the actual physical dimensions required for installation within the available space. Figure 8.45 illustrates the composite drawing preparation and approval procedure.

8.6.6.3.1.1.3 METHOD STATEMENT The contractor has to execute the works as per the method statement specified in the contract. Contractor has to submit the method statement to consultant for their approval as per the contract documents to ensure compliance with the contract requirements. The method statement shall describe the

Quality Management

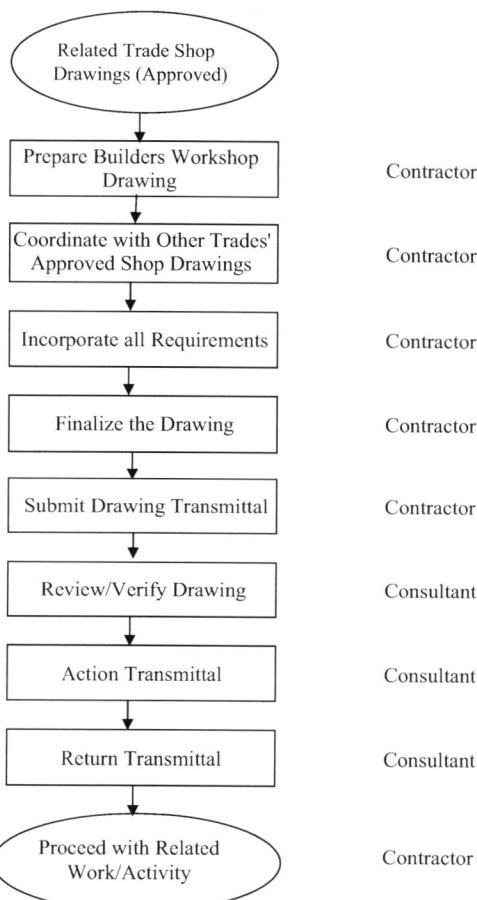

FIGURE 8.44 Builder's workshop drawing preparation and approval procedure.

steps involved for execution/installation of work by ensuring safety at each stage. It shall have the following information:

1. Scope of work: brief description of work/activity
2. Documentation: relevant technical documents to undertake this work/activity
3. Personnel involved
4. Safety arrangement
5. Equipment and plant required
6. Personal protective equipment
7. Permits/authorities' approval to work
8. Possible hazards
9. Description of the work/activity: detailed method of sequence of each operation/key steps to complete the work/activity

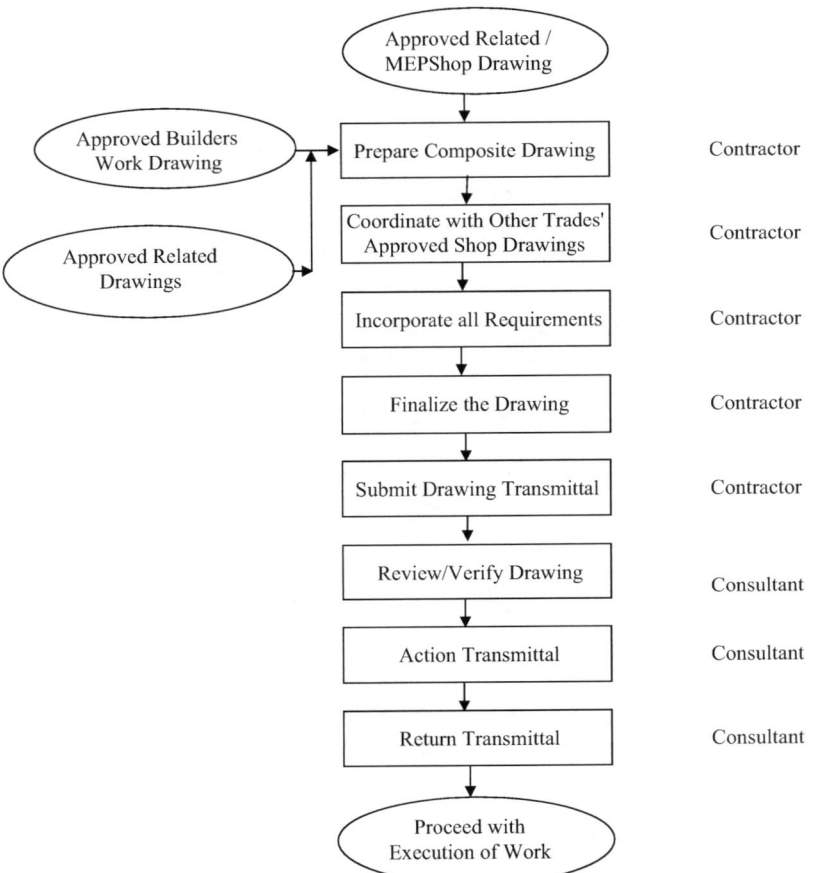

FIGURE 8.45 Composite/coordination shop drawing preparation and approval procedure.

8.6.6.3.1.2 Manage Scope Changes It is common that during the construction process, there will be some changes to the original contract. Even under most ideal circumstances, contract documents cannot provide complete information about every possible condition or circumstance that the construction team may encounter. The causes for changes are listed in Table 8.33.

These changes help to construct the project to achieve the project objective. These changes are identified as the construction proceeds. Prompt identification of such requirements helps both the owner and contractor to avoid unnecessary disruption of work and its impact on cost and time.

Contractor uses RFI form to request technical information from the supervision team. These queries are normally resolved by the concerned supervision engineer. However, it is likely that the matter has to be referred to the designer as RFI has many other considerations to be taken care which may be beyond the capacity of supervision team member to resolve. Normally, there is a defined period to respond RFI.

Quality Management

TABLE 8.33
Causes of Changes in Construction Project

Serial Number		Causes
I	Owner	
	I-1	Delay in making the site available on time
	I-2	Change of plans
	I-3	Financial problems/payment delays
	I-4	Change of schedule
	I-5	Addition of work
	I-6	Omission of work
	I-7	Project objectives are changed/modified
	I-8	Different site conditions
	I-9	VE
II	Contractor	
	II-1	Process/methodology
	II-2	Substitution of material
	II-3	Nonavailability of specified/approved material
	II-4	Unavailability of equipment and machinery
	II-5	Material not meeting the specifications
	II-6	Charges payable to outside party due to cancellation of certain items/products
	II-7	Delay in approval
	II-8	Contractor's financial difficulties
	II-9	Unavailability of manpower
	II-10	Workmanship not to the mark
IV	Miscellaneous	
	III-1	New Regulations
	III-2	Safety considerations
	III-3	Weather Conditions
	III-4	Unforeseen circumstances
	III-5	Inflation
	III-6	Fluctuation in exchange rate
	III-7	Government policies

Such queries may result in variation to the contract documents. It is in the interest of both the owner and contractor to resolve RFI expeditiously to avoid its effect on construction schedule.

Figure 8.46 illustrates RFI form which contractor submits to the consultant to clarify differences/errors observed in the contract documents, change in construction methodology, change in the specified material, etc., and Figure 8.47 illustrates the process to resolve request for variation (contractor-initiated).

Figure 8.48 illustrates the process to resolve scope change (owner-initiated).

Figure 8.37 discussed earlier illustrates a Site Works Instruction (SWI) form. It gives instruction to the contractor to proceed with the change(s). All the necessary

Project Name
Consultant Name

REQUEST FOR INFORMATION (TECHNICAL)

CONTRACT NO. : _____	R.F.I. NO. _____
CONTRACTOR. : _____	DATE : _____

To: Resident Engineer _____

REF:

SUBJECT:

<u>REQUEST FOR INFORMATION (Technical)</u>

This form is used by the contractor to request information and is normally sent to the A/E who responds on the same form.

SAMPLE FORM

CONTRACTOR: _____

| DISTRIBUTION: | Owner ☐ | Consultant/PMC ☐ | R.E. ☐ |

RESPONSE BY R.E.:

Signature of R. E. _____ Date _____

RESPONSE RECEIVED:
FOR CONTRACTOR: _____ DATE: _____

| DISTRIBUTION: | Owner ☐ | Consultant/PMC ☐ | R.E. ☐ |

FIGURE 8.46 RFI.

documents are sent along with the SWI to the contractor. SWI is also used to instruct contractor for owner-initiated changes.

8.6.6.3.1.2.1 RESOLVE CONFLICT   It is essential that changes in the project are managed as quickly as possible in accordance with the conditions of contract. However, the disputes and conflicts in the construction projects are inevitable due to the fact

Quality Management

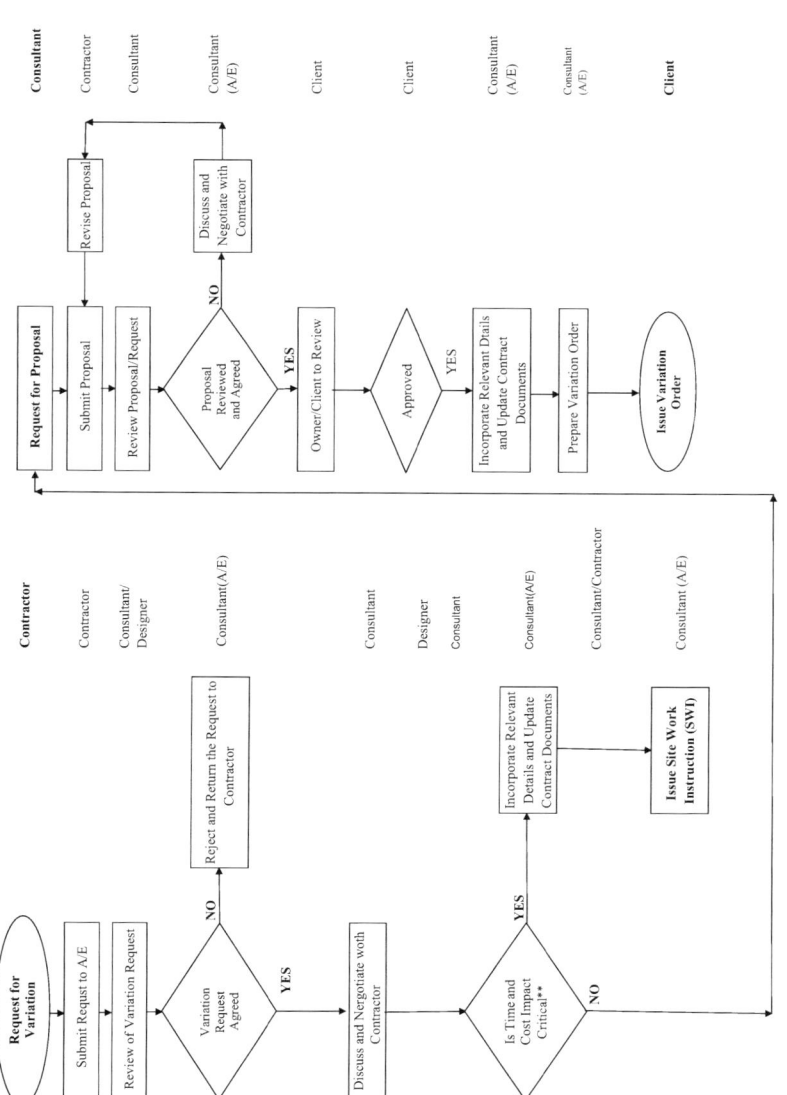

FIGURE 8.47 Process to resolve request for variation.

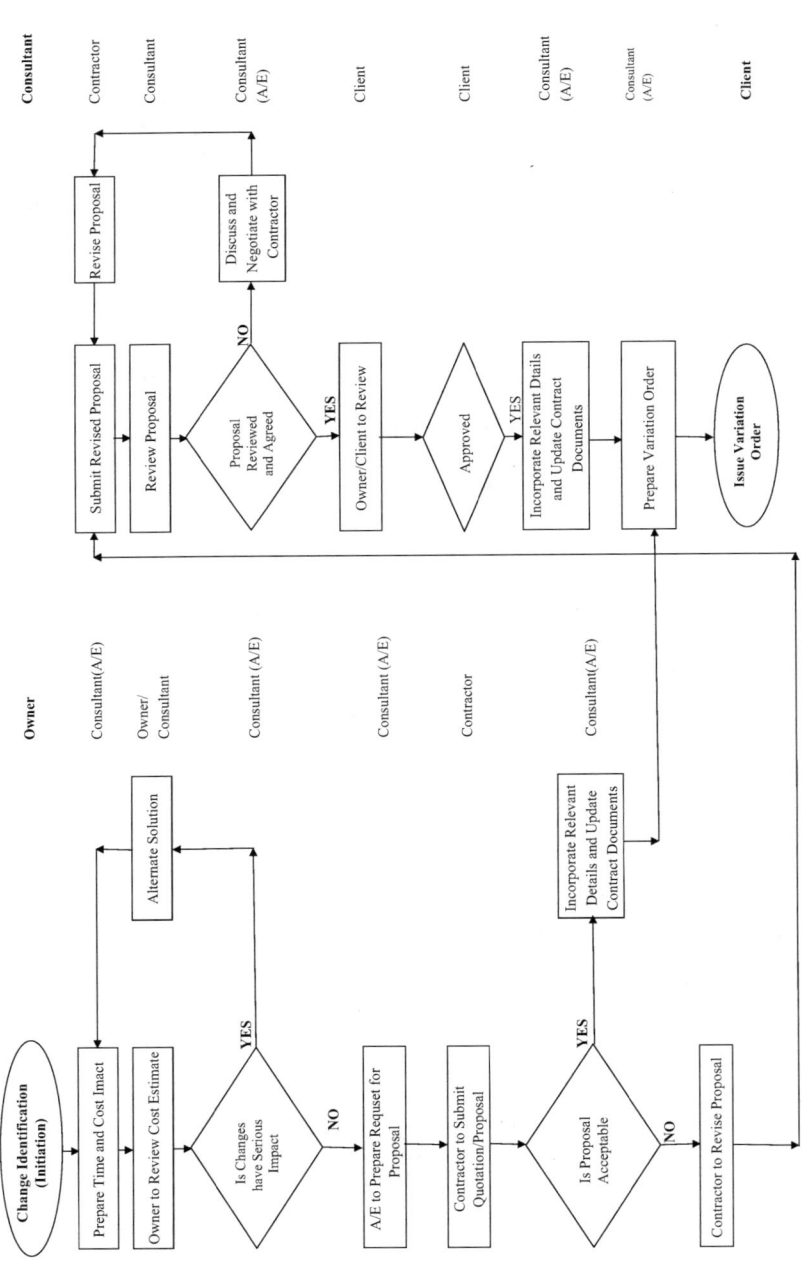

FIGURE 8.48 Process to resolve scope change (owner-initiated). (Abdul Razzak Rumane. (2016). *Handbook of Construction Management: Scope, Schedule, and Cost Control*. Reprinted with permission from Taylor & Francis Group.)

that with all the precautionary steps, the discrepancies or errors do occur in the contract documents. Following methods are normally followed to resolve the conflict:

1. Negotiation
   - The economical method to resolve the conflict is negotiation. Negotiation involves compromise. Both parties should discuss the issue by arranging the meetings of all the project team members involved and whose input to the issue will help resolve the conflict. The issue to be analyzed with the help of related documents, substantiation for claim, and justification for claim. If agreement is not reached, then involve senior representatives from both the parties.
2. Mediation
   - Mediation is a process in which a neutral third party group is involved to assist the parties to a dispute in reaching an amicable agreement that resolves the conflict.
3. Arbitration
   - Arbitration is the voluntary submission of a dispute to one or more impartial persons for final and binding determination. There are certain agencies that certify arbitrators.
4. Litigation
   - Litigation means to apply to the court to resolve the dispute.

8.6.6.3.1.3 Manage Construction Quality The construction project quality control process is a part of contract documents which provide details about specific quality practices, resources, and activities relevant to the project. The purpose of quality control during construction is to ensure that the work is accomplished in accordance with the requirements specified in the contract. Inspection of construction works is carried out throughout the construction period either by the construction supervision team (consultant) or the appointed inspector agency. Quality is an important aspect of construction project. The quality of construction project must meet the requirements specified in the contact documents. Normally, contractor provides on-site inspection and testing facilities at the construction site. On a construction site, inspection and testing are carried out at three stages during the construction period to ensure quality compliance.

1. During construction process: This is carried out with the checklist request submitted by the contractor for testing of ongoing works before proceeding to the next step.
2. Receipt of material, equipment, or services: This is performed by a material Inspection Request submitted by the contractor to the consultant upon receipt of material.
3. Before final delivery or commissioning and handover.

Quality management in construction is a management function. In general, quality assurance and control programs are used to monitor design and construction conformance to established requirements as determined by the contract specifications.

Instituting Quality Management programs reduce costs while producing the specified facility. CQCP developed as per the contents discussed in Table 8.25 under Section 8.6.4.4 is to be followed throughout the construction project. Table 8.34 illustrates the contractor's responsibilities to manage construction quality.

8.6.6.3.1.4 Manage Construction Resources The success of the construction project depends largely on availability, performance, and utilization of resources. In the construction project, the resources are linked with duration of project and each activity is allocated a specific resource to be available at the specific time. The construction resource mainly consists of the following:

1. Construction workforce
    i. Contractor's own staff and workers
    ii. Subcontractor's staff and workers
2. Construction equipment and machinery
3. Construction material, equipment, and systems to be installed on the project.

In most construction projects, the contractor is responsible to engage all types of human resources to complete the project, subcontractors, specialist installers, suppliers, arrange equipment, construction tools, and materials as per contract documents and to the satisfaction of owner/owner's appointed supervision team. Workmanship is one of the most important factors to achieve the quality in construction; therefore, it is required that the construction workforce is fully trained and have full knowledge of all the related activities to be performed during the construction process.

8.6.4.3.1.4.1 CONSTRUCTION WORKFORCE Once the contract is awarded, the contractor prepares a detailed plan for all the resources he needs to complete the project. Contract documents normally specify a list of minimum number of core staff to be available at site during the construction period. Absence of any of these staff may result in penalty to be imposed on contractor by the owner.

A typical list of contractor's minimum core staff needed during the construction period for execution of work of the oil and gas project is specified in the contract documents.

Contractor's human resources mainly consists of two categories:

1. Contractor's own staff and workers
2. Subcontractor's staff and workers

The human resources required to complete the projects are based on resource-loading program. It is necessary that all the construction resources are coordinated and brought together at the right time in order to complete on time and within budget.

The main contractor has to manage all these personnel by

1. assigning the daily activities
2. observing their performance and work output (productivity)
3. daily attendance
4. safety during construction process

TABLE 8.34
Contractor's Responsibilities to Manage Construction Quality

	Areas of Quality Control							
	Main Contractor		Subcontractors					
Serial Number / Activity	Head Office/ Quality Manager	Project site/Project Manager	Process Engineering	Piping/ Mechanical	Electrical/ Instrumentation	Civil/ Architectural	Landscape	External
1. Prepare quality control plan	■	□	□	□	□	□	□	□
2. Construction schedule	□	■	□	□	□	□	□	□
3. Mobilization	□	■	□	□	□	□	□	□
4. Staff approval	□	■	□	□	□	□	□	□
5. Prepare material submittal		■	■	■	■	■	■	■
6. Submit material transmittal		■	□	□	□	□	□	□
7. Prepare shop drawings		■	■	■	■	■	■	■
8. Submit shop drawing transmittal		■	□	□	□	□	□	□
9. Material sample	□	■	■	■	■	■	■	■
10. Off-site material inspection/testing		■	■	■	■	■	■	■
11. Receiving material inspection		■	■	■	■	■	■	■
12. Material testing		■	■	■	■	■	■	■
13. Site work inspection		■	■	■	■	■	■	■
14. Quality of work		■	■	■	■	■	■	■
15. Prepare checklist		■	□	■	■	■	■	□
16. Submit checklist		■	■	■	■	■	■	■
17. Corrective/preventive action		■	■	■	■	■	■	■

(Continued)

TABLE 8.34
Contractor's Responsibilities to Manage Construction Quality

	Main Contractor		Areas of Quality Control					
					Subcontractors			
Serial Number	Head Office/ Quality Manager	Project site/Project Manager	Process Engineering	Piping/ Mechanical	Electrical/ Instrumentation	Civil/ Architectural	Landscape	External
18 Daily report		■	□	□	□	□	□	□
19 Monthly progress report		■	□	□	□	□	□	□
20 Progress payment	□	■	□	□	□	□	□	□
21 Site safety		■	■	■	■	■	■	■
22 Safety report		■	□	□	□	□	□	□
23 Waste disposal		■	■	■	■	■	■	■
24 Reply to job site instruction		■	□	□	□	□	□	□
25 Reply to nonconformance report		■	□	□	□	□	□	□
26 Documentation		■	□	□	□	□	□	□
27 Testing and commissioning		■	■	■	■	■	■	■
28 Project close-out documents	□	■	□	□	□	□	□	□
29 Punch list		■	■	■	■	■	■	■
30 Request for issuance of substantial completion letter	□	■	□	□	□	□	□	□

■ Primary Responsibility.
□ Advise/Assist.

Source: Modified from, Abdul Razzak Rumane. (2013). *Quality Tools for Managing Construction Projects*. Reprinted with permission from Taylor & Francis Group.

Quality Management

8.6.6.3.1.4.2 SUBCONTRACTOR In most construction projects, main contractor engages subcontractors, specialist contractors to execute certain portion of the contracted project work. The main contractor has to monitor the performance of the subcontractor throughout the project to ensure project success. In order to achieve project objectives, it is essential that main contractor and subcontractor maintain partnering relationship. In order for smooth execution of the project, both parties should have cooperative, collaborative, and joint problem-solving attitude. Their aim should be to achieve a successful project and maintain long-term business relationship.

The main contractor–subcontractor relationship starts the moment the subcontractor is selected and a contract/agreement is signed to execute the project.

It is important that necessary precautions and care are taken to prequalify and select the subcontractor. Table 8.22 discussed under Section 8.6.2.6 lists questionnaires to prequalify the subcontractor.

In certain cases, the subcontractor is involved with the main contractor from the tendering stage. The main contractor takes into consideration the prices quoted by the subcontractor while submitting the proposal. The contractor price between main contractor and subcontractor can be as follows:

1. Back to back on the main contractor prices keeping agreed upon margin for the main contractor (or)
2. Negotiated prices with subcontractor in which case the main contractor's contract awarded prices are not known to the subcontractor

In order for a successful execution of the project, it is necessary to have subcontractor management plan in place to ensure that each of the subcontractors executes the project as specified without affecting the project quality and schedule.

The following is the typical content of the subcontractor management plan:

1. Introduction
2. Organization
    i. Organization chart
    ii. Roles
    iii. Responsibilities
3. Scope of work
4. Quality management plan
5. Project coordination method
6. Submittals
    i. Material
    ii. Shop drawings
7. Resource management
    i. Training
8. Change management
9. Communication
    i. Meetings
10. Risk management
11. HSE management

12. Documentation
13. Invoicing and payments
14. Close-out contract

8.6.6.3.1.4.3 CONSTRUCTION EQUIPMENT Likewise, the contract documents specify that a minimum number of equipment are to be available on site during the construction process to ensure smooth operation of all the construction activities. Please refer Figure 8.34 discussed earlier under Section 8.6.5.2 that illustrates equipment list and utilization schedule for the oil and gas project.

8.6.6.3.1.4.4 CONSTRUCTION MATERIAL The contractor also prepares a procurement log based on the project completion schedule. The contractor has to ensure that construction material is available at project site on time to avoid any delay. Delivery of long-lead items have to be initiated at an early stage of the project and monitored closely. Late order placement for materials results in delayed delivery of material, which in turn affects the timely completion of the project, is a common scenario in the construction projects. Hence, these logs have to be updated regularly and prompt actions have to be taken, to avoid delays. The contractor is required to provide twice a month or at any time requested by the owner/consultant, full and complete details of all products/systems procurement data relating to all the approved products, systems which have ordered and/or procured by the contractor for using in the construction project. The contractor maintains contractor's procurement log E-2 to keep track of the material status. Figure 8.49 is a sample procurement log. The log E-2 is normally submitted by the contractor along with monthly progress report.

8.6.6.3.1.4.5 SUPPLY CHAIN MANAGEMENT Supply chain management in construction project is managing and optimizing the flow of construction materials, systems, equipment, and resources to ensure timely availability of all the construction resources without affecting the progress of works at the site.

In construction projects, the supply chain management starts from the inception of project. The designer has to consider the following while specifying the products (materials, systems, equipment) for use/installation in the project:

1. Quality management system followed by the manufacturer/supplier
2. Quality of product
3. Reliability of product
4. Reliability of manufacturer/supplier
5. Durability of product
6. Availability of product for entire project requirement
7. Price economy/cost efficient
8. Sustainability
9. Conformance to applicable codes and standards
10. Manufacturing time
11. Location of the manufacturer/supplier from the project site
12. Interchangeability
13. Avoid monopolistic product

Quality Management

Project Name
Consultant Name

CONTRACTOR'S PROCUREMENT LOG (REPORT E-2)

CONTRACT NO. :
CONTRACTOR :

NO. :
DATE :

Activity No.	Description	B.O.Q./ Specification. No.	Estimated Quantity	Ordered Quantity	Reqd. Order Date	Date Pur. Ord. Issued	Purchase Order No.	Supplier	Method Of Shipping	Shipping Date to Kuwait	Required On Job Date	E.D.A. To Site	A.D.A. To Site	Lead Time	Remarks

FIGURE 8.49   Contractor's procurement log.

Product specifications are documented in the construction documents (particular specifications). In certain projects, the documents list the names of recommended manufacturers/suppliers. However, in order for continuous and uninterrupted supply of a specified product, the contractor has to consider the following:

1. Quality management system followed by the manufacturer/supplier
2. Historical rejection/acceptance record
3. Reliability of the manufacturer/supplier
4. Product certification
5. Financial stability
6. Proximity to the project site
7. Manufacturing/lead time
8. Availability of product as per the activity installation/execution schedule
9. Manufacturing capacity
10. Availability of quantity to meet the project requirements
11. Timeliness of delivery
12. Location of manufacturer/supplier
13. Product cost
14. Transportation cost
15. Product certification
16. Risks in delivery of product
17. Responsiveness
18. Cooperative and collaborative nature to resolve problem

In order to ensure supply chain management system, the payments are made promptly, and the cash flow system should be projected accordingly as the supply chain may affect due to interruption in payments towards supply of products.

8.6.6.3.1.5 Manage Communication For smooth implementation of the project, proper communication system is established clearly identifying the submission process for correspondence, submittals, minutes of meeting, and reports.

8.6.6.3.1.5.1 CORRESPONDENCE Normally, all the correspondence between contractor and consultant are through the submittal transmittal form (please refer Figure 8.36). Correspondence between consultant and contractor is normally done based on job-site instructions whereas correspondence between owner, consultant, and contractor is normally done through letters. Figure 8.37 discussed earlier is a sample job-site instruction form used by the consultant to communicate with the contractor.

8.6.6.3.1.5.2 TRANSMITTALS Contract documents specify number of copies to be submitted to various stakeholders. Figures 8.50a and b illustrate submittal process as paper copy and electronically.

Quality Management

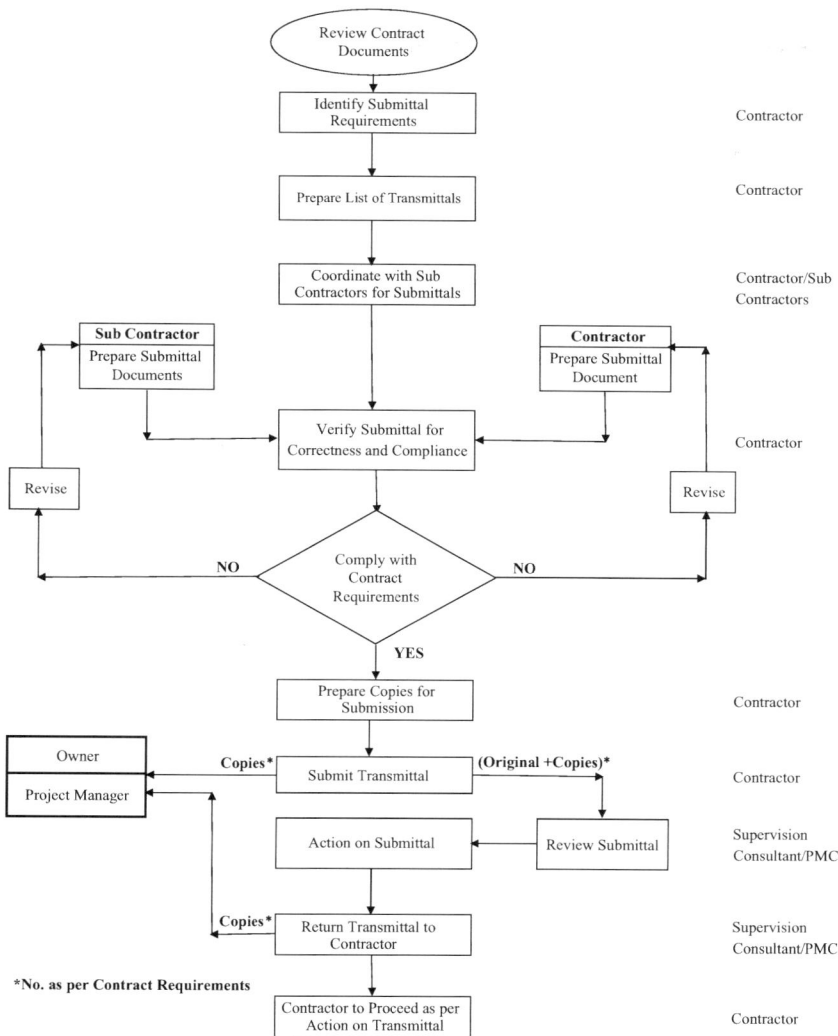

FIGURE 8.50A Submittal process (paper copy).

8.6.6.3.1.5.3 MEETINGS There are various type of meetings conducted during execution of the project. These meetings are conducted at an agreed upon frequency. The meetings during the construction phase are held for specific reasons. For example:

1. Kick-off meeting is held to acquaint project team members and discuss project objectives, procedures, and other contract information
2. Progress meetings to review work progress
3. Coordination meetings to coordinate among different disciplines and resolve the issues

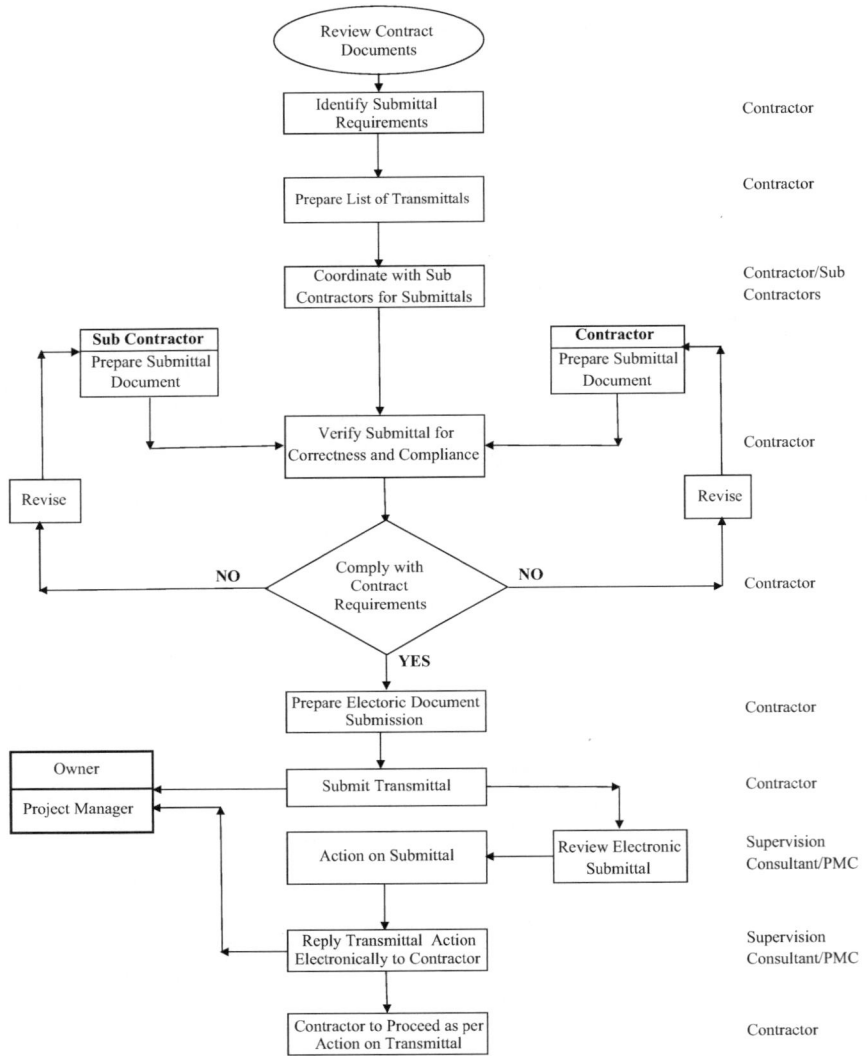

FIGURE 8.50B Submittal process (electronic).

4. Quality meetings to discuss on-site quality issues and improvements to the construction process
5. Safety meetings to discuss site safety and environmental issues

Table 8.35 lists points to be discussed during the safety meeting.

Apart from the above discussed meeting, the contractor, consultant, and owner can call for a meeting to discuss any project relevant issue and information.

Prior to conducting any meeting, an agenda is circulated to stakeholders and team members to attend the meeting. Figure 8.51 illustrates sample agenda format

TABLE 8.35
Points to be Reviewed during Monthly Safety Meeting

Serial Number	Points to be Reviewed	Yes/No
1	Whether meetings are held as planned and attended by all the concerned persons?	
2	Whether all the items from the previous meetings addressed and appropriate actions taken?	
3	Whether all the accidents/incidents/near misses recorded and reviewed to identify common issues and learning points?	
4	Whether training needs identified?	
5	Whether records of training programs maintained?	
6	Whether actions/feedback from weekly meetings reviewed and appropriate action taken?	
7	Whether suggestions/comments from safety tours are implemented?	
8	Whether safety awareness programs are regularly held?	
9	Whether safety warning signs are displayed and operative?	
10	Whether sirens and alarm bells are functioning properly?	
11	Whether temporary firefighting system in action and all the equipment are update?	
12	Whether regular check and audits are performed?	
13	Any latest regulation is introduced by authority and informed to all concerned?	
14	Was there any visit by competent authority and their observations are taken care?	
15	Whether escape route and assembly points are displayed?	
16	All the workforce have personal protective equipment?	
17	Whether First Aid box has all the required medicines?	

for meeting. The proceedings of meeting are recorded and minutes of meeting are circulated among all attendees and others per the approved form for minutes of meeting. Figure 8.52 illustrates minutes of meeting format.

8.6.6.3.1.5.4 REPORTS During the construction phase, the following reports are submitted by the contractor:

1. Progress reports such as:
   - Daily report
   - Weekly report
   - Monthly progress report
2. Safety report
3. Risk report

8.6.6.3.1.6 Manage Construction Risks During the construction project, the assigned project team members have to identify the construction risk and also to estimate likely occurrence of risk and develop response plan to mitigate and avoid the risk. Contractor has to maintain risk register and update all the time if any risk is noticed.

PROJECT NAME

Contract Number:			
Type of Meeting:		Date of Meeting:	
Place of Meeting:		Time of Meeting:	
Owner:			
Project/Construction Manager			
A/E (Consultant)			

AGENDA

SAMPLE FORM

1. Points to be discussed:

   1.1 ....

   1.2 ....

   1.3 ....

   1.4 ....

   ....

   ....

2. Any other Issues:

Signed by: ....   Po sition: ....

Deate: ....

FIGURE 8.51 Agenda format for meeting.

8.6.6.3.1.7 Manage Contract Contract management during the construction phase is an organizational method, process, and procedure to manage all contract agreements involved between the owner, contractor, subcontractor, manufacturers, and suppliers. During the construction phase, contracts are managed mainly by the following parties who are directly involved for execution of project:

1. Supervision Consultant/PMC
2. Contractor

PROJECT NAME
Minutes of Meeting

Contract NO.:	
Owner Name:	
Project/Construction Manager:	
Contractor Name:	

Meeting Type:		Minutes Number:	
		Date:	
Meeting Location:		Time:	

Attendees *SAMPLE FORM*

Number	Name	Position	Company

ITEM	DESCRIPTION OF DISCUSSION	STATUS	PRIORITY	ACTION			
				By	Due	Started	Completed

Distribution Original Copies: Owner ☐ PMC ☐ Consultant (Supervision) ☐ Contractor ☐ Other ☐

FIGURE 8.52 Minutes of meeting format.

Apart from these two parties, subcontractor and vendors also have their own contract management system.

Contract management process starts once the contract is signed. The supervision consultant/PMC is responsible to manage contract on behalf of owner. The consultant monitor scope, schedule, cost, and quality of the construction to ensure contract conditions are met. The contractor is responsible to ensure that all project works are executed within the agreed upon time and cost in accordance with the contract conditions and specification.

For successful contract management, the contractor as well as Consultant/CM/PM have to consider the following points while executing the project:

1. Use of RFI to get clarification on some aspects of the project. There are two parts in RFI. These are as follows:
   i. "Question" by the contractor
   ii. "Answer" by the Owner (Consultant)
2. Cooperating with all team members to fulfill their contractual obligations
3. Developing project execution plan considering realistic duration for each activity
4. Execution of contracted works in a timely manner in accordance with agreed upon schedule
5. Execution of project works using specified and approved materials, equipment, and systems
6. Installation of equipment, machinery, and system using manufacturer's recommended method of installation
7. Managing errors, omissions, and additions strictly in accordance with contract terms and avoiding any delays to the project
8. Providing resources to ensure timely availability of competent workforce as per resource schedule
9. Conducting meetings to monitor progress and clarify prevailing project issues
10. Dealing variations to the specified product, method, work in accordance with related specification, contract clauses and by providing substantiation and justifications that are resulted proposing alternative or substitute material
11. Resolving disputes in an amicable way by adopting cooperative approach
12. Taking action on all the transmittals within agreed upon period
13. Timely reply to all correspondences and queries
14. Communicating issues and problems well in advance
15. Not to ignore problems/issues with the hope that they might go away
16. Resolving risk by taking proper action to eliminate/mitigate the risk
17. Arranging payment of monthly progress payment as per contractual entitlement within stipulated time
18. Maintaining list of claims on monthly basis
19. Settling claims in accordance with contract terms
20. Maintaining proper logs and records

8.6.6.3.1.8 Manage HSE Requirements Contractor has to follow the HSE management plan. Contractor has to develop procedure for Project HSE review and evaluate the implementation of HSE activities.

8.6.6.3.1.9 Manage Project Finance In construction projects, maximum amount is expended during the construction phase. During this phase,

1. Owner has to make payments to:
   a. Main contractor
   b. Supervisor (Consultant)

c. Construction/Project Manager, if applicable
   d. Specialist consultant
   e. Specialist contractor
   f. Any other party such as direct appointments
   g. Owner supplied items, if any
2. Main contractor has to make payments to;
   a. Subcontractors
   b. Suppliers (material procurement)
   c. Designer, if any design work is involved
   d. Workforce
   e. Rent (equipment rent and rental vehicles)
3. Subcontractor has to make payment to:
   a. Suppliers
   b. Specialist

Owner's payment is mainly related to progress payment claimed and approved by the consultant, advance payments (if any as per contract) to the contractor, monthly fees to the consultants and construction/project manager. Contractor and subcontractor's payments are linked to the approved executed works. The contractor submits the payment on monthly basis. Figure 8.53 illustrates the progress payment submission format generally used by the contractor. Figure 8.54 illustrates the process for progress payment approval process.

8.6.6.3.1.10 Manage Claims Most claims in construction projects are due to the following:

1. Errors and omissions in the contract documents
2. Incomplete design
3. Design changes
4. Delay in payment by the owner
5. Delay in transfer of site
6. Delay in approval of submittals

The contractor during the construction phase has to identify the errors or omissions in the contract, if any, and follow the contractual procedure to resolve the issues.

Any claim submitted by the contractor has to be resolved by the project team members amicably as per the condition of contract.

8.6.6.4 Monitor and Control Project Works

Monitoring and controlling the project is an ongoing process. It starts from the inception of the project and continue till handover of the project. Monitoring and control of construction project are operative during the execution of the project, and its aim is to recognize any obstacles encountered during execution and to apply measures to mitigate these difficulties and to ensure that the goals and objectives on the project are being met.

Monitoring is collecting, recording, and reporting information concerning any and all aspects of project performance that the project manager or others in the

Contractor Name: Project Number			Project Name														
Serial Number	BOQ Reference	Description	Original Reference				Work Completed (Quantity/Percentage)							Cummulative Work			Approved Amount
							Previous			Current				Previous+Current			
			Unit	BOQ Quantity	Unit Price	Total Amount	Quantity	Percentage (%)	Total Amount	Quantity	Percentage (%)	Check List/ Approval Refrence	Total Amount	Quantity	Percentage (%)	Total Amount	

SAMPLE FORM

FIGURE 8.53 Progress payment submission format.

Quality Management

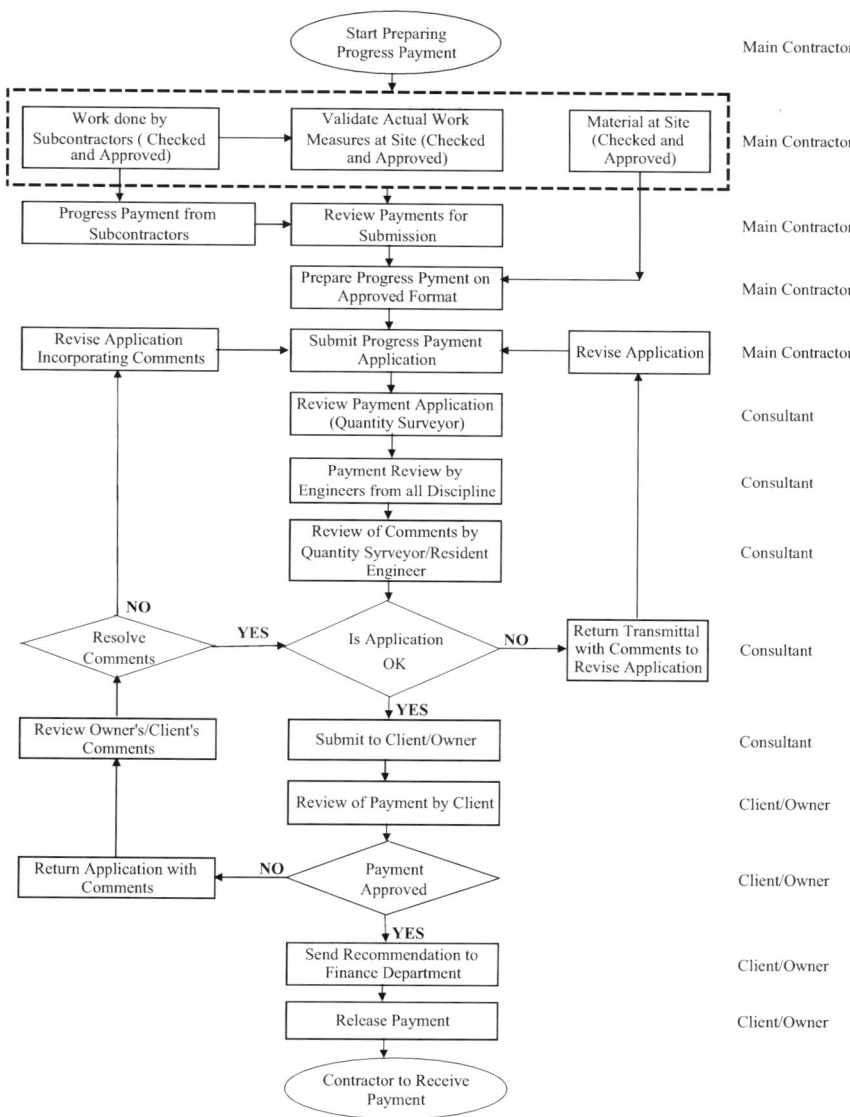

FIGURE 8.54 Progress payment approval process.

organization need to know. Monitoring of construction project is normally done collecting and recording the status of various activities and compiling them in the form of progress reports. These are prepared by the consultant, PMC, and contractor and distributed to the concerned members of project team. Figure 8.55 illustrates the logic flow diagram for monitoring and control process.

Monitoring involves not only tracking time but also budget, quality, resources, and risk. Monitoring in construction projects is normally done by compiling status

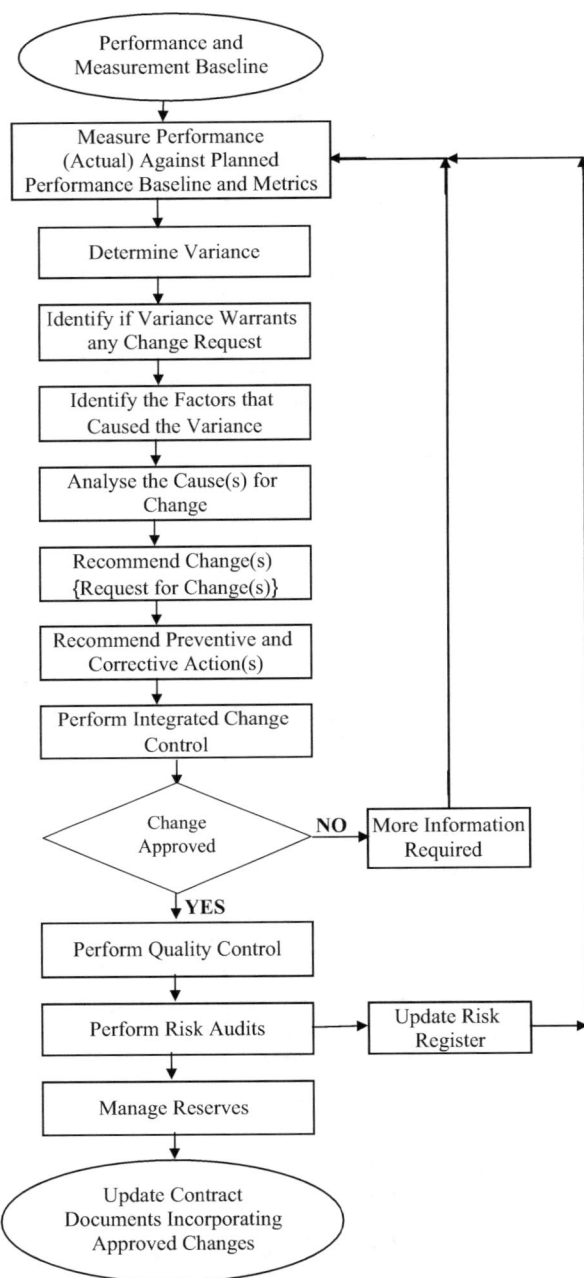

FIGURE 8.55 Logic flowchart of monitor and control process. (Abdul Razzak Rumane. (2013). *Quality Tools for Managing Construction Projects*. Reprinted with permission from Taylor & Francis Group.)

of various activities in the form of progress reports. These are prepared by contractor, supervision team (consultant), and construction/project management team. The objectives of project monitoring and control are as follows:

1. To provide an organized and efficient means of measuring, collecting, verifying, and quantifying data reflecting the progress and status of execution of project activities, with respect to schedule, cost, resources, procurement, and quality
2. To report the necessary information in detail and in appropriate form which can be interpreted by management and other concerned stakeholders to provide with the information about how the resources are being used to achieve project objectives.
3. To provide an organized, efficient, and accurate means of converting the data from the execution process into information
4. To identify and isolate the most important and critical information about the project activities to enable decision-making personnel to take corrective action for the benefit of the project.
5. To forecast and predict about future progress of activities to be performed

Project control involves a regular comparison of performance against targets, a search for the cause of deviation, and a commitment to check adverse variance. It serves two major functions:

1. It ensures regular monitoring of performance
2. It motivates project stakeholders to strive for achieving project objectives

A construction project control is exercised through knowing where to put in the main effort at a given time and maintaining good communication. There are mainly three areas where project control is required. These are

1. Quality (Scope)
2. Schedule
3. Budget

All of these areas are to be kept in balance to achieve project objectives. In order to accomplish the project objectives in construction projects, monitoring and control are done through various tools and methods. There are mainly three elements that need monitoring and control. These are

1. Quality (Scope)
2. Schedule (Work Progress)
3. Budget (Cost Control)

These are known as "Quality Trilogy" or "Triple Constraints."

In order to achieve a successful project, the supervision consultant/project manager (PMC) must handle these key attributes effectively and efficiently, and track the progress of the work from the inception to completion of construction and handover of the project for a successful completion.

8.6.6.4.1 Control Scope

Control scope is the process of monitoring the project scope and managing any changes to the scope baseline. It is common that despite all the efforts devoted to develop the contract documents (scope baseline), the contract documents cannot provide complete information about every possible condition or circumstance that the construction team may encounter.

The project requirements set out in the contract documents determine the project scope which is described in terms of project deliverables in the form of specifications and drawings. During the construction phase, the contractor has to execute/install the works as defined in the project specifications and contract drawings. The contractor has to make certain that the project changes do not result in compromising the intended project.

Figure 8.56 illustrates scope control process in the construction project.

8.6.6.4.2 Control Schedule

Monitor and control Schedule is the process to determine the current status of the schedule, identify the influencing factors that causes the schedule changes, determine that the schedule has changed, and manage the changes in the approved

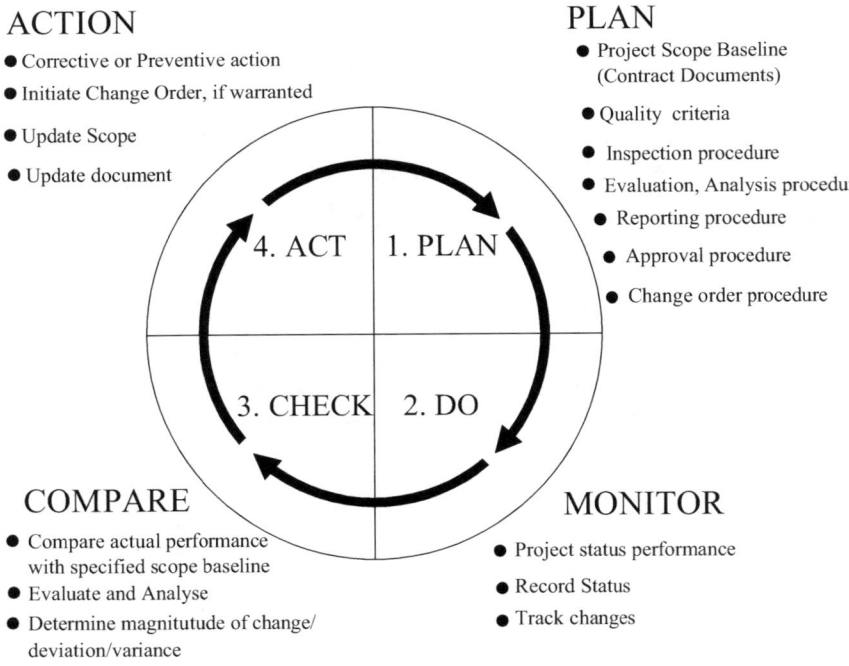

FIGURE 8.56 Scope control process.

Quality Management

project schedule baseline by updating and taking appropriate actions, if necessary, to minimize deviation from the approved schedule.

Completion of a construction project within defined schedule is most important. Time control status is prepared in different formats to monitor the project completion time. Figure 8.57 illustrates project progress status of a construction project. This chart presents overall picture of the elapsed period of the project and remaining period of the scheduled project duration, and actual progress versus planned progress.

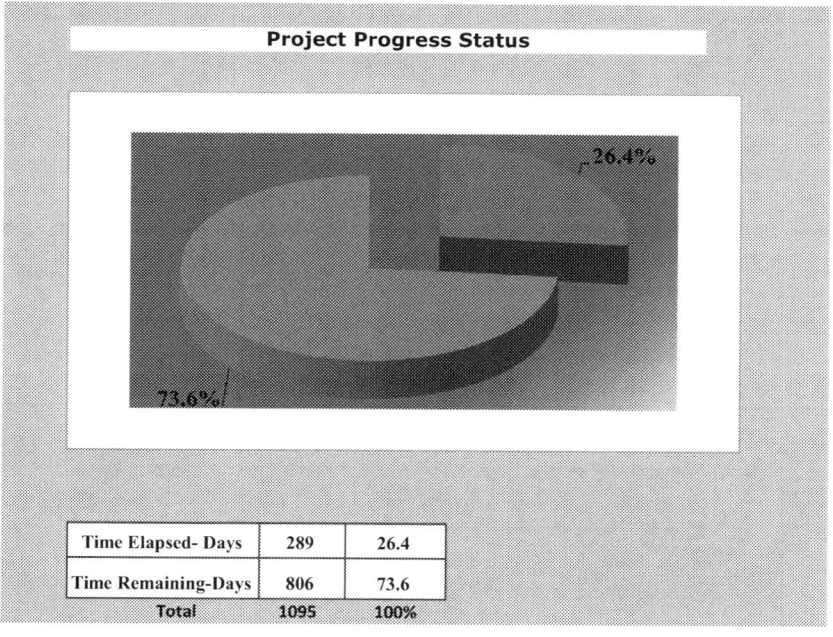

Current Project Status			
	June 15, 2020		
Completion Period (Days)	1095		
Duration Elapsed (Days)	289		
% Elapsed Duration	26.39%		
Remaining Duration (Days)	806		
% Remaining Duration	73.61%		
Total Contract Amount (KD)	258,674,521.320		
Earned Value in KD upto date	Early	Late	Actual
	49,743,110.450	45,423,245.944	43,741,861.55
Project Completion % till date	19.23%	17.56%	16.91%

FIGURE 8.57 Project progress status.

8.6.6.4.2.1 Monitor Progress Contractor's approved construction schedule is the performance baseline for construction projects which is achieved by collecting information through different methods. Work progress is monitored through various types of logs, S-Curves, reports, and meetings.

8.6.6.4.2.1.1 LOGS There are various types of logs used in construction projects to monitoring and control construction activities. The main logs used in a construction project are as follows:

1. Subcontractors' submittal and approval log
2. Submittal status log
3. Shop drawings and materials' logs – E1
4. Procurement log – E2
5. Equipment log
6. Manpower logs

These logs provide necessary information about the status of subcontractors, materials, shop drawings, procurement, and availability of contractor's resources and help determine its effects on project schedule and project completion.

8.6.6.4.2.1.2 S-CURVES Figure 8.58 illustrates S-Curve for planned work versus actual work.

8.6.6.4.2.1.3 REPORTS Different types of reports are used to monitor work progress.

8.6.6.4.2.1.3.1 PROGRESS REPORTS Apart from different types of logs and submittals, progress curves, time control charts, contractor's progress is monitored through various types of reports and meetings. These are

1. Daily report
2. Weekly report
3. Monthly report

Contractor's daily progress is monitored through daily progress report submitted by the contractor on the morning of the working day following the day which the report relates. It gives the status of all the resources available at site for that particular day. It shows the details of contractor's staff and manpower, contractor's plant and equipment, and material received at site. Details of subcontractor's work and resources are also included in the report. Along with the daily report, contractor submits work in progress report.

Monthly report giving details of all the site activities along with the photographs is submitted by the contractor to the consultant/owner for their information to know the progress of work during the month. Table 8.36 illustrates contents of contractor's monthly progress report and Table 8.37 illustrates contents of consultant's monthly progress report.

Quality Management

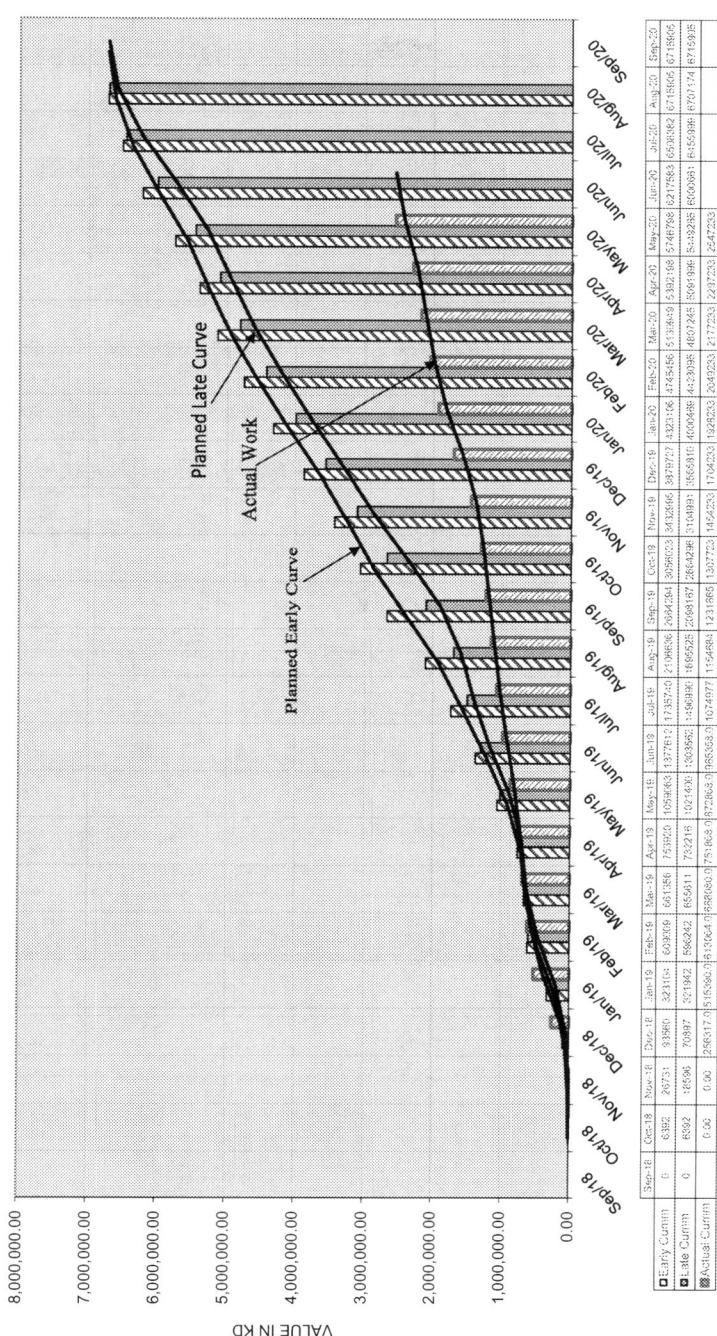

FIGURE 8.58 Planned versus actual.

TABLE 8.36
Monthly Progress Report

Serial Number	Contents		Description
1	Executive summary – Tabular		
	1.1	Summary status report	Brief description of the project status up-to-date, i.e., manpower, cash, activities
	1.2	NOC's report	No objection certificate report
	1.3	Project manager narratives	Narratives description of project status up-to-date
2	Progress layouts		
	2.1	Updated milestone table	Comparison of Planned vs. actual for contractual milestones per construction unit/design, you track the delays for major trades through the color theme
	2.2	Updated major of events table	Comparison of planned vs. actual for major trades per construction unit/design, track the delays for major trades through the color theme
	2.3	Updated layouts	Same as (2.1) above but presentation per milestone phase (drawing)
3	Updated execution program.		
	3.1	Updated milestone schedule – Roll-up "Update versus latest Target"	To indicate the status of the control & key milestones comparing the current status with the baseline
	3.2	Updated detailed schedule	All activities in details
	3.3	One-month look-ahead program	Same as (3.2), but includes only the detailed activities for the coming month BUT ON EXCEL FORMAT
4	Submittal status report (E1 Log).		Updated status of submittals
	4.1	Submittal status report	Briefly describe the project submittal status up to date
	4.2	Detailed E1 log	Detailed describe the project submittal status up to date
5	Procurement status report (E2 Log).		Updated procurement status
6	Status of information requested.		
	6.1	RFI status report	RFI
	6.2	NCR report	Nonconformance request (NCR)
	6.3	PCO & NOV log	Potential change order and notice of variation summary
7	Updated cost-loaded program.		
	7.1	Updated status report	Money progress = Physical * Budgeted cost for each running or completed activity
	7.2	Updated cost loaded schedule	Same as 7.1 but money values not percentages
	7.3	Updated work in place (%) report	Histogram and cumulative curve
	7.4	Updated cash flow report	Histogram and cumulative curve

(Continued)

TABLE 8.36 (*Continued*)
Monthly Progress Report

Serial Number	Contents	Description
8	Updated manpower histogram.	Histogram and cumulative curve
9	Updated schedule of construction equipment & vehicles	Tabular report
10	Updated critical indicators.	
	10.1 Shop drawings status report	Histogram and cumulative curve
	10.2 Material status report	Histogram and cumulative curve
	10.3 Construction: leading indicators	Each trade alone
	10.4 Line of balance diagram	It indicates all progress of the major trades as line cumulative chart
11	Updated progress photographs.	–
12	Updated safety inspection checklist.	Tabular report
13	Contractor information	Organization chart, tabular report

Source: Abdul Razzak Rumane. (2013). *Quality Tools for Managing Construction Projects.* Reprinted with permission from Taylor & Francis Group.

8.6.6.4.2.1.3.2 DIGITIZED MONITORING OF WORK PROGRESS The work progress is normally monitored through daily and monthly progress reports. Monthly progress report consists of progress photographs to document physical progress of work. These photographs are used to compare compliance with the planned activities and actual performance. Figure 8.59 illustrates traditional monitoring system.

With the advent of technology, it is possible to monitor and evaluate construction activities using cameras and related software technologies. In this process, digital images are captured through use of cameras. These photographs are processed using photo modeler software and developing 3D model view of the digital picture captured from the site. The captured as-built data are compared to the planned activities by interfacing through Integrated Information Modeling System. The use of the system

1. improves the accuracy of information
2. avoids delays in getting the information
3. improves communication among all parties
4. improves effective control of the project
5. improves document recording
6. helps reduce claims

Figure 8.60 illustrates schematic for Digitized Monitoring System.

8.6.6.4.2.1.4 SAFETY REPORT Contractor submits safety report every month listing important activities with photographs.

TABLE 8.37
Contents of Progress Report (Consultant)

1.0 Contract particulars
    1.1 Project description
    1.2 Project data
2.0 Construction schedule
3.0 Progress of works
    3.1 Temporary facilities and mobilization
    3.2 Summary of construction progress
        3.2.1 Status
        3.2.2 On-shore progress
        3.2.3 Off-shore progress
4.0 Time Control
    4.1 CPM schedule-level one (target vs. current) – summary by building/marine.
    4.2 CPM schedule-level two (target vs. current) – summary by building/division.
    4.3 30 days look-ahead schedule
    4.4 Time control conclusion
5.0 Cost control
    5.1 Financial progress
    5.2 Cash flow curve and histogram
    5.3 Work-in-place S-Curve and histogram
    5.4 Cost control conclusion
6.0 Status of contractor's submittals
    6.1 Material status
    6.2 Shop drawing status
7.0 Sub-contractors
8.0 Consultant's staff
9.0 Quality control
10.0 Meetings
11.0 Site work instructions
12.0 Variation orders
13.0 Construction photographs
14.0 Contractors resources
15.0 Others matters
    15.1 Safety
    15.2 Weather conditions
    15.3 Important developments/proposals/submissions

8.6.6.4.2.1.5 RISK REPORT Identify occurrence of new risk and report the same if this will have any effect on project progress and performance.

8.6.6.4.2.1.6 MEETINGS
8.6.6.4.2.1.6.1 PROGRESS MEETINGS Progress meetings are conducted at an agreed upon interval to review the progress of works and discuss about the problems, if any, for smooth progress of construction activities. Contractor submits

Quality Management

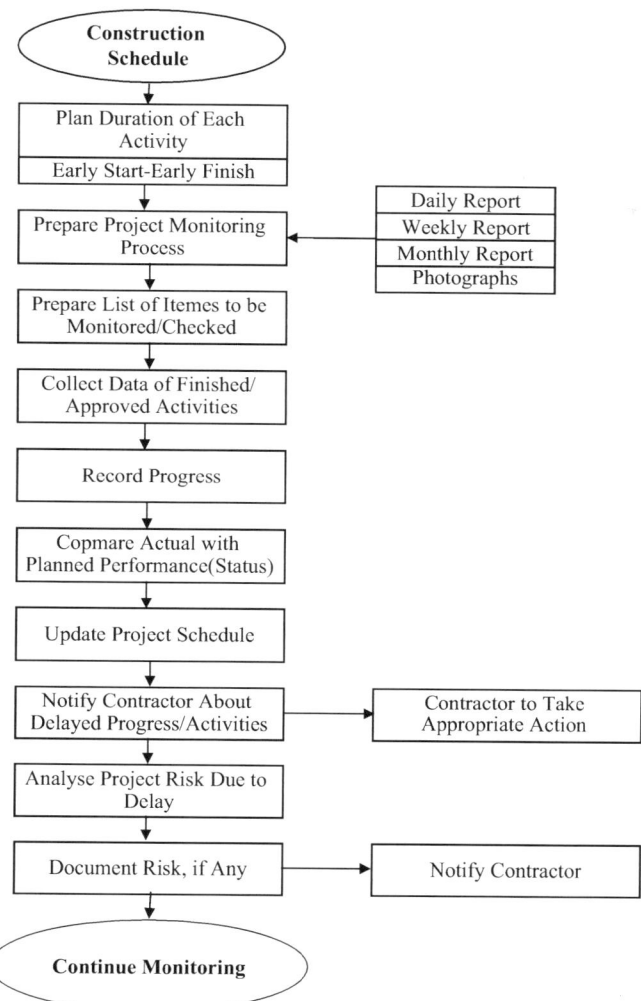

FIGURE 8.59 Traditional monitoring system.

premeeting submittal to project manager/consultant normally two days in advance of the scheduled meeting date. The submittal consists of

1. list of completed activities
2. list of current activities
3. two weeks look ahead
4. critical activities
5. materials submittal log
6. shop drawings submittal log
7. procurement log

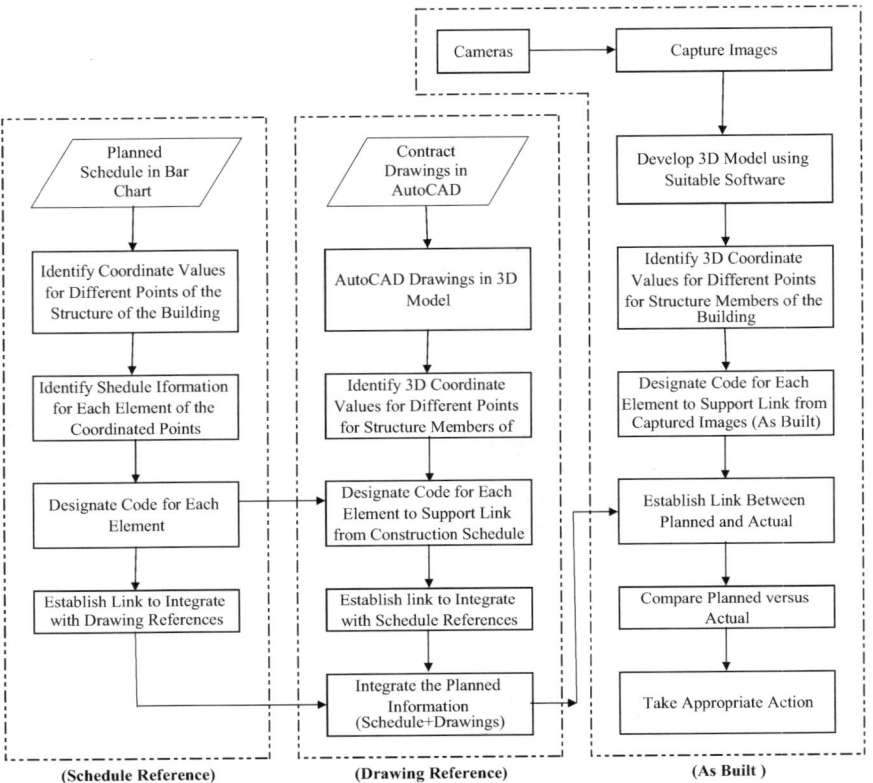

FIGURE 8.60 Digitized progress monitoring. (Abdul Razzak Rumane. (2013). *Quality Tools for Managing Construction Projects*. Reprinted with permission from Taylor & Francis Group.)

Apart from the issues related to progress of works and programs, site safety and quality-related matters are also discussed in these meetings. These meetings are normally attended by owner's representative, designer/consultant staff, contractor's representative, and subcontractor's responsible personnel.

8.6.6.4.2.1.6.2 COORDINATION MEETINGS Coordination meetings are held from time to time to resolve coordination matters among various trades.

8.6.6.4.2.1.6.3 QUALITY MEETINGS Quality meetings are conducted to discuss quality issues at site and how to improve the construction process to avoid/reduce rejection and rework.

8.6.6.4.2.1.6.4 SAFETY MEETINGS Safety meetings are also held to discuss related health, site safety, and environmental matters.

Frequency of conducting meetings is agreed between all the parties at the beginning of construction phase. Normally, the Construction Manager/Resident Engineer prepares the agenda for the meeting and circulates to all the participants.

Quality Management

Contractor informs Resident Engineer in advance about the points contractor would like to discuss, which are included in the agenda. The minutes of meetings are recorded and circulated among all the attendees and others per the approved responsibility/site communication matrix.

8.6.6.4.3 Control Cost

Cost control is the process of monitoring the status of the project to update the project cost and managing changes to baseline. This process provides the means to recognize variance from the approved plan, evaluate possible alternative, and take corrective action to minimize the risk. In order to have a successful cost control, it is essential to have the necessary information and data to take appropriate action. If the necessary information and regular updates are not available or if the action is inefficiently executed, then the risk to cost control on a project is raised considerably.

The purpose of cost control is to manage the project delivery within the approved budget. Regular cost reporting will facilitate the following:

1. Establish project cost to date
2. Anticipate final budget of the project
3. Cash flow requirements
4. Understanding potential risk to the project

Cost Control process focused on

5. Identifying the factors that influence the changes to the cost baseline
6. Determining the cost baseline has changed
7. Ensuring that the changes are beneficial for the project
8. Establishing the cost control structure and policy
9. Managing the actual changes as and when they occur
10. Monitoring cost performance to detect cost variance from the actual budget
11. Preventing unauthorized changes to the cost baseline
12. Recording all appropriate changes
13. Informing/reporting concerned stakeholders about the approved changes
14. Working to bring cost overruns within acceptable limits

Monitoring and control of project payment are essential with the budgeted amount. This is done through monitoring cash flow with the help of S-Curves and progress curves which gives exact status of payment and also identifies if it is exceeding the budget. Uninterrupted cash flow is one of the most important elements in the overall success of the project. Figure 8.61 illustrates planned and actual cost S-Curve.

8.6.6.4.4 Quality Control

In order to achieve quality in construction projects, all the works have to be executed as per approved shop drawings using approved material and fully coordinating with different trades. Proper sequencing and method statement for installation of work must be followed by the contractor to avoid rejection/rework. Figure 8.62 illustrates the sequence of execution of work.

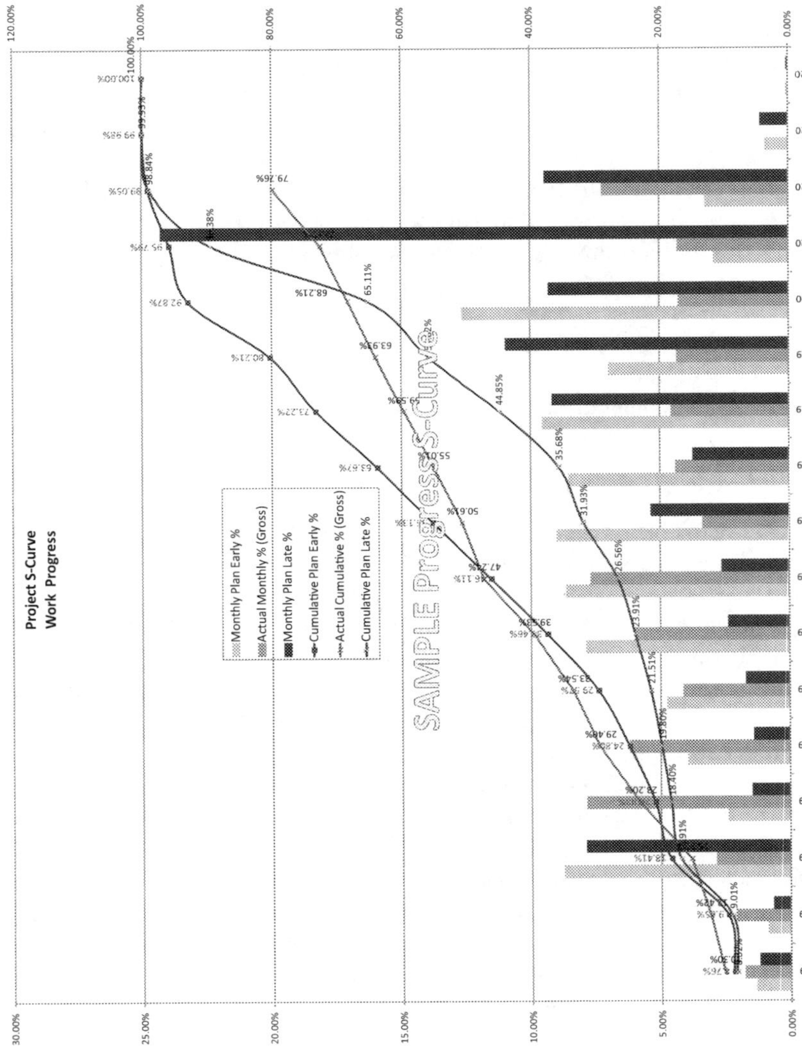

488 Quality Management in Oil and Gas Projects

FIGURE 8.61 S-Curve (work progress).

Quality Management

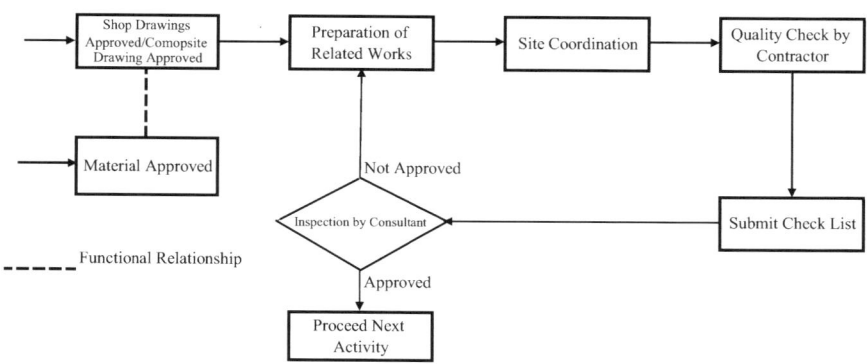

FIGURE 8.62 Sequence of execution of works. (Abdul Razzak Rumane (2010). *Quality Management in Construction Projects*. Reprinted with permission from Taylor & Francis Group.)

Following this sequence will help the contractor to avoid the rejection of works. Rejection of checklist will result in rework, which will need time to redo the works and cost implication to the contractor. Frequent rejection of works may delay the project ultimately affecting the overall completion schedule.

8.6.6.4.4.1 Inspect Executed Works The inspection of construction works is performed throughout the execution of project. Inspection is an ongoing activity to physically check the installed works. Checklists (inspection notice) are submitted by the contractor to the consultant, who inspects the executed works/installations. If the work is not carried out as specified then it is rejected and the contractor has to rework or rectify the same to ensure compliance with the specifications. During construction, all the physical and mechanical activities are accomplished on the site. The contractor carries final inspection of the works to ensure full compliance with the contract documents.

During construction process, the contractor has to submit the checklists to the consultant to inspect the works. Submission of check or request for inspection is an ongoing activity during the construction process to ensure proper quality control of construction. Concrete work is one of the most important components of building construction. The concrete work has to be inspected and checked at all the stages to avoid rejection or rework. Necessary care has to be taken right from the control of design mix of the concrete till the casting is complete and cured. Contractor has to submit checklist at different stages of concrete work and has to make certain tests, specified in the contract, during casting of concrete. In order to ensure that structured concrete works are executed without any defects or rejection and achieved concrete strength as specified, proper sequencing of works is important.

Figure 8.63 illustrates checklist for general works to be inspected by the consultant.

If the consultant finds that an item has been executed at site not as per contract documents, specification, or general code of practice, then a remedial note is issued by the consultant. Figure 8.64 illustrates a remedial note.

	Owner Name
	Project Name
	Consultant Name
	CHECK LIST

CONTRACTOR : _____   CHECK LIST No. : _____

CONTRACT No. : _____   QC Code No. _____

TO   : Resident Engineer

CCS ACTIVITY NO : _____   SPECIFICATION DIVISION : _____   SECTION : _____

AREA :
- [ ] Process Work   [ ] Civil Works   [ ] Information and Communication
- [ ] Piping Work   [ ] Electrical Works   [ ] Low Voltage System
- [ ] Mechanical Works [ ] Instrumentation Works [ ] Other (Specify)

Please inspect the following :-

Location : _____

Work : _____

SAMPLE FORM

Sketch(es) attached { } No. _____

The work to be inspected has been coordinated with all related subcontractors.

Estimated Quantity of Work : _____   Date & Time Inspection Required : _____

Contractor Signature : _____   Date & Time _____

Received By : _____   Date & Time _____

C.C.: Owner Rep: _____   Date & Time _____

(All request must be submitted at least 24 hours prior to the required inspection)

Reply: The above is Approved/Not approved for the following :-

_____
_____
_____
_____

Inspected by _____ Date & Time _____ Resident Engineer _____ Date & Time _____

Received by Contractor _____   Date & Time _____

C.C.: Owner Rep: _____   Date & Time _____

FIGURE 8.63   Checklist.

The contractor is required to reply on the same form after taking necessary action. Upon finalization of the issue, the withdrawal notice is issued by the consultant/supervision staff.

Consultant's supervision staffs always make a routine inspection during the construction process. A nonconformance report is prepared and sent to the contractor

Quality Management

Project Name
Consultant Name

REMEDIAL NOTE (RN)

Contractor: _____   R.N. No.: _____

Contract No.: _____   DATE: _____

Your attention is drawn to the following works which have not been carried out in accordance with the Contract and are therefore not acceptable. Failure to carry out remedial works within a reasonable period of time may result either in additional work at your expense, or the Employer may elect to invoke Clause ---of the General Conditions of Contract.

LOCATION:

DEFECTS:

SAMPLE FORM

Signed: _____   Received by: _____
         Resident Engineer   Contractor/Date

Distribution: ☐ Owner   ☐ Consultant   ☐ Contractor   ☐

FIGURE 8.64 Remedial note.

to make corrective/preventive action toward this activity. Figure 8.65 illustrates a nonconformance report used for a building construction project.

8.6.6.4.5 Control Resources

The challenge for the construction management is to bring together all the required resources (material, manpower, and equipment) in the correct quantity at the correct time. In the construction network programs, in addition to detailing how the project will be assembled, the resources required for each activity are also prepared. This process is called resource-loading. It is necessary that all the construction resources be coordinated and brought together at the right time in order to complete

Project Name
Consultant Name

NON-CONFORMANCE REPORT NO. [ ]

CONTRACTOR: **DATE :** / /

CONTRACT NO:

Location of Non-Conformance :

Drawing / Specification :

SAMPLE FORM

Description of Non-Conformance :

Resident Engineer : ____________________

Corrective / Preventive Action : (Proposed by Contractor)

Quality Control Engineer : ________________

Contractor Manager: ____________________ **Date:** ____________________

Comments:

Resident Engineer: ____________________ **Date:** ____________

FIGURE 8.65 Non-conformance report.

the project on time and within the budget. All construction projects track equipment and manpower employed on-site by updating the respective logs.

Figure 8.66 illustrates manpower monitoring histogram and Figure 8.67 illustrates equipment status.

Quality Management

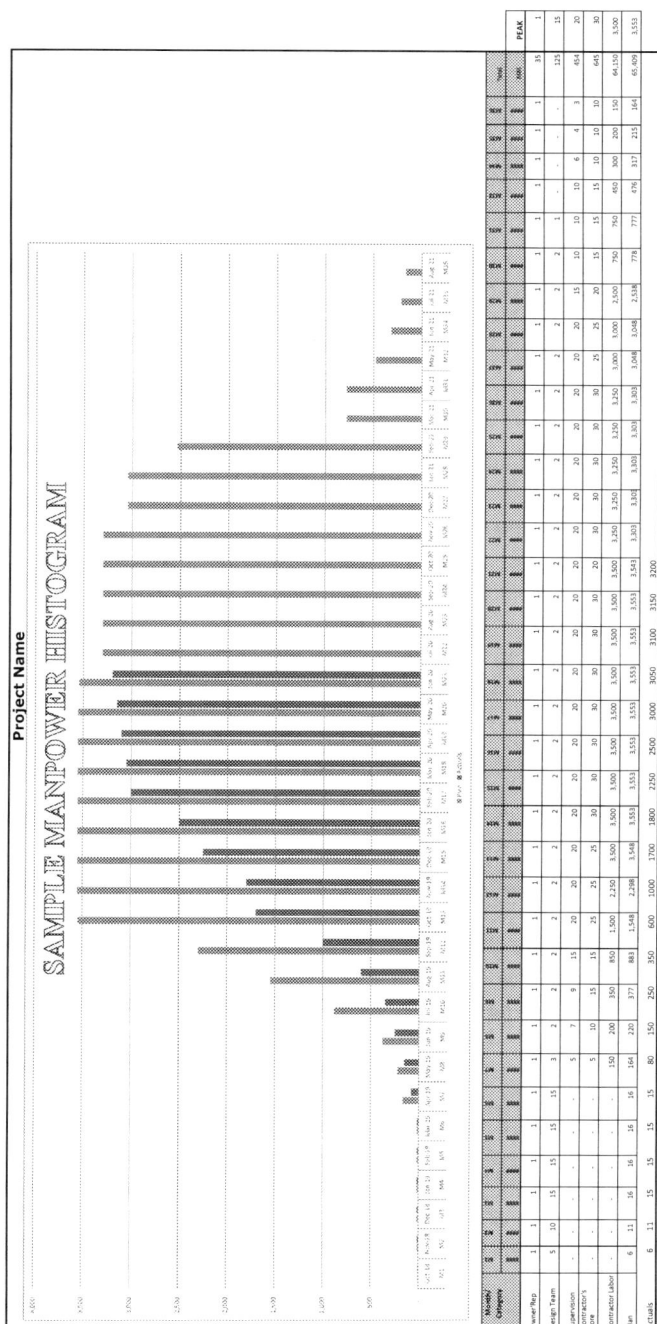

FIGURE 8.66 Project manpower monitoring histogram.

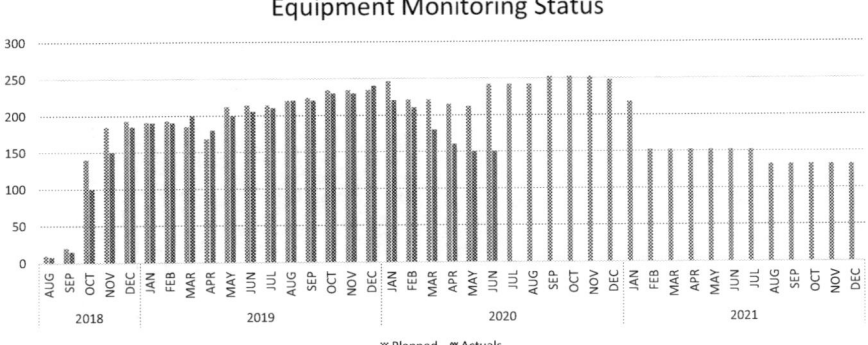

FIGURE 8.67 Equipment monitoring status.

8.6.6.4.6 Control Risk

Probability of occurrence of risk during the construction phase is very high compare to design phases. During the construction phase, uncertainty comes from various sources as this phase has involvement of various participants. The contractor has to develop risk management plan which identifies how risk associated with the project will be identified, analyzed, managed, and controlled. Risk management is an integral part of project management as the risk is likely to occur at any stage of the project. Therefore, the risk has to be continually monitored and response actions to be taken immediately.

8.6.6.4.7 Monitor Site Safety

The construction industry has been considered to be dangerous for a long time. The nature of work at site always presents some dangers and hazards. There are a relatively high number of injuries and accidents at construction sites. Safety represents an important aspect of construction projects. The construction site should be a safe place for those working there. Necessary measures are always required to ensure safety of all those working at construction sites. Effective risk control strategies are necessary to reduce and prevent accidents.

Contract documents normally stipulate that the contractor, upon signing the contract, has to submit a safety and accident prevention program. It emphasizes that all personnel have to put in efforts to prevent injuries and accidents. In the program, the contractor has to incorporate safety and health requirements of local authorities, manuals of accident prevention in construction, and all other local codes and regulations.

8.6.6.5 Validate Executed Works

The inspection of construction works is performed throughout the execution of project. Inspection is an ongoing activity to physically check the installed works. Checklists are submitted by the contractor to the consultant to inspect the executed works/installations.

If the work is not carried out as specified then it is rejected and the contractor has to rework or rectify the same to ensure compliance with the specifications. During construction, all the physical and mechanical activities are accomplished on the site. The contractor carries final inspection of the works to ensure full compliance with the contract documents.

8.7 TESTING, COMMISSIONING, AND HANDOVER PHASE

Testing, commissioning, and handover phase is the sixth phase of oil and gas construction project life cycle. Testing and commissioning has to be carried out on installed equipment, machinery, and systems to ensure that they are safe and meet the intended requirements of the project to the satisfaction of owner/end user. The testing is normally undertaken to prove the quality and workmanship of the installation. It is also known as static testing. Upon completion of static testing, dynamic testing can be undertaken; this is called "commissioning". Commissioning is carried out to prove that the systems operate and perform to the design intent and specification.

Commissioning is the orderly sequence of testing, adjusting, and balancing the equipment and system, and bringing the equipment, systems, and subsystems into operation and starts when the construction and installation of works are complete. Commissioning is normally carried out by contractor or specialist in the presence of consultant and owner/owner's representative and user's operation and maintenance personnel to ascertain proper functioning of the systems to the specified standards.

The commissioning of the construction project is a complex and intricate series of start-up operations and may extend over many months. Testing, adjusting, and operation of the equipment and systems are essential in order to make the project operational and useful for the owner/end user. It requires extensive planning and preparatory work, which commences long before the construction is complete. All the contracting parties are involved in the start-up and commissioning of the project as it is required for the project to be handed over to the owner/user. Figure 8.68 illustrates testing and commissioning process.

8.7.1 IDENTIFY TESTING AND START-UP REQUIREMENTS

Testing and start-up requirements are specified in the contract documents. It is essential to inspect and test all the installed/executed works prior to handover the project to the owner/end user. Generally, all works are checked and inspected on regular basis while the construction is in progress; however, there are certain inspection and tests to be carried out by the contractor in the presence of owner/consultant. These are especially for rotating equipment, systems, conveying systems, electrical works, low-voltage systems, information- and technology-related products, emergency power supply system, and electrically operated equipment which are energized after the connection of permanent power supply. Testing of all these equipment and systems start after the completion of installation works. By this time, facility is connected to permanent electrical power supply and all the equipment are energized.

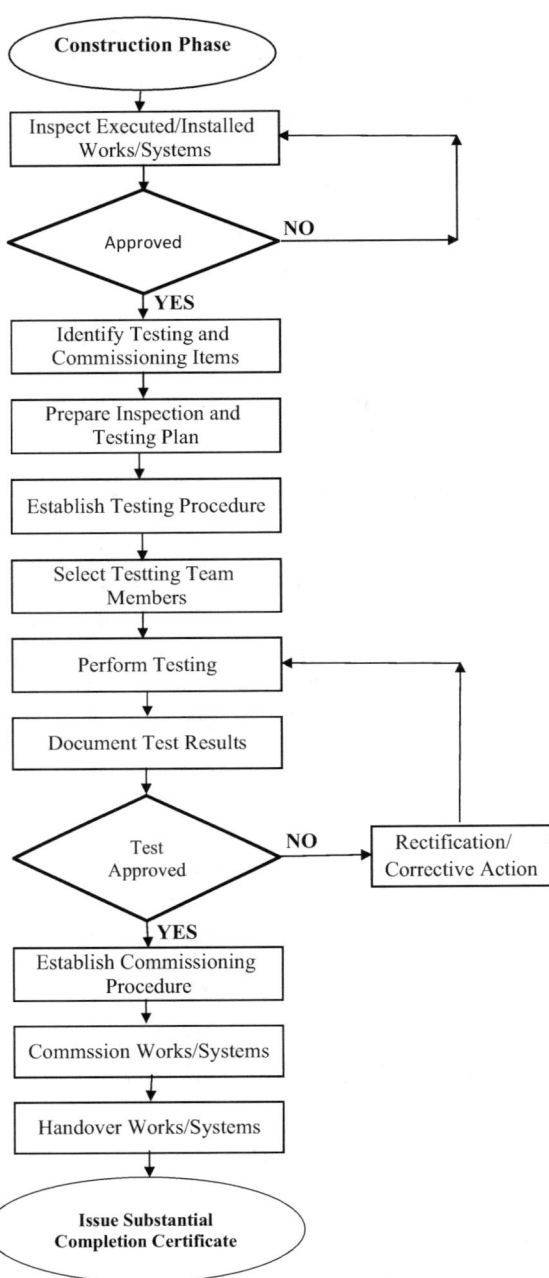

FIGURE 8.68 Logic flowchart for testing, commissioning, and handover phase. (Abdul Razzak Rumane. (2016). *Handbook of Construction Management.* Reprinted with permission from Taylor & Francis Group.)

Quality Management

8.7.2 Identify Stakeholders

The following stakeholders are involved during testing, commissioning, and handover phase:

1. Owner
    i. Owner's representative/Project Manager
2. Construction supervisor
    i. Construction supervisor (Consultant)
    ii. PMC
    iii. Specialist contractor
3. Contractor
    i. Main contractor
    ii. Subcontractor
    iii. Testing and commissioning specialist
4. Regulatory authorities
5. End user
6. Third-party inspection agency

8.7.2.1 Select Team Members

During the testing and commissioning phase, it is essential to select team members having experience in testing and commissioning of major projects. The team members can be from the same supervision team which was involved during execution of the project, if they have experience to carry out testing and commissioning. In most cases, the manufacturer's representative is involved to test the supplied equipment/systems. The owner may engage specialist firm(s) to perform start-up activities and commission the project.

8.7.2.2 Develop Responsibility Matrix

Table 8.38 lists responsibilities of consultant during the close-out phase and Table 8.39 illustrates contribution of various participants during testing, commissioning, and handover phase.

8.7.3 Develop Scope of Work

The contract documents specify the testing and commission works to be performed by contractor, subcontractor, and specialist supplier of equipment/systems. Following is the main scope of work to be carried out during this phase:

1. Testing of all equipment
2. Commissioning of all equipment
3. Testing of all systems
4. Commissioning of all systems
5. Performance testing of all equipment, systems
6. Obtaining authorities' approvals
7. Submission of as-built drawings

TABLE 8.38
Typical Responsibilities of Supervision Consultant during Project Testing, Commissioning, and Handover Phase

Serial Number	Responsibilities
1	Ensure that all the equipment, systems are functioning and operative
2	Ensure that performance tests are carried out on all the equipment, systems and equipment, systems are performing as per intended/design requirements
3	Ensure that job site instruction (JSI), nonconformance report (NCR) are closed
4	Ensure that site is cleaned and all the temporary facilities and utilities are removed
5	Ensure as-built drawings handed over to the client/end user
6	Ensure that operation and maintenance manuals handed over to the client
7	Ensure that record books are handed over to the client
8	Ensure that guarantees, warranties, bonds are handed over to the client
9	Ensure that test reports, test certificates, inspection reports handed over to the client
10	Ensure that spare parts are handed over to the client
11	Ensure that snag (punch) list prepared and handed over to the client
12	Ensure that training for client/end user personnel completed
13	Ensure that substantial completion certificate issued and maintenance period commissioned
14	Ensure that all the dues of suppliers, subcontractors, contractor paid
15	Ensure that retention money is released
16	Ensure that supervision completion certificate from the owner is obtained
17	Lesson learned documented

Source: Modified from; Abdul Razzak Rumane. (2013). *Quality Tools for Managing Construction Projects.* Reprinted with permission from Taylor & Francis Group.

TABLE 8.39
Responsibilities of Various Participants during Testing, Commissioning, and Handover Phase

	Responsibilities		
Phase	Owner	Consultant/PMC	Contractor
Testing, Commissioning, and Handover	Acceptance of project Take-over Substantial completion certificate Training Payments	Witness tests Check close-out requirements Recommend take over Recommend issuance of Substantial completion certificate	Testing Commissioning Authorities approvals Documents Training Handover

8. Submission of technical manuals and documents
9. Submission of record books
10. Submission of warranties and guarantees
11. Training of owner/user's personnel

Quality Management

12. Handover of spare parts
13. Handover of the project to owner/end user
14. Preparation of punch list
15. Issuance of substantial certificate
16. Lessons learned

8.7.4 Develop Testing, Commissioning, and Handover Plan

Figure 8.69 illustrates the logic flowchart for development of inspection and testing plan

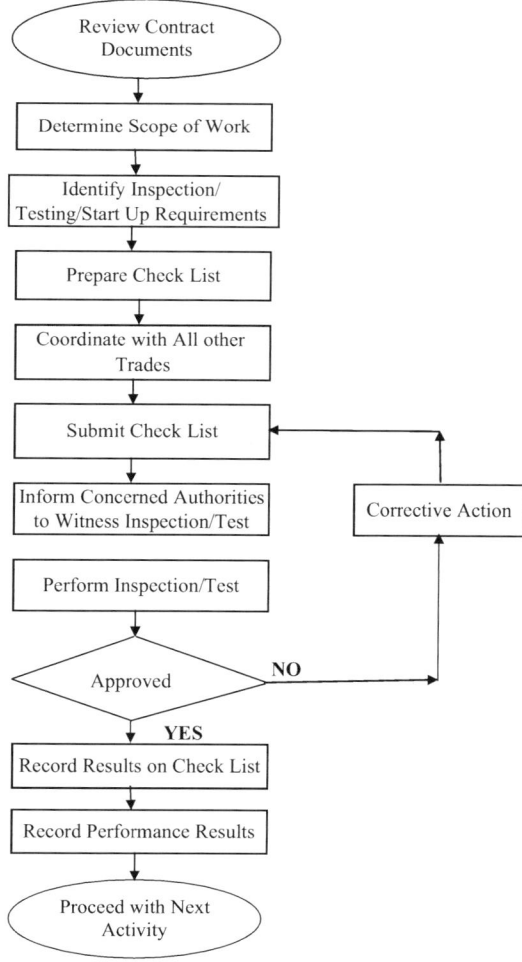

FIGURE 8.69 Logic flowchart for development of inspection and test plan. (Abdul Razzak Rumane. (2016). *Handbook of Construction Management*. Reprinted with permission from Taylor & Francis Group.)

8.7.5 Execute Testing and Commissioning Works

The testing has mainly carried out on electromechanical works/systems and electrically operated equipment/systems and rotating equipment which is energized after the connection of permanent power supply to the facility. These include the following:

1. Pumps (all types)
2. Piping works (pressure test)
3. Valves
4. Compressors
5. Heat exchangers
6. Separators
7. Generators
8. Coolers
9. Fire-fighting system
10. Flare and incinerator system
11. Leak-detection system
12. Cathodic protection system
13. Corrosion monitoring system
14. Process control and metering
15. Process automation system
16. Instrumentation and control system
17. Supervisory control system
18. Integrated automation system (building automation system)
19. HVAC
20. Water supply, plumbing, and public health system
21. Fire suppression system
22. Conveying system
23. Electrical switchgear
24. Electrical lighting and power system
25. Grounding (earthing) and lightning protection system
26. Fire alarm system
27. Information and communication system
28. Electronic security and access control system
29. Data security system
30. Public address system
31. Parking control system
32. Material handling equipment
33. Emergency power supply system
34. Electrically operated equipment.

The testing of these works/systems is essential to ensure that each individual work/system is fully functional and operates as specified. The tests are normally coordinated and scheduled with specialist contractors, local inspection authorities, third-party inspection authorities, and manufacturer's representatives. Sometimes, owner's representative may accompany the consultant to witness these tests.

Quality Management

Test procedures are submitted by the contractor along with the request for final inspection. Standard forms, charts, and checklists are used to record the testing results.

8.7.6 Manage Testing and Commissioning Quality

The contract document identifies testing and commissioning procedure to be followed on the installed equipment.

In order to follow the specified procedure and smooth functioning of testing and commissioning, the contractor/consultant has to plan quality (planning of design work), perform quality assurance, and control quality for managing testing and commissioning activities. This will mainly consist of following:

1. Plan Quality
    i. Identify testing and commissioning requirements
    ii. Identify codes and standards to be followed for testing and commissioning
    iii. Identify contract specifications to be followed for testing and commissioning
    iv. Identify regulatory requirements to be followed for testing and commissioning
    v. Identify (list) the items to be tested
    vi. Identify (list) the items to be commissioned
    vii. Establish testing and commissioning criteria
    viii. Identify manufacturer recommendation procedure
    ix. Establish detailed procedure for testing of equipment
    x. Ensure mechanical completion of the plant/equipment
    xi. Identify precommissioning requirement
    xii. Establish schedule for testing and commissioning
    xiii. Establish sequence of commissioning of equipment
    xiv. Identify and prequalify testing and commissioning agency, if mandated by the contract
    xv. Identify stakeholders to witness the testing and commissioning
    xvi. Calibration/certification of testing equipment to be used for testing and commissioning
    xvii. Acceptance criteria
    xviii. Establish corrective action plan
2. Perform Quality Assurance
    i. Witness by the stakeholders
    ii. Ensure Inspection of all installed equipment is finished
    iii. Manufacturer's recommended method statement
    iv. Ensure Operating and maintenance manual requirements are followed
    v. Sequence of testing and commissioning
    vi. Testing Test points
    vii. Approval of third party inspection agency
    viii. Approval of specialist commissioning agency

  ix. Records of testing and commissioning
  x. Calibration Certificate of testing equipment used for Testing and Commissioning
  xi. Ensure availability of "Punch List"
3. Quality Control
  i. Check line by line as per flowchart
  ii. Test points (location)
  iii. Pressure test (pipe, vessel)
  iv. Flow test
  v. Control system
  vi. Valves
  vii. Temperature
  viii. Storage capacity
  ix. Installation safety
  x. Environmental compatibility

8.7.7 Develop Documents

Figure 8.70 illustrates list project close-out documents to be submitted for project close-out.

Description	Testing and Commissioning	As-Built Drawings	O & M Manuals	Guarantees	Warranties	Authorities Approvals	Record Documents	Test Certificates	Performance Results	Spare Parts	Punch Lists	Final Cleaning	Training	Taking Over Certificate	Remarks
PROCESS WORKS															
PIPING WORKS															
MECHANICAL WORKS															
ELECTRICAL WORKS															
INSTRUMENTATION WORKS															
LOW VOLTAGE SYSTEMS															
AUTOMATION & CONTROL															
EXTERNAL WORKS															

FIGURE 8.70 Project close-out report.

8.7.7.1 As-Built Drawings

Most contracts require contractor to maintain a set of record drawings. These drawings are marked to indicate the actual installation where the installation varies appreciably from installation shown in the original contract. Revisions and changes to the original contract drawings are almost certain for any construction project. All such revisions and changes are required to be shown on the record drawings. As-built drawings are prepared by incorporating all the modifications, revisions, and changes made during the construction. These drawings are used by the user/operator after taking over the project for their reference purpose. It is the contractual requirement that the contractor handovers as-built drawings along with record drawings, record specifications, and record product data to the owner/user before handing over of the project and issuance of substantial completion certificate. In certain projects, contractor has to submit field records on excavation and underground utility services detailing their location and levels.

8.7.7.2 Technical Manuals and Documents

Technical manuals, design and performance specifications, test certificates, warranties and guarantees of the installed equipment are required to be handed over to the owner as part of contractual conditions.

Systems and equipment manuals submitted by the contractor to the owner/end user generally consist of

1. Source information
2. Operating procedures
3. Manufacturer's maintenance documentation
4. Maintenance procedure
5. Maintenance and service schedules
6. Spare parts list and source information
7. Maintenance service contract
8. Warranties and guarantees

Procedure for submission of all these documents is specified in the contract document.

8.7.7.3 Record Books

1. Manufacturing record book
2. Project record book
3. Engineering record book
4. Construction record book

8.7.7.4 Warranties and Guarantees

Contractor has to submit warranties and guarantees in accordance with the contract documents. Normally, the guarantee for water-proofing woks varies from 15 to 20 years. Similarly, warranty for diesel generator is set as 5 years.

8.7.8 MONITOR WORK PROGRESS

Schedule for testing, commissioning and handover phase to be prepared, and all the activities to be performed as per agreed upon plan.

8.7.9 TRAIN OWNER'S/END USER'S PERSONNEL

Normally, training of user's personnel is part of contract terms. The owner's/user's commissioning, operating, and maintenance personnel are trained and briefed before commissioning starts in order to familiarize the owner's/user's personnel about the installation works and also to ensure that the project is put into operation rapidly, safely, and effectively without any interruption. Timings and details of training vary widely from project to project. Training must be completed well in advance of the requirement to make the operating teams fully competent to be deployed at the right time during commissioning. This needs to be planned from project inception, so that the roles and activities of the commissioning and operating staff are integrated into a coherent team to maximize their effectiveness.

8.7.10 HANDOVER THE PROJECT

Once the contractor considers that the construction and installation of works have been completed as per the scope of contract, and final tests have been performed and all the necessary obligations have been fulfilled, the contractor submits a written request to the owner/consultant for handing over of the project and for issuance of substantial completion certificate. This is done after testing and commissioning is carried out and it is established that the project can be put in operation or owner can occupy the same. In most construction projects, there is a provision for partial handover of the project.

The contractor starts handing over of all completed works/systems which are fully functional and the owner has agreed to take over the same. A handing-over certificate is prepared and signed by all the concerned parties. Figure 8.71 illustrates sample handing-over certificate.

8.7.10.1 Obtain Authorities' Approval

Necessary regulatory approvals from the respective concerned authorities are obtained so that owner can occupy the facility and start using/operating the same. In certain countries, all such approvals are needed before electrical power supply is connected to the facility. It is also required that the building/facility is certified by the related fire department authority/agency that it is safe for occupancy.

8.7.10.2 Handover Spare Parts

Most contract documents include the list of spare parts, tools, and extra materials to be delivered to the owner/end user during the close out stage of the project. The contractor has to properly label these spare parts and tools clearly indicating manufacturer's name and model number if applicable. Figure 8.72 illustrates spare parts handing-over form used by the contractor.

	Project Name	
	Project Name	
	HANDING OVER CERTIFICATE	
CONTRACTOR : _____		CERTIFICATE No. : ☐
SUBCONTRACTOR: _____		DATE ☐

SPECIFICATION NO : _____   DIVISION : _____   SECTION : _____

DRAWING No.   BOQ REF : _____

AREA : ☐ Process Works   ☐ Electrical Works   ☐ Mechanical Works

        ☐ Piping Works   ☐ Instrumentation Works   ☐ Low Voltage Systems

Description of Work/System: *SAMPLE FORM*
_____
_____

The work/system mentioned above is completed by the contractor as specified and has been inspected and tested as per contract documents. The work/system is fully functional to the satisfaction of owner/end user. The contractor hand over the said work/system to the owner/end user as on ---------------. The guarentee/warranty of work/system shall start as on ------------- and shall be valid for a period of -----------------years(duration) from the date of issuance of substantial completion certificate. The contractor shall be liable contractually till the end of warranty/guarentee period.

SIGNED BY:

OWNER/END USER: _____   CONTRACTOR: _____

CONSULTANT: _____   SUBCONTRACTOR: _____

FIGURE 8.71 Handing-over certificate.

8.7.10.3 Accept/Takeover the Project

Normally, a final walk through inspection of the project is carried out by the Committee which consists of owner's representative, design and supervision personnel, and contractor to decide the acceptance of work and that the project is complete enough to be put in use and is operational. If there are any minor items remain to be finished then such list is attached with the certificate of substantial completion for conditional acceptance of the project. Issuance of substantial completion certifies acceptance of work. If the remaining work is minor, then the contractor has to

Project Name								
Project Name								
HANDING OVER OF SPARE PARTS								

CONTRACTOR: _____ CERTIFICATE No. : _____

SUBCONTRACTOR: _____ DATE _____

SPECIFICATION NO : _____ DIVISION _____ SECTION : _____

DRAWING No. _____ BOQ Ref. _____

AREA : ☐ Process Works ☐ Electrical Works ☐ Mechanical Works ☐

☐ Piping Works ☐ Instrumentation Works ☐ Low Voltage System ☐

Following Spare Parts have been handed over to the owner/end user

Description of Spare Parts

Sr.No.	Description	BOQ Reference	Spec. Ref.	Manufacturer	Specified Qty	Delivered Qty

(Attach additional sheet, if required)

SIGNED BY:

OWNER/END USER: _____ CONTRACTOR: _____

CONSULTANT: _____ SUBCONTRACTOR: _____

FIGURE 8.72 Handing-over of spare parts.

submit a written commitment that he shall complete said works within the agreed upon period. A memorandum of understanding is signed between the owner and the contractor that the remaining works will be completed within an agreed upon period.

8.7.11 CLOSE CONTRACT

Table 8.40 illustrates a list of activities that need to be considered for project close-out.

TABLE 8.40
Project Close out Checklist

Serial Number	Description	Yes/No
	Project Execution	
1	Contracted works completed	
2	Site work instructions completed	
3	Job site instructions completed	
4	Remedial notes completed	
5	Noncompliance reports completed	
6	All services connected	
7	All the contracted works inspected and approved	
8	Testing and commissioning carried out and approved	
9	Any snags	
10	Is project fully functional?	
11	All other deliverable completed	
12	Spare parts delivered	
13	Is waste material disposed?	
14	Whether safety measures for use of hazardous material established?	
15	Whether the project is safe for use/occupation?	
	Project Documentation	
16	Authorities approval obtained	
17	Record drawings submitted	
18	Record documents submitted	
19	As-Built drawings submitted	
20	Technical manuals submitted	
21	Operation and maintenance manuals submitted	
22	Equipment/material warrantees/guarantees submitted	
23	Test results/test certificates submitted	
	Training	
24	Training to owner/end user's personnel imparted	
	Payments	
25	All payments to subcontractors/specialist suppliers released	
26	Bank guarantees received	
27	Final payment released to main contractor	
	Handing over/Taking over	
28	Project handed over/taken over	
29	Operation/maintenance team taken over	
30	Excess project material handed over/taken over	
31	Facility manager in action	

8.7.11.1 Prepare Punch List

Owner/consultant inspects the works and inform contractor of unfulfilled contract requirements. A punch list (snag list) is prepared by the consultant listing all the items still requiring completion or correction. The list is handed over to the contractor for rework/correction of the works mentioned in the punch list. Contractor rcsubmit inspection request after completing or correcting previously notified works. A final snag list is prepared if there are still some items which need corrective action/completion by the contractor; such remaining works are to be completed within the agreed period to the satisfaction of the owner/consultant. Table 8.41 is sample form for preparation of punch list.

8.7.11.2 Prepare Lesson Learned

Construction/Project Manager, consultant, and contractor have to prepare lesson learned and document the same for future references to improve the processes and organizational performance. This includes

TABLE 8.41
Punch List

Owner name
Project name
Punch list

Punch list number:		Date:
Area		Type of work
Zone		
Serial number	Item	Remark
1	Piping	
2	Storage tank	
3	Pumps	
4	Vessel	
5	Fire alarm detectors	
6	Sprinklers	
7	Communication system devices	
8		
9		
10		
11		
12		
13		
14		
15		
16		
17		
18		
19	Any other item	

1. Reasons for delay
2. Resigns for rejection/rework
3. Preventive/corrective actions
4. Cost overrun
5. Causes for claims
6. Reasons for conflict

8.7.11.3 Issue Substantial Completion Certificate

A substantial certificate is issued to the contractor once it is established that the contractor has completed works in accordance with the contract documents and to the satisfaction of the owner. Contractor has to submit all the required certificates and other documents to the owner before issuance of the certificate.

The certificate of substantial completion is issued to the contractor, and the facility is taken over by the owner/end user. By this stage, owner/end user already takes possession of the facility and operation and maintenance of the facility commences. The project is declared complete and is considered as the end of the construction project life cycle.

The defect liability period starts after issuance of substantial certificate.

During this period, the contractor has to complete the punch list items and also to rectify the defects identified in the project/facility.

8.7.11.4 Terminate/Demobilize Team Members

The project team members are demobilized from the project and their services are either terminated or shifted to other projects, if positions are available to accommodate. In certain case, the project members join the member of functional team at the project office.

8.7.12 Settle Payments

The owner has to settle all the dues toward consultant, contractor, and other parties involved. Similarly, the contractor has to settle due payments to their subcontractors and suppliers.

8.7.13 Settle Claims

The entire project-related claims have to be amicably settled as per contract conditions to close the project.

9 Operation and Maintenance of Oil and Gas Projects

9.1 OPERATION MANAGEMENT

Operation management is a system that manages the operational activities of equipment, machinery, material, system, and resources to operate together smoothly and safely to produce desired product(s) to satisfy the customer's need. Operation management deals with the activities to produce quality product(s) that satisfy customer requirements and to be cost-efficient achieving competitive advantage for the organization. An operations management department ensures that processes are planned for and executed in a safe, reliable, efficient, and effective way to satisfy the needs of the organization and its customers.

Once the project is completed, testing and commissioning is finished, and the project is functional, the project is handed over to the owner/end user, and then it becomes part of the operation system. The operation and maintenance (O&M) of a completed project are associated with the production of products and the quality of the end products.

The operation of the project is managed by specialist operation and maintenance team to operate and maintain the project/plant. The following documents are required to operate and maintain the plant. These documents are normally submitted by the contractor during the testing, commissioning, and handover phases of the construction project.

1. As-built drawings
2. Operation and maintenance manuals
3. Operating procedures
4. Record books
5. Source information for availability of spare parts of the installed works

Figure 9.1 illustrates logic flowchart for operation and maintenance.

9.2 MAINTENANCE MANAGEMENT

Maintenance management is defined as the process of maintaining a company's assets and resources while controlling time and costs, ensuring maximum efficiency of the manufacturing process. **Maintenance** is performed on equipment, machinery, systems, or facilities to minimize the possibility of damage or the lowering of performance, operating quality because of corrosion, contamination, or deterioration for the continuous operation.

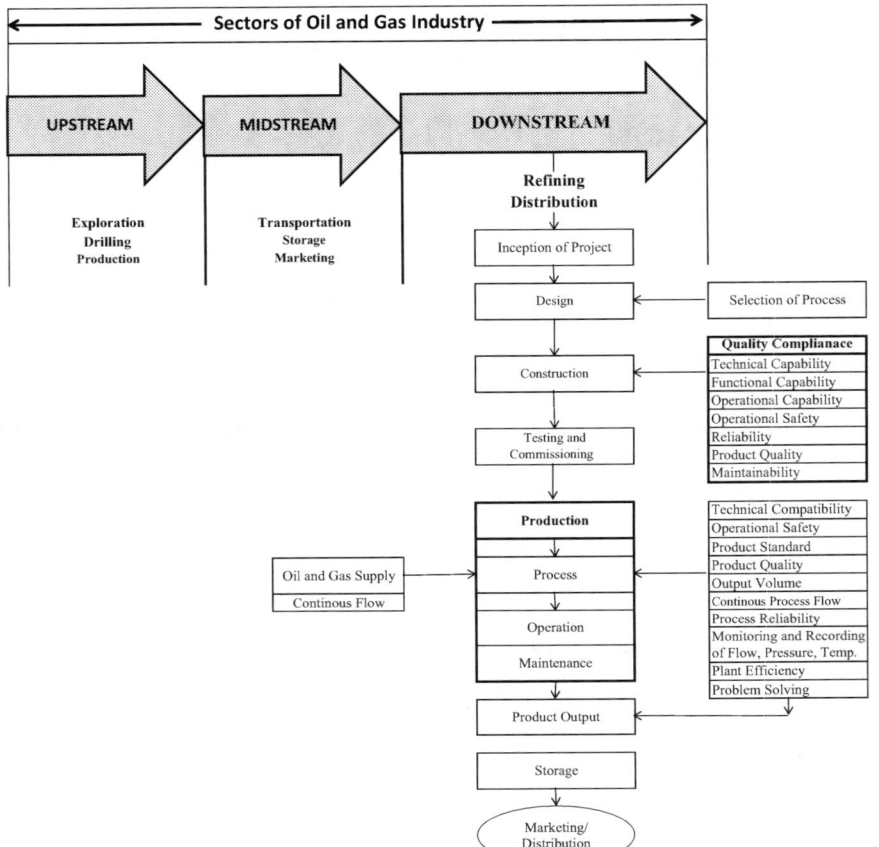

FIGURE 9.1 Logic flowchart for operation and maintenance.

9.2.1 Categories of Maintenance

The following are different types of maintenance normally being carried out, depending on the nature of activity, system, or equipment, to avoid interference/interruption with essential installation operations, endanger life or property, or involving high cost for replacement of the item.

9.2.1.1 Scheduled Maintenance

Scheduled maintenance includes those maintenance tasks whose cycles exceed a specified period. This type of maintenance relates to finish items, such as painting, polishing, parking and road markings, roof maintenance, and testing of fire alarm systems. Scheduled maintenance is aimed at avoiding breakdowns. Scheduled maintenance practice incorporates (in it) inspection, lubrication, repair, and overhaul of certain equipment, which if neglected can result in breakdown.

9.2.1.2 Preventive Maintenance

Preventive maintenance is the planned, with scheduled periodic inspection, adjustment, cleaning, lubrication, parts replacement of equipment, and systems in operation. Its aim is prevention of breakdown and failure. Preventive maintenance consists of many activities required to be checked and performed to ensure that the operations are safe and can be performed without any danger to life or facility and will not warrant high cost or long lead time for replacement. Preventive maintenance is the cornerstone of any good maintenance program.

Preventive maintenance is regularly performed on equipment/system to lessen the likelihood of it failing. Preventive maintenance is performed while the equipment is still working, so that it does not break down unexpectedly. A system of scheduled, planned, or preventive maintenance tries to minimize the problems of breakdown maintenance.

9.2.1.3 Corrective or Breakdown Maintenance

Corrective or breakdown maintenance for systems or equipment is scheduled (planned) and unscheduled (unplanned). Unscheduled (unplanned) is unanticipated maintenance because of system or component failure. If the problem relates to an essential service or has any hazardous effect, then an emergency response is necessary. If the problem is not critical, then routine response is adequate. Corrective maintenance is performed to identify, isolate, and rectify a fault so that the failed equipment, machine, or system can be restored to an operational condition within the tolerances or limits established for in-service operations.

Corrective or breakdown maintenance implies that repairs are made after the equipment is out of order and it cannot perform its normal function any longer, e.g., an electric motor will not start and a belt is broken. Under such conditions, production department calls on the maintenance department to rectify the defect.

9.2.1.4 Predictive Testing and Inspection

Predictive testing and inspection refers to testing and inspection activities that involve the use of sophisticated means to identify maintenance requirements. For example, specialized tests are to be carried out:

- To locate wearing problems on bus bar bends of an electrical main's low tension panel
- To detect insulation cracks
- To locate thinning of pipe walls and cracks
- To detect problem in rotating equipment due to unusual sound
- To identify vibration problems for equipment such as chillers, diesel generator sets, and pumps
- To locate heat buildup (rise in temperature)
- To predict troubles before the equipment such as pressure gauge, temperature gauge, and resistance strain gauges fails

Predictive maintenance techniques are designed to help determine the condition of in-service equipment in order to predict when maintenance should be performed.

Predictive maintenance extends the service life of equipment without fear of failure and thus promises cost savings over routine or time-based preventive maintenance, because tasks are performed only when warranted.

9.2.1.5 Routine Maintenance

Routine maintenance is an action taken by the maintenance team to restore a system or piece of equipment to its original performance, capacity, efficiency, or capability.

9.2.1.6 Replacement of Obsolete Items

Replacement of obsolete items refers to the work undertaken by a maintenance team to bring a system or component into compliance with new codes of safety regulations or to replace an item that is unacceptable, inefficient, or for which spare parts are no longer available or have had technological changes. Early detection of problems may reduce repair and replacement costs, prevent malfunctioning, and minimize downtime as, for example, with communication systems, audio visual systems, and equipment.

9.2.1.7 Shutdown Maintenance

Shutdown maintenance is performed while equipment is not in use. Shutting down machinery can be costly, but sometimes due to the nature of the defective part/machine, shutdown maintenance is the only viable maintenance procedure.

9.3 MANAGE OPERATION AND MAINTENANCE QUALITY

In order for smooth functioning of plant operation and maintenance, the company has to plan quality (planning of design work), perform quality assurance, and control quality for managing operation activities. This will mainly consist of the following:

1. Plan quality:
    - Identify operation and maintenance requirements
    - Identify codes and standards of the product(s) to be produced
    - Identify operational safety of the plant
    - Identify (list) the items to be produced
    - Identify (list) the quantity (volume) of each item to be produced
    - Confirm the capability/ability of the process to produce the required quantity (volume)
    - Identify production procedure to be followed
    - Establish detail procedure for operation of the plant
    - Ensure the process equipment are commissioned and are ready for start-up
    - Identify whether predefined operational requirements are complete and are safe and reliable
    - Establish schedule for production
    - Ensure availability of operations team
    - Ensure availability of maintenance team
    - Confirm continuous availability of feedstock

- Identify and prequalify testing and commissioning agency, if mandated by the contract
- Ensure control systems are operative
- Acceptance criteria of output product(s)
2. Perform quality assurance
   - Ensure all the equipment are operational and safe for usage
   - Ensure safety of team members
   - Ensure all equipment are operative and functional
   - Manufacturer's recommended operation methods are followed
   - Ensure operating and maintenance manual requirements are followed
   - Ensure processing sequence is followed
   - Testing of samples at the agreed-upon frequency
   - Calibration of processing equipment at agreed-upon frequency
3. Quality control
   - Recording flow rate
   - Recording pressure at various points
   - Recording temperature
   - Monitoring control systems
   - Product quality within control limit (statistical quality control)
   - Product quantity (volume)

9.3.1 Total Quality Management in Operation and Maintenance

The concept of total quality management can be applied to operation and maintenance function.

Figure 9.2 illustrates concept of total quality management in operation and maintenance.

9.4 OPERATION AND MAINTENANCE PROGRAM

Addressing operation and maintenance considerations at the start of the project contributes greatly to improved working environments, higher productivity, and reduced energy and resources costs. During the design phase of the project and up to handover of the facility, O&M personnel should be involved in identifying maintenance requirements for inclusion in the design. The goal of effective O&M is to achieve the intent of the original design team, i.e., the systems and equipment delivery services to the user enhancing a comfortable, healthy, and safe environment. O&M should also include long-term goals such as economy, energy conservation, and environmental protection. To create an effective O&M program, the following procedures should be followed:

- Ensure that up-to-date as-built drawings for all the systems are available.
- Ensure that operational procedures and manuals for the installed equipment are available.
- Prepare a master schedule for operation and preventive/predictive maintenance.

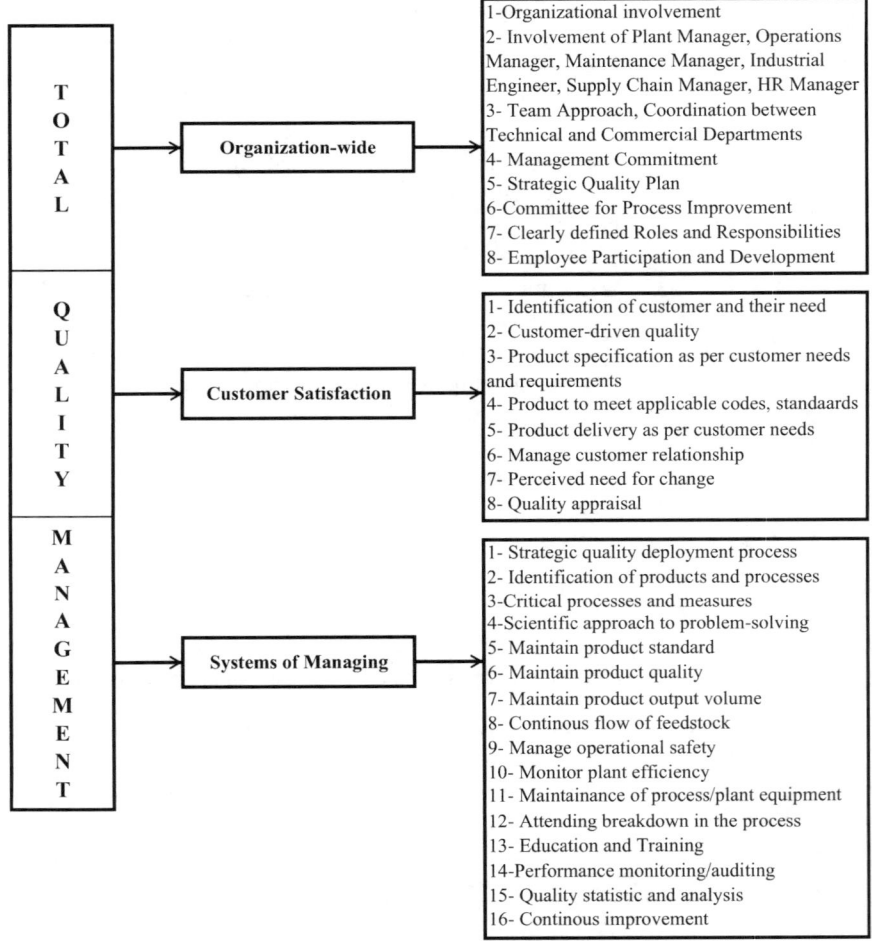

FIGURE 9.2 Concept of total quality management in operation and maintenance.

- Implement preventive maintenance programs complete with maintenance schedules and records of all maintenance performed for all the equipment and systems installed in the project/facility.
- Follow manufacturer's recommendations for maintenance of equipment and systems.
- Ensure the maintenance personnel have full knowledge of the equipment and systems installed and for which he/she is responsible to operate and maintain.
- Ensure that O&M personnel are provided training during the testing, commissioning, and handover phase.
- Offer training and professional development opportunities for each of the O&M team members.

Operation and Maintenance

- Implement a monitoring program that tracks and documents equipment performance to identify and diagnose potential problems.
- Implement a monitoring program that tracks and documents systems performance to identify and diagnose potential problems.
- Perform predictive testing and inspection for critical and important items.
- Use preventive maintenance and standbys, so that the failed components can be isolated and repaired without interrupting system performance, thus minimizing equipment failures.

For an effective O&M program, the following specific items should be considered:

1. Process equipment
2. Pumps
3. Compressors
4. Generators
5. Control systems
6. HVAC (heating, ventilation, and air conditioning) equipment
7. Electrical systems and equipment
8. Public health items
9. Life safety (fire suppression systems)
10. Cleaning equipment and products
11. Landscape maintenance

A guideline to prepare a preventive maintenance program, taking into consideration the most important equipment and systems of an oil and gas plant, is illustrated in Table 9.1.

TABLE 9.1
Preventive Maintenance Program for Oil and Gas Plant

Sr. No.	Activities	Maintenance Frequency				
		Daily	Weekly	Monthly	Quarterly	Yearly
Accumulators						
Automation system						
Boilers pressure vessels						
Building automation system						
Chiller units						
Compressors						

(*Continued*)

TABLE 9.1 (*Continued*)
Preventive Maintenance Program for Oil and Gas Plant

Sr. No.	Activities	Maintenance Frequency				
		Daily	Weekly	Monthly	Quarterly	Yearly
	Coolers					
	Cooling towers					
	Diesel generator set					
	Electrical switchgears					
	Electrical works					
	Electrical motors					
	Filters					
	Firefighting system					
	Flare system					
	Nitrogen generators					
	Steam generators					
	Heat exchangers					
	Heaters					
	Instrumentation and control					
	Material handling equipment					
	Packages					
	Piping					
	Process control and metering					
	Pumps					
	Fire pumps					
	Jockey pump					

(*Continued*)

Operation and Maintenance

TABLE 9.1 (*Continued*)
Preventive Maintenance Program for Oil and Gas Plant

Sr. No.	Activities	Maintenance Frequency				
		Daily	Weekly	Monthly	Quarterly	Yearly
	Sump pump (drainage pump)					
	Public health items					
	Separators					
	Scrubbers					
	Storage tanks					
	Supervisory control system					
	Water tanks					
	Transformers					
	Turbines					
	Valves					
	Automatic transfer switch					
	Access control system					
	Audio visual system					
	Cathodic protection system					
	Cleaning equipment					
	Dimmer system					
	Elevator					
	Escalator					
	Fire suppression system					
	Information and communication system					

(*Continued*)

TABLE 9.1 (*Continued*)
Preventive Maintenance Program for Oil and Gas Plant

		Maintenance Frequency				
Sr. No.	Activities	Daily	Weekly	Monthly	Quarterly	Yearly
	Kitchen equipment					
	Parking control system					
	Public address system					
	Security system					
	Uninterrupted power supply					
	Landscape					
	Plants					

Bibliography

American Society of Civil Engineers. (2012). *Quality in the Constructed Project: A Guide for Owners, Designers, and Constructors*, Second Edition. Reston, VI: American Society of Civil Engineers.

Ashford, J L. (1989). *The Management of Quality in Construction*. London: E & FN SPON.

Chase, R., Aquilano, N., and Jacobs, F. (2001). *Operations Management for Competitive Advantage*, Ninth Edition. New York: Irwin, McGraw-Hill.

Chung, H.W. (1999). *Understanding Quality Assurance in Construction: A Practical Guide to ISO 9000 for Contractors*. London: E & FN SPON.

Construction Industry Institute. (April 1990). *Total Quality Management: The Competitive Edge*. CII Publication 10-4. Austin, TX: University of Texas.

Construction Industry Institute. (June 1992). *Guidance for Implementing Total Quality Management in Engineering and Construction Industry*, CII Source Document 74. Austin, TX: University of Texas.

Construction Industry Institute. Source Document 51. Austin, TX: University of Texas.

Crosby, P.B. (1979). *Quality Is Free: The Art of Making Quality Certain*. New York, McGraw-Hill.

Deming Institute. (2000). Out of the Crisis. www.deming.org.

Feigenbaum, A.V. (1991). *Total Quality Control*, Third Edition. New York: McGraw-Hill.

Gryna, F.M. (2001). *Quality Planning and Analysis*, Fourth Edition. New York: Irwin, McGraw-Hill.

International Organisation for Standardisation, from ISO, http://www.iso.org.

Ishikawa, K. (Translated by David J. Lu). (1985). *What Is Total Quality Control? The Japanese Way*. Englewood Cliff, NJ: Prentice-Hall, Inc.

Juran, J.M., and Godfrey, A.B. (1999). *Juran's Quality Handbook*, Fifth Edition. New York: McGraw-Hill.

Kerzner, H. (2001). *Project Management: A Systems Approach to Planning, Scheduling and Controlling*, Seventh Edition. New York: John Wiley & Sons.

MasterFormat. (2016). The Construction Specifications Institute and Construction Specification, Canada. http://www.csi.com.

Oakland, J.S. (2014). *Total Quality Management*. Abingdon, UK: Taylor & Francis Group.

Project Management Institute. *A Guide to the Project Management Body of Knowledge*. PMBOK® Guide-Fifth Edition. Newtown Square, PA: PMI.

Pyzdek, T. (1999). *Quality Engineering Handbook*. Boca Raton, FL: CRC Press/Taylor & Francis Group.

Rumane, A. R. (2013). *Quality Tools for Managing Construction Projects*. Boca Raton, FL: CRC Press.

Rumane, A. R. (2016). *Handbook of Construction Management: Scope, Schedule, and Cost*. Boca Raton, FL: CRC Press/Taylor & Francis Group.

Rumane, A. R. (2017). *Quality Management in Construction Projects*, Second Edition. Boca Raton, FL: CRC Press/Taylor & Francis Group.

Rumane, A. R. (2019). *Quality Auditing in Construction Projects*. Abingdon, UK: Routledge/Taylor & Francis Group.

Tague, N.R. (2005). *The Quality Toolbox*. Milwaukee, WI: American Society for Quality Press.

Thorpe, B., Sumner, P., and Duncan, J. (1996). *Quality Assurance in Construction*. Surrey, UK: Gower Publishing Ltd.

Turner, J.R. (2003). *Contracting for Project Management*. Surrey, UK: Gower Publishing Ltd.

Index

A/E 102, 103, 260, 285, 367, 386, 456, 457, 458, 470
accessibility 319, 344, 368, 436
acoustic 8, 69
activity network 137, 138
addendum 297, 307, 373, 374, 377, 379
aesthetic 204, 319, 321
affinity diagram 137, 140, 143, 144
agency 42, 57, 99, 115, 192, 195, 197, 275, 277, 278, 279, 280, 297, 314, 318, 354, 372, 374, 443, 459, 497, 501, 504, 515
agenda 381, 382, 468, 470, 486, 487
align 31, 145, 178, 179, 181, 452
Altshuller, G. 187, 188, 189, 190
appraisal 7, 172, 175, 208, 371, 516
arbitration 459
as built 486
Ashford, J. L. 13
ASQ 8, 10, 12, 29
assumptions 316, 320, 339, 354, 404, 435, 441
attribute control chart 20, 159
audit, first party 99
audit, second party 99
audit, third party 99
authority 9, 57, 110, 111, 119, 192, 194, 202, 208, 222, 223, 226, 243, 258, 277, 278, 297, 372, 383, 384, 469, 504
automation 7, 170, 199, 200, 206, 321, 325, 330, 351, 352, 353, 357, 358, 363, 368, 400, 430, 437, 447, 500, 502, 517
avoidance 6, 45, 50, 249

backfilling 155, 157, 170, 405, 406, 447
bar chart 333, 336, 337, 338, 358, 404, 486
Bard, J. F. 25
benchmarking 45, 147, 149, 150, 151, 326
bill of material 283
bill of quantity 33, 149, 259, 266, 283, 362, 364, 365, 396
Black Belt 177, 180, 182
black box 80, 82, 266, 267
brainstorming 24, 45, 140, 160, 161, 189, 318, 367
breakthrough 30, 180
builder's workshop drawings 450
business case 194, 195, 214, 304, 306, 312, 314, 315, 316

categories of costs 172
cause and effect 23, 130, 150, 352, 358, 430
cellular 164

certification, ISO 84, 94, 95, 285, 291, 294, 298, 393, 442
champion 177, 181
change management 52, 86, 88, 90, 93, 186, 269, 463
changing views of Quality 15
Chase, R. 31
check sheet 23, 129, 130, 131
classic quality tools 129, 130, 135, 210
CLIPSCFM 79, 80, 264, 267
close out 406, 502, 507
compatibility 19, 60, 204, 208, 219, 321, 340, 343, 347, 356, 359, 433, 445, 502, 512
competence 81, 83, 86, 88, 91, 93, 223, 226, 280, 284, 301
composite drawing 56, 187, 385, 421, 450, 452, 454
compressor 73, 74, 309, 415, 439, 500, 517
constructability 19, 37, 117, 204, 206, 207, 208, 272, 321, 325, 342, 344, 347, 348, 355, 359, 360, 368, 391, 396, 399, 429, 436
construction quality 36, 38, 109, 113, 173, 257, 261, 263, 302, 305, 306, 409, 459, 460, 461, 462
constructor 33, 34, 82, 259, 260, 261
contract management 52, 109, 112, 127, 257, 395, 402, 409, 423, 426, 427, 470, 471, 472
contractor's quality control plan 37, 39, 40, 79, 91, 109, 112, 257, 265, 268, 304, 401, 402, 409, 411
controllability 49
cost efficient 62
cost of project 123, 287
cost of quality 6, 16, 17, 129, 150, 155, 172, 173, 188, 203, 208
cost loaded 407, 408, 421, 482
cost reimbursement 195, 283
critical path method (CPM) 137, 140, 142, 358, 404
critical to quality (CTQ) 150, 152, 183, 327
Crosby, P. B. 5, 16, 17, 27, 29, 30, 250
corrosion 67, 199, 325, 331, 351, 353, 357, 401, 431, 432, 440, 447, 500, 511
customer focus 31, 79, 86, 88, 90, 93, 264, 265
customers need 5, 32, 217, 255, 262, 268, 511
customized 34, 104, 105, 118, 129, 247, 251, 254, 262, 264, 271, 309

daily report 414, 462, 469, 480, 485
data security 199, 206, 325, 347, 352, 353, 358, 401, 440, 500

523

data sheet 326, 339, 348, 351, 353, 358, 399, 432
definite 101, 327, 329
Delphi techniques 45, 160, 162, 318
Deming, W. E. 5, 13, 16, 17, 18, 20, 21, 22, 29, 30, 212, 250, 284
demobilize 509
design-build 115, 117, 118, 119, 319
design management 93, 287, 300, 301, 302, 386, 387
design quality 108, 109, 113, 117, 258, 291, 324, 339, 348
dewatering 190, 405, 406, 447
DFSS (Design for Six Sigma) 182, 327
digitized 483, 486
DMADDD 187, 188
DMADV 160, 164, 177, 182, 183, 203, 327
DMAIC 157, 158, 177, 183, 184, 187, 210
downstream 105, 106, 110, 111, 112, 113, 114, 203, 253, 255, 256, 257, 258, 309, 512
drainage 145, 383, 405, 406, 431, 439, 447, 519

earthwork 341, 363
ecology, ecological 317, 348, 355
economy 10, 13, 62, 204, 254, 318, 321, 328, 344, 356, 429, 439, 464, 515
enhancement 30
executed works 114, 139, 210, 258, 266, 304, 401, 446, 450, 473, 489, 494, 495
external works 199, 200, 206, 352, 354, 358, 393, 394, 401, 405, 406, 431, 447, 502

fabrication 10, 7, 70, 102, 136, 151, 159, 170, 323, 349, 351, 357, 417, 422, 436, 449, 451
facility management 326, 344
factual approach 32, 80, 264, 266
Feigenbaum, A. V. 4, 5, 9, 12, 16, 20, 21, 29, 30, 249, 250
fence, site 383
fire alarm system 71, 74, 352, 401, 405, 406, 440, 447, 500, 508, 512
fire and gas 199, 200, 204, 206, 309, 325, 351, 431
fire protection 67, 74, 213, 321, 347
firm fixed price 118, 195, 282, 319
Flare 430, 432, 439, 500, 518
flowcharting 150, 152, 158
FMEA 150, 152, 153, 154, 158, 186
Ford Global 8D Tool 182
foreman 5, 22, 248, 295, 299, 394
formwork 57, 141, 190, 239, 241, 364, 365
function, functional 11, 22, 29, 31, 33, 38, 42, 52, 124–126, 175, 180, 231, 260, 266, 271, 274, 310, 315, 326, 346, 448, 449, 477, 495

geotechnical 203, 328, 348, 355
globalization 42, 79, 94, 247
goodness 5, 16, 177
grading 173, 400, 430

green belt 177, 180, 184
grounding 430, 500
Gryna, F. M. 9, 15, 20
Guaranteed maximum price (GMP) 283
guarantees 413, 498, 502, 503, 507
guilds 3, 14, 247
gurus 5, 16, 21, 22, 24, 27, 250

HAZID 348, 352, 357, 358, 391, 432, 434, 435
HAZOP 348, 352, 354, 391, 432
heavy engineering 103, 252, 285, 286
histogram 23, 130, 132, 134, 361, 416, 482, 483, 484, 492, 493
house of quality 175, 176, 326, 327
Hydrology 348, 355

Industrial Revolution 3, 5, 7, 8, 14, 102, 247, 248
Information and Communication (ICT) 200, 206, 325, 352
innovation 13, 129, 160, 163, 182, 201, 319
integrated systems 115, 276
integration 6, 42, 43, 125, 126, 127, 184, 220, 249, 333
interdisciplinary 24, 109, 170, 207, 253, 257, 340, 341, 348, 398, 432, 434, 435
interrelationship diagraph 137, 143, 145
involvement of people 31, 79, 264
IoT 398
IQMS 42, 43, 44
Ishikawa, K. 5, 6, 21, 23, 24, 25, 29, 30, 130, 155, 250
ISO 9004 78, 267
ISO 14000 42, 44, 75, 84, 94, 95, 219
ISO 18000 68, 219
ISO 22000 76
ISO 27000 68
ISO 28000 76
ISO/TS 29001 97
ISO 45000 68, 216, 219, 220
ITB (Invitation to Bid) 348, 350, 366, 371
ITP (Inspection and Testing Procedure) 443

Jacob, F. 31
Juran, J. M. 4, 5, 8, 9, 13, 16, 25, 26, 28, 29, 30, 250, 259, 260
JUSE 28
just in time 166

Kaizan 165, 167
Kanban 165, 166
Kerzner, Herold 6, 9, 11, 15, 22
Kick off meeting 111, 205, 256, 385

landscape 171, 199, 200, 206. 295, 323, 325, 331, 352, 354, 358, 369, 393, 394, 401, 405, 406, 422, 431, 437, 447, 461, 462, 517, 520
leak detection 199, 200, 204, 206, 235, 309, 331
lean tool 129, 164, 165, 204

Index

lesson learned 498, 508
lifecycle 128, 167, 311, 319
litigation 54, 60, 286, 287, 290, 293, 396, 427, 459
LNG 105, 253
load, live 203, 328, 355
load, seismic 203, 328, 355
load, wind 203, 328, 355
logistic 46, 54, 290, 317, 443
loss prevention 199, 200, 237, 331, 352, 357
low bid 57, 118, 193, 209, 282, 283, 373, 376
low voltage 147, 176, 323, 406, 490, 502, 505, 506
lump sum 116, 118, 123, 195, 282, 319

machines 24, 25, 104, 238, 251, 309
maintainability 204, 321, 347, 429, 512
manageability 49
manufacturing plant 101, 103, 104, 146, 252, 285, 286, 309
MasterFormat® 362, 363, 364
material handling 59, 216, 236, 319, 325, 328, 351, 440, 500, 518
matrix, diagram 137, 145
measurable 66, 197, 317, 326
mediation 459
meeting, coordination 304, 348, 361, 467, 486
meeting, safety 237, 240, 428, 468, 469, 486
MEP 141, 143, 206, 325, 332, 351, 353, 357, 368, 401, 431, 454
metrological 355
middle ages 3, 14, 247
midstream 105, 213, 253, 511
mind mapping 160, 163, 164
mistake proofing 165, 167, 170, 204, 434
mitigation 43, 45, 47, 50, 55, 56, 57, 58, 195, 218, 232, 234. 344, 368, 436
mobilization 53, 56, 108, 111, 141, 143, 209, 210, 234, 256, 378, 380, 381, 382, 383, 385, 403, 406, 426, 461, 484
Motorola 177
MTO List 353, 354, 376, 379
mutual beneficial 32, 80, 264, 266

narrative 324, 331, 339, 352, 354, 369, 482
National Quality Award 30
need analysis 191, 193, 194, 197, 279. 306, 307, 313, 314
need statement 191, 194, 196, 198, 306, 313, 314, 315
negotiation 53, 56, 280, 288, 289, 403, 307, 338, 371, 373, 389, 459
NFPA 67
nominal group technique 160, 163
nominated 382, 390
notice to proceed 108, 131, 209, 256, 380, 381, 385, 413, 421

Oakland, John 5, 16, 26, 27, 250
obsolete 514

OHSAS 18000, 43, 220, 231
optimization 347
organization-wide 6

Pareto analysis 314
Personal protective equipment (PPE) 216, 236, 239, 243, 453, 469
Poka-Yoke 27, 165, 171
power plants 103, 120, 252, 285, 286
predesign phase 312, 333
predictive 237, 513, 514, 515, 517
Pre-FEED 321
preferred alternative 107, 112, 192, 195, 196, 257, 312, 313, 319, 321, 340
preliminary schedule 107, 112, 113, 194, 205, 257, 322, 329, 342
principles, quality 34, 52, 249, 262, 264, 265
prioritization 45, 47, 49, 137, 147, 148, 183, 191, 194, 333, 397
process analysis 29, 129, 147, 150, 203, 210
process approach 31, 79, 80, 264, 265
process mapping 150, 152, 156
process type projects 103, 285
programming phase 312
project brief 87, 117, 315, 320
project delivery system 110, 115, 273, 276
project phases 124, 125, 127, 274, 309
proximity 49, 63, 466
punch list 391, 462, 499, 502, 508, 509
purchasing 58, 59, 60, 51, 66, 104, 252, 309
Pyzdek, T. 4, 6, 8, 13, 65

QFD 129, 175, 176, 190, 326, 327
qualification based 193, 209
quality based system 288
quality circles 21, 22, 23
quality cost, please *see* cost of quality
quality definition 20, 255
quality engineering 7, 12, 249
quality forms 28, 77, 79, 266, 267
quality guru 5, 16, 21, 22, 24, 249, 250
quality in construction 32, 33, 34, 35, 118, 217, 251, 259, 262, 264, 272, 487
quality in manufacturing 32
quality inspection 3, 7, 14, 217, 247, 249
quality management plan, planning 36, 37, 268, 310, 348, 382, 387, 463
quality manager 59, 322, 323, 347, 349, 387, 461, 461
quality matrix 36, 38
quality planning 12, 25, 28, 173, 268
quality policy 12, 76, 77, 80, 83, 84, 86, 88, 90, 93, 94, 96, 267, 269
quality principles, please *see* principles of quality
quality trilogy 25, 35, 124, 274, 477
quantitative 17, 18, 23, 25, 28, 29, 45, 46, 48, 129, 158, 175, 318, 319

radiography 8
realistic 197, 317, 328, 428, 472
record books 110, 113, 257, 498, 503, 511
refining 105, 253, 512
register, risk 45, 46, 47, 49, 469, 476
reliability 6, 9, 60, 62, 63, 66, 97, 195, 204, 206, 318, 321, 347, 429, 464, 466, 512
request for information (RFI) 55, 156, 208, 282, 278, 290, 292, 397, 456
request for proposal (RFP) 193, 209, 284, 287, 457
request for qualification (RFQ) 290, 291, 292, 293
residential 101, 103, 252, 285, 286
responsive 49, 63, 283, 378, 466
retention 45, 50, 498
roadmap 79, 177, 267
roles and responsibilities 84, 99, 234, 235, 263, 269, 276, 281, 322, 412, 420, 426, 427, 516
run chart 130, 133, 134, 136

sanitary 103, 383
scatter diagram 123, 130, 135, 136
schematic 119, 120, 206, 266, 325, 330, 331, 348, 351, 353, 358, 429, 431, 444, 483
scientific management 7, 14, 27, 29, 248
S-curve 404, 480, 487
seismic load, please see load, seismic
seven point action plan 22
Shewhart, W. A. 6, 8, 11, 14, 22, 29, 158, 248
Shingo, Shigeo 16, 27, 171, 250
shoring 170, 203, 328, 355, 405, 406, 447
short listing 281
Shtub, A. 25
shutdown 192, 206, 430, 514
single stage 283, 297
SIPOC 155, 157, 159
site facilities 111, 256, 378, 383
site safety 41, 163, 240, 278, 413, 462, 468, 486, 494
site survey 352, 354, 355, 358, 383, 399
site work instruction 413, 455
solid process plants 103, 252, 285, 294
source information 503, 511
statistical process control 9, 14, 17, 18, 20, 23, 31, 33, 42, 130, 131, 157, 158, 159, 356
statistical quality control 6, 11, 248, 515
statutory 46, 53, 57, 78, 194, 319, 326, 344, 354, 400
stone age 3, 247
subconsultant 37, 89, 104, 308
supportability 204, 321, 347, 429
sustainability 60, 62, 106, 188, 194, 204, 206, 282, 309, 317, 318, 321, 344, 347, 368, 429, 436, 464
switchgear 71, 430, 431, 440, 500, 518
system approach 31, 79, 264, 265
systems engineering 80, 106, 253, 254, 310

Taguchi, Genichi 16, 28, 230
Tague, Nancy 18, 24
target price 195, 283
Taylor, Fredrick 27
Taylor's system 7, 248
TC 176 80, 267
team approach 15, 28, 33, 218, 250, 262, 263, 265, 270, 516
team leader 33, 177, 180, 181, 261, 322, 347, 349, 387
technical manual 110, 113, 257, 498, 503, 507
technological 52, 102, 176, 192, 195, 272, 310, 316, 514
technology 7, 12, 213, 216, 231, 273, 398, 483, 495
terminate 11, 509
terms of reference (TOR) 37, 38, 87, 170, 195, 198, 199, 200, 201, 272, 277, 306, 318, 320, 342, 360
thermography 8
Thorpe, Brian 11
tie-in material 326, 353, 432
topography 355
total quality control 6, 14, 30, 249
total safety management 216, 218
traffic studies 203, 328, 330
transcendent 4
tree diagram 137, 147, 149
triple constraints 124, 125, 274, 477
Triz 129, 160, 164, 187, 188, 189, 190
Turnkey 115, 116, 195, 276, 283, 319
two stage 283, 297

ultrasonic 8
unit price 195, 282, 474
universe 253
upstream 105, 213, 253, 512
user's requirement 35, 36, 52, 110, 124, 260, 262, 268, 278, 310, 317
utilization 30, 95, 219, 319, 334, 404, 417, 460, 464

validate 114, 185, 186, 258, 342, 348, 367, 401, 443, 475, 494
value engineering 37, 43, 44, 108, 114, 119, 200, 202, 206, 258, 277, 279, 286, 287, 302, 341, 346, 348, 362, 366
value stream mapping 165, 171
variable control charts 20
vendor list 250

warranties 414, 498, 502, 503
waste reduction 165, 172, 219
wind load, please see load, wind
work breakdown structure (WBS) 106, 107, 254, 255, 310

zero defects 16, 17, 27, 30, 61, 171, 174, 218, 441, 449
Zoning 429

Printed in the United States
By Bookmasters